9/23/93

SCIENCE, AMERICAN STYLE

D1065598

Library of Congress Cataloging-in-Publication Data

Reingold, Nathan, 1927–
    Science, American style
    p.cm.
    Includes bibliographical references and index.
    ISBN 0-8135-1660-9 (cloth)   ISBN 0-8135-1661-7 (paper)
    1. Science—United States—History.   2. Science—United States—
History—19th century.   I. Title.
    Q127.U6R397   1991
    509.73—dc20   90-9068
                                                            CIP

To Ida who was there, waiting for me, when it all started

SCIENCE, AMERICAN STYLE

NATHAN REINGOLD

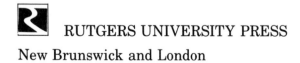

RUTGERS UNIVERSITY PRESS

New Brunswick and London

CONTENTS

SCIENCE, AMERICAN STYLE

Introduction

Faced with the daunting task of explaining why I wrote this array of pieces over a long span of years, I must reluctantly fall back on autobiography. "Reluctantly" because a great deal of my professional work has involved pointing out errors of omission and commission in past individuals' confident assertions about themselves. Perhaps a note of uncertainty is just the right way to a start. I am against historical writings asserting or implying closure. Histories should be heuristic and frankly have an element of incompleteness. The historian should have the good sense to convey a sense of humility, no matter how strong his beliefs. (I have not always succeeded in this.) And the sense of humility should spring from awareness of the problems of sources and the limitations or deficiencies of the research process.

What follows are seventeen of my essays arranged by rubrics not necessarily mutually exclusive. Each has a preface (all but one being newly written) explaining the origins of the pieces and, sometimes, my present views. Although the selected items are largely on events, individuals, and institutions in the United States, often European references are present, particularly those relevant to issues in both the history of science and in American history. From time to time readers will encounter discussions of ideas about the nature of history and of historical concepts specific to the topics of the essays. Finally, this collection has much on institutional histories, often specifically in terms of individuals and groups of individuals. These are not, to use archaic terms, "Sketches for a Natural History of Science in the United States." Rather, the articles are responses to specifics of my career and to the particular development of a historical specialty, the study of the growth and the role of the pure and applied sciences within the United States of America. I will try to explain the origins of at least some of my various historical preoccupations in the pages immediately following. All of these writings are incomplete in the sense of being not final destinations but signposts indicating the directions of future research for myself and my fellow historians.

A few points on my background are in order. I entered the history of the sciences from American history. My intellectual allegiance is not to particular scientific disciplines, to philosophy, or to the social sciences but to history—not even to history of science as something autonomous in its specialized splendor. Although an academic, I am not part of academe but a civil servant. I take great pride in being the first person to have worked at all three of our great national cultural institutions, the National Archives, the Library of Congress, and the Smithsonian Institution. The effect on my work as a historian is quite great. Most obvious are the occasions, the opportunities for specific research forays. Less obvious, but perhaps more fundamental, is how the three institutions have influenced both my attitudes and my methods. A sense of my indebtedness sometimes emerges in what follows. I am also grateful to them for aid and encouragement at particular points in my career. (I must also add my acknowledgment of financial and other support from the American Philosophical Society, the National Academy of Sciences, the National Science Foundation, the National Endowment for the Humanities, and the Lounsbery Foundation.)

Perhaps the most obvious consequence of my employment history is the recurring concern with the behavior of humans with respect to formal organizations. More fundamental, in my view, is my exposure in the National Archives, the Library of Congress, and the Smithsonian Institution to an immensity of sources disclosing the staggering complexity of past and present science and technology. The professional literature barely skims the surface in the topics covered and in the depth of analysis. In no way could I even comprehend all of this immensity. Being aware of the extreme selectivity of what was covered by the existing literature yielded a skepticism about confident generalizations of historians, philosophers, and sociologists.

In my own practice of history, I early adopted a strategy different from that of conventional history of science, which tended to a top-only (not even top-down) fixation on great ideas, great scientists, and great revolutionary events. In contrast, I looked for the routine, the lowest common denominator. The routine, the lowest common denominator interested me for its own sake but also because I did not believe one could really characterize deviations from the norm as great without a good sense of the routine of each situation.

In dealing with the vast bodies of sources, I was strongly influenced by the perceptions dating from my early work in the National Archives of the existence of structures within and between collections. If I could clone myself, the other Nathan Reingold would spend all his time on investigations of the histories of bodies of unpublished records of institutions and individuals. Analyzing collections is just great fun, especially where the boxes in serried ranks assembled obscure a complex history made even more opaque by the passion for order of curators and archivists. Superficially, my reaction is quite like that of devotees of other puzzles (crossword, jigsaw, mathematical, etc.). Not having a clone to indulge in this fascinating pleasure, I often reluctantly leave collections with questions unresolved.

As a supposedly serious, high-minded historian, my puzzling over collections is not purely self-indulgence. Analysis, even incomplete, gives a history illuminating the structure and composition of bodies of institutional records and personal papers. Even small, apparently simple bodies can surprise the aware historian. At best, understanding a body of sources explains what is or is not present. That, in turn, helps in the understanding of the writings on the sheets of paper. A second possible benefit is to suggest where related materials might exist.

A number of the essays that follow are really detective stories. By chance or otherwise I came across interesting unexpected documents. Next, there is a hunt through collections, sometimes scattered widely in depositories. Then, having assembled a large sample of sources suggesting convincing patterns, I display my skills as a historical artisan. Details are selected to convey a texture of the past and arrayed to present what I hope is a cogent account and explanation. The best description of my feelings during this process is given by a French historian, Jean Meuvert, quoted in an article I accidentally came across: "a feeling of humility, certainly, but also of mounting excitement and often of pure joy." I try to keep the excitement and pure joy in perspective as I savor big pictures and wonderful details. Try as I might even with weighty topics I cannot evade seeing a Cecil B. De Mille epic with a grandeur tinctured by the absurd. Often the specifics of the past, no matter how important, remind me of the world of the great comedian, W. C. Fields. Quite properly, only occasionally do I let on to this and indulge in a little tongue in cheek.

Before considering the course of my specialty, the most important influence on my historic work, I want to acknowledge two other factors. To this day I have a great admiration for and happy memory of Richard H. Shryock. At several points below I specifically discuss his role in shaping my outlook. Although best known as a historian of medicine, Shryock had a wider range of concerns which still influence me. First and by far the most important was the sense of the need to add the sciences to the body of American history. Second, that interest in the history of science in the United States included applied fields. Not for Shryock nor for me the snobbish fixation on purity. Third, to understand U.S. events required at best a comparative prospect. And fourth was the resolve to view pure and applied fields in a broad cultural prospect. The sciences cannot be understood historically without an understanding for each epoch of the nonscientific.

Conferences were a very important feature of the early growth of the history of the sciences within the United States. Some of the papers that follow might never have been written without the stimulus of a conference invitation. The indifference theme paper was presented at a meeting at Northwestern University in the late 1960s. It startled me that so many cared about science in nineteenth-century America to come to Evanston. Many of us had a sense of being alone or part of a rather tiny band. Later conferences like those sponsored by the American Academy of Arts and Sciences at Newagen, Maine, reinforced a growing sense of excitement and self-confidence. References to this development appear at several points in my essays. Comparison with another historical specialty may clarify what was at issue.

The winter 1984 number of *Art Journal* is on the history of American art. The issue editor, Professor Jules David Prown of Yale, contributed an introductory statement entitled, "Art History vs. the History of Art." The parallels with the study of the sciences within the United States were striking.

In both, a formerly marginal specialty, often regarded with disdain by the mainline practitioners, had experienced a rapid growth in a short time period. Prown pointed to an increase in posts, major exhibitions, a rising tide of collecting activity, and an outpouring of research. From being the interest of a very few, American art now attracts many scholars. Its study, if not central in the history of art, is a bustling and recognized specialty. I could point to analogous developments best summed up by two publications: Marc Rothenberg's bibliography, *The History of Science and Technology in the United States* (1982), and the first volume of the new series of *Osiris* edited by Sally G. Kohlstedt and Margaret W. Rossiter, *Historical Writing on American Science* (1985). Both publications stimulated me to introspective reflections. For both fields two obvious questions are posed: Why the former neglect? And why the recent upsurge?

Having written an essay published in 1972 on the alleged American indifference to basic research, I could not miss Prown's dwelling on the analogous concept for his field:

> The second-class status of American art has been based on a fundamental and literal discrimination. American art has been perceived as generally of lower aesthetic quality than European art and, by virtue of a logical fallacy that equates the quality of scholarship with the perceived quality of the art being studied, it has therefore been largely snubbed by leading art historians.

Aesthetic quality and basic (or pure) research are terms alike in being laden with assumptions about the ideal nature of art and of science, respectively. Although these ideal attributes are presumed to exist independently of the contingencies of time, place, and person, there is an increasing trend to think and to act otherwise among students of art and of science, and not simply in the American context. Why that is so and how the changes occurred is one of the recurring themes to many of my writings. I was there; I participated; I contributed and have strong feelings on many issues.

Prown's analysis of the historiographic issues reinforced the sense of parallelism but only up to a point. Art and science are not identical. Historical study of aesthetic objects and of scientific research must diverge. He distinguishes between the "history of art" and "art history." According to him, the "history of art" is "the study of the causation . . . of the population of objects we call art." The objects are both the occasion for and the subject of the historian of art's labors. In contrast, Prown describes "art history" as a field where the focus is on history and the objects are a means to that end, "especially social and cultural history." Analogous but not identical arguments occur in some of my articles. But one difference exists between Prown's account and what occurred in the history of the sciences in the United States. Prown's "history of art" is concerned, among other things, with national styles. That idea is a recurrent concern of many, like myself, involved in the new bustling field. The nature of the former reigning orthodoxy in the history of science is invoked by me at points in my writings to explain and to argue for a new best practice in research, one involving a differing perspective on the concept of national style. Since written history is the result of an interaction between the evidences from the past and the concepts and work habits of present-day historians, in the absence of convincing substantial evidence to the contrary, national styles are best seen as transitory, like styles in fashion, not something fixed and immutable.

Prown does not really explain why his upsurge occurred, nor is he aware, apparently, of parallels with my upsurge and even others. In both science and art the rise in scholarly concern followed quickly upon recognition both of a considerable increase in national activity in both domains and of attaining world class status, if not a temporary hegemony. In art the growth of contributions during this century culminated in the recognition that a "New York School" had supplanted the "School of Paris" after World War II. In the sciences a steady growth since the late nineteenth century became grossly apparent in the research and development feats of World War II and in the outpouring of diverse results in the following fifteen years. While the coincidences are striking, I doubt there is a single hidden cause for the sequence of events in both the sciences and the arts in the United States. I incline, rather, to a belief in chance simultaneous culminations of long-term trends in the growing support of culture and learning in the United States. As I more than hint at a few points in my writings, I see the sciences in the United States achieving world class (not hegemony) earlier than World War II in a number of scientific disciplines.

Another, quite different relationship links Prown's field and mine. Prown is primarily writing about the visual arts. Similar patterns— the study of the thing itself in contrast to seeing the thing as part of a historical environment—exist in other arts such as literature and music. Besides science, much the same occurs with philosophy, political theory, and religion. The literal contents of texts and objects are reified at the expense of the contexts within which contents came into being and had consequences for humans. Although persisting, the dichotomy is rather foolish. Contents and contexts do not exist the one without the other. Boundaries between them are often very hazy. What we have in all these areas of human creativity is a reaction of awe to the greatness of cultural monuments—triumphs of intelligence, emotional striving, and human sensitivity. Both concerns, with contents and with contexts, are different ways of expressing awe and wonder in the presence of what are the heights of human history to so many of us.

Whether one elects to see the monument existing for its own sake or existing in creative tension with a larger society—national or otherwise—is a matter of personal commitment often taking a distinctive functional path. In each dispensation there is a tendency to think in terms of essences in the sense of defining characteristics persisting for all time. The essences supposedly defining "science" and "art" each then justifies, even sanctifies, an intense myopic fixation on the objects of science and of art. The essences supposedly defining "France," the "United States," or "England" establish the boundaries and ground rules of historical discourse. Although extensive literatures define, discuss, and elaborate on all these essences, many others are barely explicit in either written or oral discourse, providing numerous opportunities for scholarly employment in teasing them out.

Essences can often serve another purpose. Whether in art or in science, from them derive a tendency to determinisms. If science is defined, let us say, as a certain body of texts each generated by an internal logic, each being a successive approximation to Truth, then the historian's path is not so much indicated as rigidly circumscribed. If a national style is supposedly enthusiastically antitheoretical, practical, and hostile to the undemocratic emergence of great innovators, history is almost like a train that cannot even jump the tracks during a collision. The determinisms "work" because so little is admitted to the scholar's discussion and so much is excluded by definition.

I will have nothing to do with the whole apparatus of cultural monuments, essences, determinisms, and idealized national styles. I am a believer in and even sometimes a practitioner of the social history of the sciences in the United States. Social history of science is concerned with the "social structure of scientific work and the social conditioning of the intellectual history" of the various disciplines. A second definition somewhat amended from the original is: "Our task is to understand the nature, place, role and impact of these technical branches of knowledge in the general American culture."

"Technical branches of knowledge" are clearly an object of interest of the first definition of social history of "sciences." But the term "social" makes a crucial distinction between the "scientist" and the "science," that is, between the creators and practitioners of scientific work and the body of knowledge itself. "Social," as used here, refers to what people do, and scientific thinking or investigation (i.e., creating knowledge) is an action of people—that is, a form of human behavior. A true social history is neither obsessed with successive cognitive achievements nor confined necessarily to a particular discipline. Above all, a social history shuns the tacit identification of the knowledge and the knower endemic to so much of the history of science.

If we examine a body of scientific and technical knowledge solely in rational and intellectual terms—as sets of symbols organized consistently according to specified rules and standards—then the body of knowledge is indeed properly abstracted from humans and their environments. If we consider how and why humans devised the symbols in time, how and why the symbols spread in space and time, becoming accepted, rejected, or modified, in no way can we limit the history of the symbols to the workings of supposedly pure rational processes. All this amounts to is the assertion that the history of the sciences is a branch of history, not a branch of science or philosophy.

What I am accepting as my working hypothesis is by no means universally held. Nevertheless, it is increasingly accepted in many forms of intellectual discourse. The knowledge we refer to as science and technology is not a replication in symbolic form of some ultimate reality, nor does this knowledge simply correspond, as best as possible, to this ultimate reality. The data and concepts are human artifacts, constructions reflecting beliefs and values of all kinds. They are meant to be useful in generating more data and concepts or useful in the sense of serving a particular need—whether technological, medical, ideological, or psychological. In this sense we can say that "scientists" are adding to human culture in three ways: first, as makers of conceptual artifacts; second, as makers of material artifacts; the third way is less obvious but perhaps crucial for a social history—they are forever generating traditions which are painfully retrievable from evidences in writing and from structured interviews. A discipline like physics is a continually evolving tradition, not a fixed body of certified truths and methods.

The social history of science is not limited to great and new ideas nor to great scientists, engineers, and physicians. It is not limited to the best practice of a field at a given time nor to the most commonly held views. Nor does it evade older forms of theory and practice nor those of very modest impact. Neither does it scorn the archaic nor the blatantly false. If held by a portion of the population in our historical epoch, creationism and astrology are valid problems for the labors of social historians of science.

A whole generation of research has resulted in a considerable degree of assent on these points. Let me give two examples. On the same page of the issue of *Science* magazine of March 2, 1984, the following appeared in two book reviews by authors of widely different backgrounds: (1) "the important role that society at large plays in scientific development" and (2) "almost totally ignores the recent reevaluations by historians of science who attempt to place leading thinkers in their social and intellectual milieux, the better to understand the interplay of empirical discovery, new ideas, research techniques, and personal rivalries." My second example is from a 1984 essay review in *Isis* by Lynn White, the great historian of medieval technology who is the only person to have served as President of the American Historical Association, the History of Science Society, and the Society for the History of Technology. He ascribes the attenuation of the old view of science to "the emergence of a profound interest in the ecology of science: that is, in how theoretical science in any period and place has shaped its total context, and how reciprocally it has been shaped by its environment, cultural and other. The history of modern science is not a triumphant process of chalking up an endless series of discoveries of absolute truth achieved by Galilean methods. It is integral to all the rest of history and is in no way different in kind from all the other sorts of human experience."

A social history of science is a history of the behavior of a population identified as "scientists," however we define that term. While an important aspect of that history is the behavior of members of the population to one another, it is by no means limited to that. On the contrary, a social history of science should primarily place its subjects within larger contexts, particularly the varied relations, interactions, and processes involving individuals, institutions, and social groups outside the definition of "scientist." While much remains undone along these lines, in the last generation much was accomplished furthering that aim both for the United States and for Western Europe.

I view myself both as an American historian specializing in the history of the pure and applied sciences and as a historian of the sciences in the nineteenth and twentieth centuries. I want to place scientists within specific past environments—perhaps *ecological* is as good a term as *anthropological* to give my bent. I have a strong antipathy against abstracting the science and the scientists from the historic specifics of their milieus. I am interested in the complexity of the past world inhabited by my subjects, not in oversimplifications tacitly assuming either the autonomy of knowledge, the hegemony of supposed social laws, or a presumed progression to an ultimate reality or to a more glorious discipline. My primary obligation is to explain scientists' behavior within their social settings, including their cognitive behavior, not to analyze scientific findings as components of bodies of knowledge.

To summarize my position: *All history is local history*—somehow or other the historian has to explain specifics of particular past milieus;

generalizations must simultaneously arise from and explain specifics. *All local history is social history*—there is no intellectually valid way of escaping the necessity of dealing with individuals and groups interacting. *All social history is intellectual history*—social interactions are only understandable in terms of the general views of those involved and/or of the concepts historians bring to the surviving evidences of the past. *All intellectual history is local history*—while we all have the hope that our concepts are valid, if not timeless universals, it is much more likely they are tribal behaviors peculiar to local groups and destined to erode away. Imbued with the sense of the internationalism of their labors and the timeless universality of their contributions, the scientists are a particular, very important tribe whose local histories I find engrossing.

In preparing this particular volume, I was aided by the comments of Reese Jenkins, Marc Rothenberg, Seymour Cohen, Jeffrey Stine, Paul Theerman, and James Fleming. They are not responsible for the final selection nor my own comments. Over the years I have run up debts of various kinds to many in the course of my research. There are simply too many archivists and librarians to remember, let alone to list by name. I am inordinately fond of archivists and librarians and still identify with them. (I often still use the first person plural when talking about the National Archives and the Library of Congress.) Best of luck and much thanks to each and every one who has helped in the past. I look forward to seeing you again on future research trips.

These essays, with one exception, are reprinted and appear with permission of the original publishers. I am greatly appreciative of their cooperation, now and in the past. *Nature* printed "Reflections on Two Hundred Years of Science in the United States" in its issue of July 1, 1976 (copyright 1976 by Macmillan Magazines Ltd.). "Definitions and Speculations" appeared in A. Oleson and S. Brown, eds., *The Pursuit of Knowledge in the Early American Republic,* Johns Hopkins University Press, 1976. "American Indifference to Basic Research" is from George N. Daniels, ed., *Nineteenth Century American Science, A Reappraisal,* Northwestern University Press, 1972.

"Science in the Civil War" is from the September 1958 *Isis.* "Cleveland Abbe at Pulkova" is from the April-June 1964 *Archives Internationales d'Histoire des Sciences.* "Alexander Dallas Bache" originally appeared in *Technology and Culture,* April 1970. "Theorists and Ingenious Mechanics" is from the October 1973 issue of *Science Studies.* "Joseph Henry on the Scientific Life" is a previously unpublished paper presented at the meeting of the American Association for the Advancement of Science, March 1, 1974.

"Graduate School and Doctoral Degree" is from N. Reingold and M. Rothenberg, eds., *Scientific Colonialism: A Cross-Cultural Comparison* published by the Smithsonian Institution Press in 1987. "National Science Policy" first appeared in A. Oleson and J. Voss, eds., *The Organization of Knowledge in Modern America, 1860–1920*. Johns Hopkins University Press, 1979. "The Case of the Disappearing Laboratory" is from the Spring 1977 *American Quarterly* (copyright 1977 by the American Studies Association).

"Refugee Mathematicians in the United States" appeared in the May 1981 *Annals of Science*. A slightly revised version of "Vannever Bush's New Deal" was printed in *Historical Studies in the Physical and Biological Sciences* in 1987. *Expository Science: Forms and Functions of Popularisation*, edited by T. Shinn and E. Whitley Reid, 1985, contained "Metro-Goldwyn-Mayer Meets the Atom Bomb."

"Uniformity as Hidden Diversity" is from the November 1986 *British Journal for the History of Science*. *History of Science*, 1981, contained "Science, Scientists, and Historians of Science." "Through Paradigmland" originally appeared in the November 1980 *Social Studies of Science*.

THE NATIONAL STAGE

1 □ Reflections on Two Hundred Years of Science in the United States

This paper was an unexpected by-product of the bicentennial of U.S. independence. In 1976 I organized a conference on the history of sciences in the United States for the annual meeting of the American Association for the Advancement of Science at Boston. Shortly afterward, an editor of *Nature*, the British journal, visited me. When he asked about the conference, I groused about the European neglect of the history of the sciences within the United States. In 1964/65, I had lived in London while studying scientific contacts between Americans and Europeans of the mid-nineteenth century. On many occasions I had been vexed and even outraged by the ignorance expressed about the history of the sciences in the United States and by the smug, patronizing tone often manifested by British commentators. I informed the man from *Nature* I was a bit tired of being told the British were Greeks imparting wisdom to us, the cruder Romans across the Atlantic. We were not Romans, and the British were far from Greeks. He responded by challenging me to write a paper for their Fourth of July bicentennial issue.

Although aimed at a British audience, this freewheeling piece, naked of footnotes, summed up attitudes arising from nearly twenty years of work. Social history and intellectual history appear in juxtaposition, not segregated in distinct compartments. Comparisons with other societies seasoned the prose. Written in a spirit of euphoria, the essay remains a great favorite of mine. Today I read it as closing one stage of my research and announcing my increasing interest in twentieth-century topics.

When I published the conference proceedings in 1979 (*The Sciences in the American Context: New Perspectives*), I used the piece to introduce the collection.

Earlier this year a British historian of science wrote to me to say he was giving a few lectures on the institutional history of American science before 1914 in his course. I was astonished. Hardly any non-American historians do research or display interest in science in the U.S.A.; presumably their courses reflect this. An East German has a few pieces dealing with astronomy; a Dutch historian this year published an article on a nineteenth-century American physicist; a few years ago, British geologists did a multivolume history of geomorphology, a field they defined as overwhelmingly derived from U.S. geologists.

In contrast, there is no shortage of writings and courses outside the U.S.A. on science policy or sociology of science with strong emphasis on recent events in America. Nobel Prizes are counted, statistics marshaled, and very current trends analyzed. In this dispensation, science in the U.S.A. starts with a contemporary occurrence—the Mansfield Amendment, Sputnik overhead, the atom bomb, or the arrival of refugees from Hitlerism. But research and development in the U.S.A. is important on a global scale. It did not spring up overnight but was the result of a complex, fascinating historical metamorphosis. In 1776 the British North Americans, south of former French Canada, had one great investigator, Benjamin Franklin, and a handful of others we can denote as scientists. By 1900 the new nation had essentially evolved a vigorous scientific community with many of the basic structural and functional characteristics so evident today. This included strong commitments to fundamental research in various fields. In the present century these characteristics produced what is now so visible. It is a meaningful historical movement, not simply a banal success story, and deserves critical scrutiny by scholars. Rather than give a recital of great names and notable contributions, or genealogies of institutions, let us consider the sciences in America as a set of historiographic problems particularly worth studying in comparative prospect.

Influence of de Tocqueville

Americans and Europeans are still influenced by Alexis de Tocqueville's *Democracy in America* (1835). The work is best described as a hortatory essay to the French in the form of an account of a mythical country perversely given the name of a real republic on the continent of North America. Noting the absence of the usual landmarks of European history—kings, churches, nobles, castles, dynastic squabbles and sectarian wars—de Tocqueville concluded that the mechanism of American history was different, a vast arena for the smooth workings of impersonal dynamic laws. He approved this lack of romance, drama, and great trends because of the desire to avoid the disruption of revolutions. This view recurs in the influential writings of Henry Adams, the descendant of Presidents who cast a jaundiced eye both on his society and on the consequences of science. That he thought America and China were alike in this sense says much about his insight into both histories. Like the immigrant myth of gold in American streets to explain prosperity, this Newtonian line of reasoning aborted analysis; change simply occurred, obviating any need for study of the prior actions of men and institutions.

De Tocqueville noted the presence of energetic go-getters intent on acquiring wealth, and he met no savants, devotees of abstract thought who required a niche immune from daily pressures in a hierarchical aristocratic society. A democratic society was technologically oriented and indifferent to basic research. In fact there were scientists in antebellum America. Someone like Nathaniel Bowditch, the translator of Laplace, or Joseph Henry at Princeton, might have dented his certitude. Or going forward in time to a school chum of Henry Adams, the mathematical physicist turned idealist philosopher Charles Sanders Peirce was every bit a grand savant. And there were others.

De Tocqueville did not say there would be no science nor great pure scientists in the United States. Applied science would yield bits of basic science—that is, he was of an older view that did not see basic science as logically and chronologically anterior to technology. And by chance a great pure scientist might arise in the American republic. De Tocqueville believed that pure science ideally required the individualism, the drama of an aristocratic order. Although he was wrong in predicting a steady-state country (and Henry Adams in seeing the heat death of entropy), de Tocqueville had grasped an important truth. The western European ideal of a value-free science independent of society would take different forms in the U.S.A. One can argue the extent to which this ideal was realized in Europe; the structured society of its nations, at the very least, permitted the illusion of realization. Great men, ideas, great events existed, at least figuratively, in isolation. In the U.S.A., their analogues were often bathed, if not immersed, in social, political, economic, and ideological realities.

To this day, many American scientists incline to the view that their society is indifferent, if not hostile, to basic science, despite ample evidence of the growth over time of support for research, even research of the most abstruse nature. A proper perspective on this peculiar attitude is a comparison with literature, music, and art. Once the arts also were presumed blighted by the American environment. A democratic society could support popular or vernacular cultures but not high cultures. No one makes this accusation any more about the arts, which now flourish in symbiosis with the popular culture. Only in science is there uneasiness and sensitivity, obviously a phenomenon of great importance.

Early in the 1850s, the aged Alexander von Humboldt wrote approvingly of American efforts in a wide range of physical and biological topics related to geography. Because he wanted Americans to receive greater international visibility, Joseph Henry of the Smithsonian Institution suggested preparing what eventually became the Royal Society's catalog of scientific literature of the last century. Henry originally limited the work to the exact sciences, so he clearly had more than scientific geography in mind. Although the first series of this catalog, up to 1863, disclosed that roughly 5 percent of the authors were from the United States, Europeans were not impressed.

By the end of the century, American activity in the earth sciences loomed large. Though an extension of Humboldt's view, geology reeked of the romance of the Old West, so attractive to generations of manboys on both sides of the Atlantic. It did not hurt that the leader of the geological community in America was John Wesley Powell, whose hand was lost at Shiloh in the Civil War and who had first explored the untamed Colorado. Powell's other role as a pioneer ethnologist only heightened the effect. Although very creditable, the scientific exploration of the West was like the rumored gold in the streets: something waiting there. The effort did not impress those interested in mathematical and experimental fields. Only by 1951/52, did a French geologist, Emanuel de Margerie, produce his two-volume *Etudes américaines*. Earlier, a few works noted that astronomy had become a very active discipline in the United States by 1900.

Shortly after that date, a significant act of recognition occurred. Reacting to the founding of the Carnegie Institution of Washington and the Rockefeller Institute, the Germans founded the Kaiser Wilhelm Gesellschaft. In the years before World War I, a number of German observers commented favorably on the quality and scale of the sciences in the United States, some warning against the threat to German hegemony in research. Apparently, the familiar competition of Britain and France was less worrisome.

One of the Carnegie beneficiaries, astrophysicist George Ellery Hale, added Rockefeller funds and built the great 100-inch telescope at Mount Palomar. That great instrument and E. O. Lawrence's pre–World War II cyclotron symbolized science in the United States to many on both sides of the Atlantic. Such large-scale efforts reflected the nation's natural endowment; great instrumentation reflected the supposed American propensity for "practical" devices rather than the abstruse theorizing of true savants. The attributes of Hale and Lawrence reinforced these stereotypes. Hale was a fabulous promoter and organizer, a veritable Morgan or Rockefeller of science; Lawrence, also a manipulator, went on to invent a television system.

Usually overlooked were scientists of a different mold. A contemporary of Hale's, the great geneticist Thomas Hunt Morgan, used the fruit fly in a series of austere, rigorous experiments. With a passion for scientific exactitude, Morgan shunned empire building and applications. When Hale and Robert A. Millikan invited Morgan to the California Institute of Technology, they initially envisioned a program of science applied to medicine, but Morgan insisted on pure research. Although the United States was a world power in the sciences when Morgan went to Pasadena, a generation passed before sustained, serious historical research began, and it did so then largely because of World War II and its consequences.

This 200-year history is not explicable by any assumption of American singularity. Like the nation, the sciences in the United States are part of Western Civilization; the thirteen rebellious colonies were a provincial outpost of this civilization. It is not wholly accidental that the first college in this country was founded in 1636 in Massachusetts Bay Colony in a village later named Cambridge. Modern science originated in Western Europe and developed an eastern wing in Russia and a western one in the United States of America. Despite the obvious influence of British (and French) science, the most interesting comparisons and relations of the research scene in these former colonies of the United Kingdom are with Germany and the Soviet Union, a point to which I will return.

Because of its origins as a provincial outpost of western Europe, the United States was never properly described as a young country, that is, a country lacking the kind of history associated with the cultures of the European nation states. More precisely, the United States had never experienced a youth consisting of dynastic and sectarian violence, a rigid class structure, and an all-powerful established church. This country was born to adulthood in the sense of having many of the trappings of modernity. To conservatives, the lack of the usual traditions and their detritus was a basic flaw; to liberals and radicals their absence was a glorious opportunity. With some exceptions, both groups were skeptical about developing a high culture of science in a democratic environment.

In 1776, the United States was not a blank slate. Selectively, patterns of life and thought in seventeenth- and eighteenth-century Europe flourished and persisted; in time they became Americanized, diverging from a western Europe undergoing its own reactions to intellectual and economic changes. At the same time, a kind of feedback from these European developments prevented any gross cultural split between the populations on both sides of the Atlantic. In the U.S.A. there was an avid pursuit of the European printed word. Even modest cities of the Mississippi Valley in the antebellum era possessed a good number of books and journals. Young scientists visited Europe, and many went on to study at overseas universities. A Louis Agassiz came in 1846, as did many others in the last century. The German émigrés of the 1930s were simply an instance in a two-century-old process. In

1973 about 6,000 scientists and engineers emigrated to the U.S.A. At all times in its history a minority of the scientists were foreign born.

Scientific Heritage

The colonists in revolt insisted they were only asking for their just rights as Britons, even though most in the U.K. probably did not see the rights in that way. Facing western Europe, American investigators regarded science as part of their heritage; ruthlessly and cynically they brought over the ocean whatever they wanted of this scientific heritage, disregarding any plaints about the European nature of the best of science. Being provincials in the eyes of western Europe, they were often infuriated by patronizing treatment. In fact, American scientists were still vexed by patronizing Europeans as late as the years between the two world wars. Yet, from the very beginning, American investigators wanted nothing as much as the good opinions of their peers in Europe; at the same time many strongly desired to present their results as a symbolic return to Europe for the great gift of science.

Because of this ambiguous relationship with European science, the American scientific community developed a differing pattern of nationalism and internationalism than their overseas counterparts. Let us disregard the fatuous cliché that "The sciences were never at war." It is a meaningless truism applicable to sets of concepts and data, not communities of humans. Real, live scientists usually rise and salute when the flag is raised, especially during wartime. Even in peace, national differences and rivalries influence the behavior of scientists. Being outside the rivalry between the French and Germans after Sedan and not sharing the British anguish that Sir William Perkin discovered the dye but Germany garnered the industry, American scientists could work very comfortably in an international setting. At the same time, national interests were never totally absent.

Another differing pattern arising from colonial roots is the development of a cult of the people, substituting for king, nobility, and gentry. In a post–World War II movie about a hack senator trying for the presidency, William Powell in the title role makes a campaign promise to give a Harvard degree to every American at birth. Mass culture, like dreams, sometimes disclose suppressed beliefs. Just as scientists (and conservative critics) were uneasy about the lack of a proper hierarchy and widely agreed on norms, the American public had qualms about status based on education. Contrary to popular belief, the nation inherited a class system from Europe and still displays a social stratification based on wealth. Contrary to de Tocqueville, quite a number of families retained high status over many generations, in some instances even constituting a patriciate. This is in spite of a significant degree of upward mobility in the past for all groups with the notable exception of those of African descent. Yet classlessness is widely assumed in the form of crowding to a middle position.

Education

Education can provide a status countering classlessness. Traditionally, expertise is highly valued here. At the same time, it is one of the glories of American history that the learned have much power and honor but lack the certainty of an assured niche. When historians turned to the history of the sciences in America, they noted that scientists wanted an elite, meritocratic basis for their group, often rejecting the validity of public intrusion into their affairs. Filled with such historiographic problems, I spent a year in Britain in 1964/65 studying British analogues of my favorite nineteenth-century American scientists. An acquaintance in the Civil Service, now deceased, invited me to lunch at the Athenaeum Club. As I looked around, he said, "This is like your Cosmos Club in Washington." But I knew differently. Just a few days before I had read in manuscript a letter from an officer of the Royal Society to a council member, then little known. Charles Darwin was invited to the Athenaeum for lunch to discuss a matter before the Society. From my readings of many manuscripts, I had already concluded that in comparison to these British scientists, my American "elitists" were raving egalitarians. Even if they had not been infected with republican or democratic principles, a continuing establishment control was not feasible given the size and complexity of the U.S.A.

Just as wealth conferred status over generations, so did education and intellectual attainments. Family groupings of scientists appeared in the last century. For example, modern geodesy was brought to the U.S.A. by the Swiss-American Ferdinand Rudolph Hassler. A generation later the Nova Scotia–born mathematical astronomer Simon Newcomb married a Hassler. Their daughter became the wife of W. J. McGee, the geologist and anthropologist who was one of John Wesley Powell's lieutenants. Charles Sanders Peirce's father and brother were professors of mathematics at Harvard. J. Willard Gibbs's father (of the same name) was a professor of languages at Yale. William James's father was a Swedenborgian philospher, undoubtedly a surprise to his teacher, the physicist Joseph Henry. A present foreign Fellow of the Royal Society is a third-generation scientist.

These family groupings have never constituted an intellectual aristocracy because of the rapid growth of the scientific community over the past two centuries. A founder of the Institute for Advanced Study, the mathematician Oswald Veblen's father studied physics at Johns Hopkins during its golden age; he later taught at a midwestern university. Oswald's uncle Thorstein became a great social theorist. They were a family of Norwegian immigrant intellectuals. A small town in South Dakota produced two notable physicists, E. O. Lawrence and Merle Tuve. The so-called new immigration from eastern and southern Europe by now has yielded scientists of Jewish, Italian, Slavic, and other ancestries. In spite of the persistence and importance of a few families as sources for recruits, the scientific community has never been a closed group. The newest immigrants from Asia are entering the various disciplines. By now about 5 percent of science and engineering doctorates are of Asiatic origin. Fifteen years of consciousness-raising may very well result in an inflow of blacks, women, Chicanos—even more American Indians.

In the U.S.A., scientists are overwhelmingly a middle class group. In spite of individuals from the upper class and some from the lower, becoming a scientist until recently was means of going from the lower to the upper middle class. At present such social mobility seems on the wane. If true, it is all the more important to study that portion of the middle class producing the bulk of the professional classes which provide an essential element of stability in the U.S.A. Relatively little is known about such families in contrast to concern for the genteel culture of the patriciate or the pains and traumas of the poor and the outcast. As scientists are less numerous, they are more accessible to research than other professional groups. Where family influences are lacking, we still do not know why a small minority of middle class youngsters in the past opted for scientific careers, until very recently a choice of limited remuneration and power.

Perhaps the youngsters acted because they, like most Americans, adhered to a cult of knowledge. Here the comparisons between the U.S.A., Germany, and the Soviet Union are important. For more than half of its history, the U.S.A. was unknowingly in a race with Germany. Early in the eighteenth century, nearly 150 years before the U.K. did so, the Prussian state made elementary schooling compulsory and universal for males. With the later expansion of German universities, the British were fated to be overtaken. The Germans trained large numbers for particular niches in their society, each niche with a specific place in an elaborate hierarchy. The professoriat were an elite of elites, having gone beyond the doctorate in their habilitation, a point the Americans did not quite grasp when graduate schools arose under the influence of German higher education. What American scientists did note was the high status of the German professoriat. In Prussia each chair holder was a higher civil servant. American scientists often hungered for the same kind of assured high place in their own nation.

Throughout the nineteenth century, Americans expanded all levels of their education system; in retrospect, they were trying to match Germany. The statistics are not very clear but, apparently around 1900, the time Germans became conscious of an American threat, the U.S.A. had surpassed Germany in numbers at least. Quality was another question. American universities were largely viewed as an extension of the democratic thrust to provide a responsible electorate. Some institutions viewed themselves in a patrician sense, as training cultured gentlemen for service to the society. Typically, even the mass schools regarded themselves as engaged in both cultural uplift and in a service role. The sciences served both purposes. Mass plus quality tended to become the ideal.

A number of examples. In the middle of the large state University of Illinois at Urbana before a library with nearly four million books is a plot of corn planted to remind all of the university's origin as a land grant school for agriculture and mechanic arts. When the University of Chicago opened its doors in 1891, it was a great university. This oft-repeated judgment refers to its scholarly research strength. But its president, W. R. Harper, at the same time launched an extension program of correspondence and evening course. The Professor of Mathematics at Chicago, Eliakim H. Moore, the first pure mathematician elected to the National Academy of Sciences, received an honorary doctorate from Göttingen early in this century. He conceived his authority to extend from research in pure mathematics to supervising the training of elementary school teachers of arithmetic. Professor Moore, the University of Chicago, and the University of Illinois—like many other American individuals and institutions—wanted theory and practice, culture and service, elite and mass somehow combined or in juxtaposition.

The Soviet Union and the U.S.A. are probably the only remaining countries still believing in some form of the idea of progress, a Marxist version in one, an Enlightenment variant in the other. The two countries have curiously parallel histories in the sciences. Both were Humboldtian because of the need to manage their large land masses. Both turned engineering into mass professions of high status. The Russians have won the numbers race in that category, a more serious fact than mere heft of armed might. The U.S.A. lagged behind Russia in pure mathematics, and may still. Russian universities have a much different role in the sciences. Centralization is greater in the Soviet Union, whereas egalitarian instincts limit centralization in the U.S.A. Our society has a primal urge to replicate anything deemed desirable. In addition, strong anachronistic sentiment for rusticity ultimately works against the big. Rusticity is something the Russians shun.

Besides the idea of progress, the cult of knowledge in the U.S.A. has other traces of the seventeenth and eighteenth centuries now largely absent from western Europe. By establishing much of the broad form and content of what is now designated science, the Scientific Revolution defined some concerns as nonknowledge and started segregating the high culture of the sciences from the vernacular culture. But the useful arts partially escaped this exclusion until the nineteenth century. At the same time the casual amateur and the modest practitioner were increasingly displaced by the savant. (Significant scientific amateurism probably survived longer in the U.K. and the U.S.A. than in continental Europe.) The Scientific Revolution gave form to a previously rather loose, amorphous conglomeration; and the Industrial Revolution hardened the new pattern.

Much of this process had bypassed the British North American colonies. The vernacular, the nonscience, the amateur and the applied, survived and flourished after independence when the high culture of science began a slow steady growth. Europeans established a scheme of things in which there was a congruence between social, intellectual, and institutional hierarchies—not a perfect congruence, but enough to avoid many of the problems of the Americans. On the western side of the Atlantic, the heritage of the past and the thrust of historical development did not neatly separate grand savant and practitioner; theoretician and earnest mechanic; the abstruse and the vernacular. They were scrambled together. A persisting tension developed between the scientific elite and their perception of fundamental research on the one hand, and the mass community and their thrust for diffusion of knowledge (often older and sometimes applied), on the other. As a result, the cult of knowledge in the U.S.A. encompassed a research ideal, but not a basic research ideal in spite of all the exertions of many generations of scientists.

In this research ideal was a blurred boundary between theory and practice. Originally, religion provided a basis for pure research. The idea of design presupposing the existence of a Designer readily yielded a drive for research untrammeled by sordid motives. Secularization purged the theological mission, leaving a pure research ideal. At the same time the old, general notion of the utility of knowledge was elevated by this association with a designing deity. All research carried out His purpose and disclosed His will. Only overly rationalistic, sensitive scientists worried about the problems of defining discrete forms of research styles and goals.

As a happy compromise, even before the Civil War, American scientists launched a counterstrategy: to fractionate off basic research by establishing enclaves largely immune to the national tendency to conflate theory and practice. Higher education often served as the site of these enclaves. A notable recent example is the Institute for Advanced Study at Princeton. As it were, the scientists were levying an overhead charge for theoretical science against the gross national product. Many were uneasy about the more common mode of supporting theoretical science, which also appeared before the Civil War. This is to define applied missions so as to necessarily include a basic research component. In the 1920s Harvard's E. H. Hall (of the Hall effect) corresponded with C. J. Davisson (a future Nobel laureate) of Bell Telephone Laboratories. Impressed by Davisson's work, Hall sent the head of the laboratories, H. D. Arnold, a letter praising the company's enlightened support of basic research. Nothing of the sort was intended, Arnold replied; we are simply interested in ways of increasing the supply of electrons. Sweeping mandates like this are widely present in so-called mission-related research. Believing sincerely in their approach, the scientists still argue in favor of broad definitions of applied missions. But many also favor protected enclaves for basic research not subject to the vagaries of economics, international affairs, public health policies, and the like. Yet, it is naive to hope for and expect immunity from the ebb and flow of history.

In spite of $32 billion for research and development in 1974, in spite of over a half million engaged in research and development, in spite of $4 billion for basic research, U.S. scientists are still uneasy about their place in society. There is an overwhelming cult of the nation with dominating symbols—George Washington, the flag, the national anthem, Abraham Lincoln, the Statue of Liberty. Science is not represented among the symbols of the cult of the nation; even technology once had the iron horse and Thomas Edison. And our country has just gone through ten terrible years raising nagging doubts about progress. Largely because of the antiscience quirks of Lyndon Johnson and Richard Nixon, inflation has eroded the real strength of the national research budget.

From the vantage point of my ivory tower, I find prospects encouraging, starting with President Gerald Ford's increase in the budget to make up for some of the slippage from the Johnson and Nixon years. American scientists have always functioned well while loudly complaining. Even before the Civil War, they demonstrated great talents as entrepreneurs, improvisers, and artful dodgers. Nor am I nonplussed by the predicted numbers of surplus Ph.D.'s. Except for about fifteen post–World War II years and perhaps a few years before the Great Depression, there has always been a surplus of scientists on paper. A surplus can act as a stimulus, both on the nation and on individuals.

Whatever happens, the course of the sciences in the U.S.A. will remain an arena of tensions and ambiguities, a historical movement in which prosaic reality turns out to be a symbol, often exhilarating, sometimes grotesque, occasionally a black comedy. It has been a history of trying to have one's cake and eat it. But perhaps that is true of all of the past two hundred years of our national life.

$2_\square$ Definitions and Speculations: The Professionalization of Science in America in the Nineteenth Century

In 1964/65, I spent a year in Great Britain and the Continent ostensibly doing research for a projected biography of Alexander Dallas Bache. While I found plenty on Bache and on Joseph Henry, the real value of the year was elsewhere. I was an innocent abroad studying the Old World, both historical and contemporary. While not formally engaged in comparative history, like a number of my contemporaries in U.S. history I wanted to define my piece of the past in terms of similarities and differences with Europe.

In particular my British experiences influenced my perceptions of professionalization. The increasing role of certified experts was a familiar theme in U.S. history, notably in writings on the Progressive Era. Antebellum America witnessed the birth of a professional scientific community. In our work my colleagues and I savored instances of the clash of professional pride and democratic ideology. In Britain I encountered a truly hierarchical society where proper deference was both proffered and exacted. My supposedly elitist Americans, in comparison, were almost paragons of a populistic egalitarianism.

A few years later the American Academy of Arts and Sciences invited me to speak in the first of a series of conferences on the role of learning in the United States. Clearly favored was a paper on professionalization showing how Henry and Bache purged all those pesky amateurs. But a paper in the traditional vein on the amateur–professional split embroidered with juicy quotations of mutual antipathy no longer appealed to me. Instead I penned "Definitions and Speculations: The Professionalization of Science in America in the Nineteenth Century." It originally appeared in the conference proceedings, *The Pursuit of Knowledge in the Early American Republic* (1976). Omitted is an appendix which is a serious but tongue-in-cheek "numerical excursion" attempting estimates of the size of the scientific community.

One aim of the paper was to counter the anachronistic application of our present arrangements and beliefs to a far different setting in the past. That aim required some countering of the influence of sociological discourse, too often devoid of any sense of historical changes in time. Although greatly influenced by my reading of that literature, I could not accept the view of some sociologists that they were scientists dealing with universal laws invariant with respect to time and place.

By now an appreciable body of literature exists on the development of the American scientific community from its roots in colonial times into the early twentieth century. Not surprisingly, professionalization of science is a major theme in many of these writings, perhaps more so than in the comparable writings on science in other nations. It is a story of how full-time professionals necessarily and inevitably supplanted amateurs, however talented and devoted. From the increasing complexity of scientific knowledge arose the necessity and the inevitability of higher standards of training and recruitment coupled with full-time employment. Accompanying professionalization was an insistance on the self-government of the scientific enterprise under standards arising from within the scientific community. The leading scientists of antebellum America advocated this view of professionalization in their public discourse, and even more openly in their private communications.

Others, then and now, saw the matter differently: as an elite establishing a monopoly of competence, and therefore of jobs, by squeezing out sincere, perhaps promising, participants in the scientific endeavor. Because American historians turning to the study of science in their country were acutely conscious of the importance of the problem of an elite in a democracy, there has developed a unique and extensive body of literature on the American scientific community. Few countries have anything comparable. For France, it is true, there are writings on the role of the artisans in the execution of Lavoisier and the dissolution of the Academy during the Revolution. A rather skimpy historical literature exists on the professionalization of science in Britain,[1] but the replacement of gentlemen amateurs by a gentlemanly professoriat is remote from the problem of an elite in a democracy.

Why is there such a difference between the American historical literature and the writings on European science? The first and perhaps inevitable observation is that historians of European science were overwhelmingly concerned with internal history, while Americans were avidly interested in placing science and scientists in a larger national context. Even the Marxists writing on these matters could not tip the scales. Influencing the Europeans, Marxists and non-Marxists alike, was a second factor: the existence of a hierarchial social structure was usually assumed, with the scientists occupying a specific high place in the structure. Shedding tears over amateurs was then quite inappropriate. In contrast, American scientists and other intellectuals could not assume a fixed exalted position in a social hierarchy. It is one of the glories of the American nation that the learned have achieved great power and prestige without the comforting certainty of a secure position in the national scheme of things.

When historians wrote about the crucial decades, about 1830–1870, in which the American scientific community came into being, we naturally stressed the scientists' expression of a professional ideology with elitist, antidemocratic overtones. Even better, we sought out those incidents in which emerging scientific pride clashed with the misguided sincerity of the amateur, often animated by a Baconian faith. There was the struggle over the Smithson bequest involving that preprofessional body, the National Institute. The Institute itself was involved in maneuvers for primacy with the Association of Geologists and Naturalists, the latter then transmuting itself from a professional to a quasi-professional body—the American Association for the Advancement of Science. This organization, in turn, became the focus of two related incidents in which the leading scientists enforced their standards: the rather pathetic case of John Warner, and the confusion over the publication of the proceedings of the Cleveland meeting of 1853. To these we can add the lawsuit involving the criticism of Ebenezer Emmons's geological views, which smacks of a heresy trial, and the very tangled relations of Matthew Fontaine Maury with the Lazzaroni.[2]

The use of the first person plural in the comments above is deliberate; this is a tradition in which I grew up and to which I have contributed. Yet there is a need to stand back and rethink this matter of professionalization. What I propose writing is an essay of definitions and speculations. In order to produce a coherent analysis of the professionalization of American science, I have chosen not to end my discussion at 1860 but rather to consider developments up to the close of the nineteenth century.

Defining professionalization is a thankless task.[3] Many writers start out by warning that fashioning a definition is very difficult because there exists no ideal profession; one would be better advised to talk about the position of any particular profession on a kind of spectrum leading to an ideal. What definitions do emerge are based upon the three traditional professions of law, medicine, and the ministry, to which are added a number of newer professions, notably the sciences, engineering, and several other occupations requiring some particular body of knowledge. A number of near professional or quasi-professional fields are then brought in for comparison.

Further complicating any attempt at precise definition is the pervasiveness of certain common usages of the word that influence individual reactions. There is the connotation that a professional is someone in full-time employment, as in "the world's oldest profession." There is also the connotation of competence, that a professional is necessarily better than an amateur, as in sports. Finally, there is the notion of a professional being very decidedly white collar, as distinct from manual occupations, however skilled.[4] In America the tendency is to merge all these usages and call any occupation a profession if skill is required. While this usage of "profession" does not coincide with the definitions of academics, it reflects something very important in the American scene. In the United States there is a great tendency for all occupations to strive for professional status, to regard themselves so after the attainment of certain stigmata, and to obtain a measure of recognition from the general public, even from the state and local governments in the form of licensure laws—much to the dismay of the more established professions.

Conventionally, professions are defined as occupations that require special training in a particular body of knowledge or a particular technique or technology. Professionals usually form associations to advance the body of knowledge they serve and to enforce high professional standards. These are particularly important because the common definitions of a profession assume an applied component requiring a service ideal. Quite typically, writers on professionalization talk about clients—either individuals, as in the case of law or medicine, or possibly corporate organizations employing professionals. Increasingly, it is recognized that professions are not exclusively practiced individually but in bureaucratic structures, giving rise to concern over whether the profession can function autonomously. Can a professional act according to the best dictates of a body of knowledge and the ethical standards of the profession when subject to the control of a bureaucracy? Pessimists say no; optimists find hope in the professionalization of bureaucracy. As part of the high service ideal, professionals are presumably not primarily concerned with pecuniary gain but with the symbolic recognition that comes from high performance within their fields.[5]

The lack of clients or the lack of a specific application provides problems to some writers on the role of the sciences as professions. Several writers flatly deny that the sciences are professions in any meaningful sense of the word unless there is a very strong applied component. Lewis and Beer, in one of the few articles by historians on the professionalization of science in the twentieth century, categorize the nineteenth century as a nonprofessional era, when scientists were simply gentlemen interested in advancing knowledge. Professionalization comes to American science when scientists become significantly employed in industrial research laboratories and in government policy formation.[6]

In two articles notable for their parallelism, Shils and Ben-David,[7] who are most anxious to show that basic research is the apex of scientific professionalization, begin with general considerations on how a particular body of men go about finding knowledge but conclude by considering the role of the scientist in society, how and when scientists have the autonomy to practice their professions, and what their relationships with government and industry should be. Research achievements, in their view, arise from the high internal standards of the disciplines. What is assumed is that these standards are both applicable and carried over to dealings with outsiders where all the classic problems of client relations exist but on a national and international level. Specifically, Shils and Ben-David are concerned with showing that research—especially the most abstract and basic research—is *the* highest product generated by the professional scientist. Another aim is to illustrate that, despite the absence of any specific individual or corporate body as the client, the scientist maintains the very highest standards in the performance of his task. Talcott Parsons handles this dilemma by making the university and the academic disciplines rather than the traditional learned professions the paradigm of professionalism.[8]

In this interpretation, ethical problems requiring licensure and regulations are avoided in the sciences through informal monitoring by individual scientists or scientific organizations, seen primarily in the refereeing process of the scientific journals. To quote Shils:

The ethical problems which arise in the other professions where an expert deals with laymen have no equivalent at the heart of science, where many experts confront and scrutinize the results of each other's work as embodied in publications. Professional associations and science do not need committees on ethical practices because there are no defenseless laymen. The disinterestedness of scientists is a disciplined concern for truth which is imposed by the watchfulness of many others equally well disciplined.[9]

What Shils is talking about, of course, is not the scientists' relations with the public but only their relations among themselves. What is noticeable here is the high value placed upon science as a body of knowledge and upon scientific research as an endeavor made high-minded by an involved process of self-policing and moral inculcation carried out during years of training. Self-policing is seen as preferable to interference in science by the state in any way, shape, or form. As science becomes increasingly involved in applied areas and scientists become increasingly involved in public issues, the general populace, at least in America, is no longer inclined to accept the self-policing expertise of scientists without question. One writer even reports the development in the state of California of licensure for geology, an old established field of science.[10]

One of the effects of the last two decades of science policy disputes in the United States has been to convince many lay people that, whatever their intellectual expertise, scientists are simply another pressure group. It is widely understood that they have their own viewpoints and interests, and that these do not necessarily correspond to any other group interests in the population or indeed to that of the population at large. That this growing recognition occurred at the time when American historians were beginning to investigate seriously the antecedents of their scientific community is no accident. It helped shape our research. Perhaps this was unfortunate, because, in fact, the antebellum scientific community had relatively little power to exercise.

In this essay, I would like to offer two definitions of a profession— one sociological and the other operational. The first is taken from the writings of the Englishman Geoffrey Millerson: "It is a type of higher-grade, non-manual occupation, with both subjectively and objectively recognized occupational status, possessing a well-defined area of study or concern and providing a definite service, after advanced training and education."[11]

Note the complete absence of any rhetoric about ideals. I am willing to accept research and teaching as well as many other activities as being professional activities of scientists, even though this may not fully accord with Millerson's definition: What I propose to use is another rather simplified operating definition. Today, in the twentieth century, we define a scientist as anyone who meets at least two and usually three of the following characteristics: (1) he or she possesses a suitable educational degree (e.g., a Ph.D. or M.S.); (2) he or she is in an occupation or has an occupational title that requires the possession of that degree or at least of the knowledge supposedly entailed by obtaining that degree (e.g., professor of chemistry); (3) he or she belongs to a professional association of people who supposedly have the knowledge for which the degree was acquired and who, as a group, are presumably interested in advancing it (let us say, the American Chemical Society). While these three characteristics can be used quite successfully now and even as far back as 1900, they are not usable in the period prior to 1860. The situation prevailing then was so different that one simply cannot use the definitions of professionalism that appear in most of the current sociological literature. There were very few degrees one can compare to those awarded today. Professional societies were very rare. And only by stretching terms can many positions be categorized as "scientific" by today's standards.

It will be necessary for us to think of another kind of situation that will define the total scientific scene. I shall do this by trying to offer a model of what existed prior to the development of professionalism as we now know it and to suggest in very general terms what happened from then until approximately 1900 when professionalization was clearly rampant. In developing this model, a principal theme is the presence of a real confusion between certification and accomplishment. Most of the sociological writings stress the attainment of intellectual mastery over a body of knowledge, sometimes described as theoretical knowledge. If we look into the historical literature, it is usually assumed that the real professionals have a greater degree of mastery than their opponents, the amateurs. It is further assumed that this mastery is associated with genuine intellectual accomplishments, as it is, indeed, in many cases. Yet reflection on Darwin should cause some second thoughts; accomplishment in his case did not require formal certification. And to paraphrase Walter Cannon, if Darwin was not a professional in terms of accomplishment, who was?[12] What we assume today about degrees and the like was not true in the past.

What is most frequently involved in professionalization, past and present, is certification, that is, some way of determining that a particular person is indeed qualified to practice a given profession. Certification, a process beyond getting a degree or passing a qualifying examination, reoccurs at successive stages of a career. Certification is the nonintellectual aspect of professionalization, the seeking and attaining of place, income, and influence; it is part of the entire social context influencing scientists. The literature, both historical and current, implies that the object of professionalization is the accomplishment of high goals and ideals, whether it be the advancement of knowledge or staving off death and pain. In the case of science, Ben-David rightly points out that the leading scientists were originally charismatic figures, and the attainment, for example, of a professorship in Germany was to some degree an assessment of charismatic quality and thus a guarantee of accomplishment. Something like this happened in other countries of the Western world, even the United States. By now Ben-David notes, "A professorship in the United States becomes little more than the best-remunerated stage of a normal career" rather than necessarily a mark of exceptional gifts.[13]

There are great virtues in this position; most people are not leaders or geniuses. Most scientists do not publish, because most do not do research.[14] Of those who do publish, most publish very little.[15] And authorship may occur within a brief span of professional life. Judging by the recent literature, of those articles published, most are either not cited at all or are cited with extraordinary infrequency.[16] (The mere fact of noncitation, however, does not necessarily mean that an article was not seen and used.) To paraphrase the sociologist E. C. Hughes, training for and the doing of research is often for the purpose of getting a post not requiring research.[17] In sum, research results of a few cast a noble glow over the community.

If most never publish or publish little, the relationship of the leading scientists to the bulk of their fellow professionals becomes a very interesting problem. In the pre-1860 period these assertions about nonpublication and noncitation were more or less true, insofar as we can judge by rather useful but imperfect statistics.[18] What was different was the occupational structure, and indeed the rationale of the whole system. We may define our past scientific community as consisting of three groups:

1. *Researchers* are individuals characterized by a single-minded devotion to research, resulting in an expertise yielding an appreciable accomplishment by past standards certainly and in retrospect in some instances. Most, but not all, are in scientific occupations.

2. *Practitioners* are individuals wholly or largely employed in scientific or science-related occupations. Those who publish are less prolific and less significant in terms of accomplishment than researchers.

3. *Cultivators* are best described at this point rather than defined. In what follows, professionalization is viewed from their standpoint, not from the perspective of the leading scientists, the researchers.

The Cultivators

Unfortunately, the pejorative connotations associated with the word amateur prevent my using it. I say "unfortunately" because I would like to preserve the etymological implication of the word, that is, "lover of," because it says something important. "Cultivator" was chosen deliberately because of the relation of the word to the terms "cultural" and "cultivation." More specifically, it was actually used by Sir David Brewster to describe the mass of people in attendance at early meetings of the British Association for the Advancement of Science.[19] This crowd was not of the "professional" eminence of scientists like Brewster and the other principal leaders of the Association. Using "amateur" is also undesirable because over the years it has become associated with another term to form the expression "gentlemen amateur." With this goes another connotation—that in the past there was an appreciable body of men in the United States and elsewhere who were not professional scientists but who had a sincere interest in the knowledge of science, and who somehow or other participated meaningfully in the scientific endeavor. What has occurred are some very loose extrapolations from very special cases, like Thomas Jefferson—a very distinguished cultivator indeed.

People who extrapolate from the exceptional case do so because of confusion of terms more than anything else. What they are reacting to is a specific culture (if I may use that murky term) in the United States that one can describe as the culture of polite learning. This could be acquired by attendance at one of the colleges or by self-study. Polite learning included some knowledge of mathematics and various scientific subjects. Possessing polite learning in no way implied being a significant amateur or cultivator of science any more than entering today's American educational system—high school or college—and acquiring a certain standard core of information involving scientific subjects makes one a scientist. Early in the last century there existed in the United States, as well as in the countries of western Europe, what we can call "vernacular" cultures. Most of the people did not possess polite learning. Beyond the illiterates there were, of course, a substantial body of people who had acquired the basics of reading, writing, and ciphering—part of the vernacular culture.

The cultivator was a particular kind of person who possessed learned culture. For various reasons the cultivator had a more specific knowledge in the sciences. Unlike others in the learned culture, the cultivator actually applied his knowledge in some kind of activity. A man who was simply part of the learned culture might very well become a nominal member of one of the local societies with an interest in science. By the Civil War, if not earlier, a certain amount of social prestige was attached to membership in learned societies. When, prior to the Civil War, the American Association for the Advancement of Science came to his city to hold its meetings, such a man might give money to the support of the meetings, serve on the local arrangements committee, and in many other ways throughout his life show himself appreciative of science. In doing this he was simply reflecting, perhaps, the pervasive belief in progress, a progress that arose from science. Science was a good thing, and he was well aware of that, but he was not using science in any way.

A cultivator, on the other hand, would not only join one of the local societies but, indeed, come to meetings with some degree of regularity. He would help the more advanced devotees of particular disciplines. He might very well participate actively in the meetings. A cultivator might take an active role in the management of the society. He might be moved by the infectious enthusiasms of his friends and his own eagerness to deliver a paper—maybe two. Perhaps stimulated by praise or by his own feelings of satisfaction, this might even be sent on to Benjamin Silliman, Sr., in New Haven if the author's friends, whose opinions he respected, so urged him.

In discussing this learned culture, I am carefully avoiding the trap of assuming that there was once one culture. I think that there were always at least two—the learned culture and the vernacular culture. In the learned culture there were ways in which the scientific and the nonscientific interests tended to blend and perhaps even interact at times, just as today. The use of terminology shows how cultivators and noncultivators sometimes viewed the relationship. Early in the nineteenth century, for example, Benjamin Vaughan referred to people in the learned culture, scientists and others, as literary gentlemen,[20] and as late as the very brief heyday of the National Institute, Joel Poinsett referred to literature as the "vehicle of science."[21] In the early years of the Albany Institute, however, the differentiation of science and literature was quite clear.[22] When John C. Spencer asked Francis Wayland to contribute to the 1844 National Institute meeting, he described the Institute as engaged in "the effort . . . to establish a Confederacy of Science and Literature for our whole country,"[23] implying an awareness of a disjunction to be bridged. Periodicals and societies sometimes had the words "Literary and Philosophical" in their titles early in the century. Interestingly, as late as 1900 the Census Bureau had an occupational category for "Literary and Scientific Persons,"[24] so even at that date a relationship was presumed. The presence, for example, of an extensive nonmathematical review by Nathaniel Bowditch of technical astronomical works in the *North American Review* in 1825[25] might indicate a fusion of science and literature within a single learned culture before the rise of professionalism, but such instances are misleading.

By examining a number of journals in Joseph Henry's library and other titles found by association, the position of science in the learned culture is somewhat clarified. These journals are mostly from the Mid-Atlantic states before 1840, and, judging from the sample, science had an honored but minor position in the learned culture. At most, 30 percent of a volume might be scientific, as in volume 4 of the *New York Review* (1839); more common are volumes containing 5 or 10 percent scientific matter. A Philadelphia periodical, the *American Quarterly Review,* ranges from 10 to 20 percent in the years 1827 through 1830, but the percentage drops to zero or a nominal 2.5 percent when Carey and Hart cease being the publishers. The *Belles-Lettres Repository and Monthly Magazine* of New York has sections on mathematical problems and meteorological reports in 1821 but little else of a scientific nature.[26]

This unscientific sample says much about the place of science, even beyond the sparseness of content. To use modern terms, the articles were often reviews or secondary accounts, rather than original science. Natural history, surveys and explorations, and reviews of textbooks are common here. By this period the article rather than the monograph was the common vehicle for dissemination of new research, usually in a scientific periodical. A smattering of secondhand reports or literary reviews did not really convey much of scientific developments to the so-called gentlemen amateurs. If any trend is discerible, it was a diminution of scientific content in these periodicals. A cultivator would have to turn to the scientific literature to maintain any serious interests in research.

Cultivators might go on to become practitioners of science—that is, work at science full time—and, indeed, become fairly eminent research scientists. There are a few cases, one of which I will discuss later. The cultivators were quite often concerned with their own self-education, rather than the increase or dissemination of new knowledge.[27] Not being interested in publishing, the cultivators tended to regard what they did as a source of pleasure—a form of relaxation, a hobby.[28] It was even a sport.

The groups they formed were neither organizations devoted to specialized subjects nor professional associations that were implicitly and explicitly certifying bodies. What was often encountered were social groupings, even though a few members were scientifically proficient. Geoffrey Millerson refers to these groups in Britain as study associations, noting that in the course of time some evolved into truly professional subject groups.[29] Even today a few of the British organizations still retain a mixture of competent specialists and amateur devotees. Study associations, according to Millerson, were the vehicle by which men in the middle class raised their social status. I see few signs of that in America, although in time many of the old general learned societies became prestige groups in their localities. Even the more specialized local societies that were interested in natural history tended in time to have a degree of social prestige attached to them.

I do not use the term "social grouping" in any pejorative sense. I think the societies enabled certain people to do research and stimulated others to aid science. Indeed, a recent British article argues in a very ingenious and forceful manner that the Royal Society's origins are not to be found in any Baconian drive for utility, nor in any particular attribute of a Puritan, Presbyterian, or Anglican bloc, but in the desire of a congenial group of men with similar tastes to meet and to share these interests.[30]

What could a cultivator do? What did cultivators do? W. J. Reader in his book on the rise of professions in England cites a work early in the last century that advised gentlemen going to the country to make and use scientific instruments.[31] This advice was taken, as it were, by a few Americans who also made and used instruments of various kinds. We have no idea how many cultivators had telescopes to watch the heavens, but we know some actually published their observations. There were also individuals who delighted in solving mathematical problems.[32] Abiel Holmes, the father of the poet-physician Oliver Wendell Holmes, kept meteorological journals, as did countless other individuals in Britain, the United States, and on the Continent. When Joseph Henry set up his telegraph-linked meteorological service, these men manned the various stations in the network and later in the century also provided readings of barometers and thermometers to the Signal Corps's weather service and its successor, the Weather Bureau of the Department of Agriculture.

Then there were the ladies and gentlemen who filled the lecture halls to hear talks on chemistry by Benjamin Silliman, Sr., and countless itinerant lecturers like E. L. Youmans. One wonders what they actually received from all those varied talents, but some certainly were moved to buy chemical equipment and to delight in getting reactions, spectacular and otherwise. People interested in chemistry sometimes applied this knowledge to mineralogical analysis. Collecting minerals was a widespread scientific activity, and the identification of minerals was of interest both to serious scientists and to casual cultivators of the field. As one leafs through manuscript collections in natural history, there is a seemingly endless procession of collectors of flora and fauna, as well as of geological specimens. Judging by the subject distribution of the articles published by Americans and listed in the *Royal Society Catalogue,* it is safe to say that almost every branch of science practiced in the first six decades of the last century encompassed matters of concern to cultivators of science.

To be more specific, let us consider some events in the life of Joseph Henry. On November 25, 1828, in a meeting of the Albany Institute at which Henry spoke, Major General Erastus Root delivered a talk on two different topics—problems of the design and use of thermometers, and an interpretation of the solar spots of 1816.[33] General Root (1773–1846)[34] had gained his military title by virtue of his high rank in the state militia. In the year he spoke, as well as in 1827 and 1830, he was the Speaker of the New York Assembly. Presumably he attended the meeting of the Institute while the legislature was in session in his capacity as a corresponding member. Root never published these papers, nor any other paper I can find. In 1795 he did produce an introduction to arithmetic for common schools that went through additional editions early in the nineteenth century.

In 1837 Henry had another encounter with a politician who had scientific interests or pretensions. While visiting Britain, he was perturbed to discover that the American Minister, Andrew Stevenson (1784–1857), was planning to make a few remarks on the subject of light to the physical section of the British Association. As he later wrote, "I feared he would have made himself ridiculous . . . at this he appeared somewhat offended." Stevenson however made amends, as Henry put it, by a very nice address at one of the dinners.[35]

Andrew Stevenson was a former Governor of Virginia. He served as rector of the university in that state upon his return from Britain, but, as far as I have been able to tell from the usual sources and a fairly good biography,[36] he did not have any noticeable interest in science. His biographer mentions that while at William and Mary at the end of the eighteenth century he received a good standard education, which included mathematics and natural philosophy. Root may very well be a cultivator, but Stevenson hardly seems to have made the grade. One might characterize him as one of the literary gentlemen possessing polite learning who did not realize that professionalism was emerging. He might very well have been one of the men Henry and Bache complained of when they talked about ignorant politicians interfering in scientific matters.

Yet Henry's position on this is not at all unequivocal. If we look carefully at other instances in his life, we find him hobnobbing with rank cultivators. For example, as is well known, Henry had no desire to yield certain points to the National Institute, and even toward the close of his life he was making acid comments about it.[37] Nevertheless, in 1849, 1850, and 1851 he was successively elected as one of the three vice-presidents of the organization, declining reelection the following year. In 1855, when the Institute was nearing extinction, he was elected vice-president again but declined reelection.[38] I do not interpret this as a subtle move on the part of Henry to keep an eye on the National Institute to prevent it from gaining an advantage in any matter of interest to him as secretary of the Smithsonian Institution and as a strong proponent of the development of a proper scientific community in the United States. I think—and here I speculate—that he regarded the National Institute as one of those social societies that could and did serve a real purpose in providing education and stimulation in science. He objected, however, to the Institute pretending to be a proper scientific body, responsible for genuine research tasks. In a later address Henry very clearly indicated the distinction.[39] His position was neither ambivalent nor anachronistic.

In the post–Civil War years Henry actually wrote a clear and revealing statement of advice to people desiring to form local societies with scientific interests.[40] To me this indicates that he did not conceive of a situation in which cultivators would be wholly supplanted. He recognized that the Smithsonian Institution used these individuals not only in the weather service but also as aids in acquiring specimens and in the exchange of publications and information. In the 1850s—the exact date is not too clear to me as yet—Henry joined with other researchers, practitioners, and cultivators in the city of Washington to form a club that met weekly even during much of the Civil War and persisted for years after that conflict. Its members were not young men starting out on their careers but included some very eminent individuals. In the post–Civil War years, a member who participated with delight, according to his recollections, was the Secretary of the Treasury, Hugh McCulloch; his interests seemed largely social, perhaps partly educational.[41]

Consider, also, Charles Nicoll Bancker (1778[?]–1869). Originally a New Yorker, Bancker moved to Philadelphia and made a considerable amount of money first in importing and later in insurance and mortgage underwriting. He became a member of the American Philosophical Society, even serving on a few committees to evaluate submitted publications. His eulogist in 1869 referred to him as "a man of general scientific tastes and attainments" who made his instruments available to public lecturers on chemistry and natural philosophy.[42] From the surviving Bancker correspondence in the library of the American Philosophical Society we know that Henry at Princeton did borrow instruments from Bancker. His heir put these up for sale in 1871; there were 787 cataloged items, although the number of individual pieces must have been much larger. It is an extraordinary collection, and a letter of Henry's, partially quoted in the catalog, states that with the addition of a few of the most recent pieces it would be a complete set of apparatus for the physical sciences.[43] Indeed it would be. I have no idea why a Charles Nicoll Bancker acquired such an enormous body of philosophical apparatus. The catalog of his private library published two years previously contains a wide range of books from all fields, but only a modest number of scientific works.[44] For some reason Charles Nicoll Bancker simply collected all of these instruments. As in the case of Thomas Jefferson, one should not infer from this evidence that nineteenth-century insurance executives had significant interests in the sciences.

Another man who was part of the wide circle of Henry's acquaintances was Joseph G. Cogswell (1786–1871).[45] A Harvard graduate, Cogswell is perhaps best known as the first head of the Astor Library. His adult career seems devoted to nonscientific pursuits. At Harvard he was close to the natural philosopher John Farrar and apparently had considerable interests in mineralogy and botany. In 1819 he did the conventional grand tour of Europe, visiting not only Goethe but Gauss, and in many other ways displayed interest in the sciences. As he puts it, he was "foolish enough" to pick up an unearned Ph.D. at Göttingen. His memorialist plays down Cogswell's later scientific interests, but Cogswell was not only Librarian of Harvard but served as Professor of Mineralogy and Geology at the college (1821–1823).

Shortly after his return from Europe, Cogswell joined George Bancroft and taught in an experimental school at Round Hill, where, incidentally, one of the instructors was Benjamin Peirce. The 1826 prospectus for Round Hill stated: "A very considerable proportion of time is assigned to the Mathematics. We consider the study of them in connection with the languages as essential to the best discipline of the mind. The natural sciences are pursued rather as a relaxation, and to quicken the powers of observation."[46]

One student at Round Hill was a precocious but difficult youth from a wealthy New York banking family, Sam Ward. At first Sam Ward did not seem particularly promising in mathematics. But with a little study he became extraordinarily proficient—so much so that, instead of joining the family banking firm, he went to study with Nathaniel Bowditch, which meant, of course, working with Benjamin Peirce, who was then helping Bowditch with his translation of Laplace.[47]

In 1832 Ward became a member of the Examining Board at West

Point, and so impressed the faculty that two friends of Joseph Henry, Charles E. Davies and Captain Edward Ross, acting on the authority of Superintendant Sylvanus Thayer, offered him an assistant professorship in mathematics. By this time Ward had already published a little periodical containing mathematical problems that he proceeded to solve, an American edition of an algebra text, and two review articles in the *American Quarterly Review,* one on probability and the other on Locke. Faced with a concrete offer of a teaching job, the eminent status of a practitioner of mathematics, and the certainty of parental disapproval, Sam Ward convinced his father that he needed time to think, perhaps to get further training, and would therefore require a trip to Paris, which he described at the time he arrived as "the City of Sin and Science." While in that city he proceeded to buy the library of Legendre. On his return, Sam Ward abandoned any idea of teaching or full-time practice of mathematics and joined the family firm.[48]

This was the start of a very spectacular career as a bon vivant and lobbyist. Sam Ward is best known in American history as the "King of the Lobby," and for the possible role that he may have played in the "Mystery of the Public Man." And yet, one wonders about his mathematics. Ward's biographer asserts that he only used the knowledge of probability to further his favorite interest, determining what card would come up next. However I note that his papers in the New York Public Library contain collected autographs of eminent scientists. Would a careful examination indicate that he maintained his youthful concern for mathematics so we can consider him a cultivator and not merely the sinner brother of Julia Ward Howe? When J. J. Sylvester resided in New York City after leaving the University of Virginia during his first stay in America, Sam Ward was a good friend.[49]

Quite a different case is presented by Lewis Morris Rutherfurd.[50] He was clearly an amateur in the sense of pursuing an activity for the love of it and not being paid in any way for his work. Rutherfurd was a Williams graduate who came from a fairly well-known American family and married a relative of the Stuyvesants. Interestingly, he followed the suggestion of the English author cited previously about building devices in his country estate—the country estate being the traditional Stuyvesant acres in what is now the Lower East Side of Manhattan. There he became a pioneer of spectroscopy and astrophysics. In the period up to 1860 Rutherfurd published very little, and even if we add to it the publications occurring from 1860 until his death, the total is not very large. Nevertheless, his life is clearly one of single-minded devotion to research and the achievement of considerable expertise and accomplishment, so I am inclined to classify Rutherfurd as a researcher. Strictly speaking, he does not belong in a section on cultivators.

Rutherfurd had a number of interesting attributes. His friend and memorialist, Benjamin Apthorp Gould, reports that Rutherfurd was on the *America* when it won the cup now bearing its name.[51] I suspect that the spirit of sports and competition animated Rutherfurd but not simply in the sense of winning. In yachting, for example, attention must be paid to all sorts of details of the construction, design, and outfitting of the boat, as well as the details of the actual handling of the yacht. It is this attention to technical matters, to details, that characterized much of the work of Rutherfurd on diffraction gratings and on the application of photography to astronomy.

Rutherfurd had another characteristic quite often found among cultivators—the reluctance to publish. It was not important for him to publish. He did not need to publish in order to earn a living, to get a job, or advance his position in some bureaucracy. Being independently wealthy, he sought only the satisfaction of good workmanship and of knowing his own achievements. Again a point is the analogy with yachting on the scale of the America's Cup races. There are few of these events. One performs rarely and has to win rarely. If one does well, win or lose, it is very creditable.

Rutherfurds were, however, all too rare to provide the foundation for a scientific community in the mid-nineteenth century. Men of a different stripe—the practitioners—would dominate the scientific scene. But even in 1888 G. Brown Goode could report the existence of about three thousand cultivators.[52] And as late as 1940 the American Philosophical Society issued a publication for the benefit of "laymen scientists."[53] Long before that time, however, the cultivator had nearly vanished from both the research scene and the institutional fabric of science.

The Practitioners

In choosing the term "practitioner" I deliberately want to establish a parallel with the established professions of law, medicine, and the clergy. Practitioners have employment in which they, some way or other, use their scientific training, or at least achieve their positions and salaries by virtue of presumed scientific competence. There is no trouble in differentiating practitioners from cultivators. The latter are simply not usually remunerated for whatever scientific work they do. Some of the practitioners do indeed publish, while others do not, giving rise to an important problem in differentiating researchers from practitioners. The distinction on one level between a practitioner and researcher is that the latter performs more research and of a higher quality than the researching practitioners.

The Elliott statistics enable us to confirm what some people understood from less precise sources of information.[54] There were in the United States in the period up to 1860 a fair number of men who earned their living in the practice of science, and there were two leading sources of employment—the most obvious being the government, primarily the federal government and to a lesser extent the states; the other being the colleges and universities. Indeed, the central point of Daniels's book on Jacksonian science, the Beaver dissertation,[55] the Elliott dissertation, and numerous other works is to stress the role of the professoriat in the development of science. A large percentage of American scientists went to college or obtained some degree or training beyond the secondary school level, and a surprising percentage of these ended up teaching. In 1835 Henry made a distribution list for a reprint of an article of which he was particularly proud;[56] aside from a few old friends from Albany and three Princeton trustees, the list consists largely of men from the American professoriat: Farrar at Harvard, Renwick at Columbia, the elder Silliman at Yale, and Bache at Pennsylvania.

The problem with the professoriat, as well as the government practitioners, from the standpoint of the development of research as a professional activity, lies in determining how much research was done within the institutional framework. It is not enough to cite, as Daniels did,[57] that there were twenty-one scientific jobs in 1802, by which he clearly means academic posts: most of these men spent the greatest part of their time teaching rather than conducting research. Somehow or other we must convert figures on practitioners from counts of articles to something like a full-time equivalent measurement. Even today, less than 40 percent of holders of Ph.D.'s are engaged in research and development, and no more than 30 percent of all scientists employed in educational institutions are involved in such activities.[58]

The great trend in science in America from 1820 to 1920 is not the growth in accomplishment, considerable as that was, but the near extinction of cultivators and the tremendous expansion of the number of practitioners. By 1900, perhaps, the real problem that animated many of the scientists agitating for increased research output was how to differentiate themselves from practitioners and technicians, as well as from physicians and engineers. The young Linus Pauling and Warren Weaver both took engineering degrees as undergraduates because they were initially unaware of science as a career.[59] Scientists were often animated, I suspect, as much by the desire to avoid a lower place on the occupational totem pole as by any intellectual stimuli.

During the nineteenth century the United States generated a large number of posts for practitioners ranging from the marginal to the first rate. In France someone like Gay-Lussac could occupy multiple posts, a practice known as the *cumul*. This practice was quite common, and young French scientists in the pre-1860 period and afterward often found themselves waiting in the hope that an opening would come. In Britain there were very few academic or other posts for practitioners. One particular possibility that comes to mind was the Royal Military

Academy at Woolwich, which employed Sylvester and others. One had to be a gentlemen amateur to practice science in Great Britain; it was often an enforced necessity. Even some of the chairs at the universities paid little to their holders.

In Germany the situation was far more complicated because there were many more university posts than in either Britain or France, and good ones. In the United States there were many chairs at the college-university level, but they were not good posts in comparison to professorships in the leading German universities of the mid-nineteenth century. I think the German situation was probably far better than we realize because there were quite a number of other institutions employing scientists. We know something of Freiburg, the mining academy, but all too little about agricultural research stations, medical research institutes, and various kinds of other organizations. The German situation, at least in the first three-quarters of the nineteenth century, must have been far better than that in any other major country in the West.

The posts in the United States, however, made it possible for people of modest income to find employment and a lifetime occupation as scientists. They encouraged the development of a large body of trained, competent, middling-level scientists. I say this not in any pejorative sense; many of these scientists were far above the average and had quite creditable records. I use the term in contrast to words like "genius" and "charismatic." In a country like France, Germany, or Britain, it was assumed (and may still be) that major posts were awarded after a very intensive sifting out and that the award of the post was a recognition of eminence, an acknowledgment that a man had become, as Ben-David put it, a charismatic leader of science. The American assumption is quite different, namely, that genius is somehow taken care of but that society must concern itself with providing for myriad needs by widening the number of professional posts. Distinctions between professions are minimized, and gradations of rank within professions are blurred. Not that these differences are not known and felt, but public policy and the ways of the society often combine to overlook them.

This trend goes back quite far in American history. In 1827 the Board of Regents of the University of the State of New York, under the prodding of the state legislature, quietly adopted a policy that placed scientific and technical occupations on the same level with the older learned professions insofar as state support was concerned.[60] I have characterized this trend as "occupational egalitarianism," meaning that it was assumed that all vocations were to be treated equally in the eyes of the government, provided they were based on a suitable level of education and training. In the United States occupations would be encouraged by public policy and by social pressures to become professions, to acquire some kind of a learned component, and to be considered the equal of even some of the older prestige professions. This did not mean that in actual practice each profession received equal shares of power and money. On the contrary, American society has been, from its very beginning, one in which there were great differences in these shares and in the deference accorded to groups.[61]

Lest anyone still harbor a confusion with law, medicine, and clergy, let me state that, in the course of a considerable amount of reading of manuscripts and printed material in this period, I have no recollection whatsoever of seeing any scientist cite the three established learned professions as models to be followed—or to be avoided, for that matter—in the development of science on a specialized professional level.[62] This is by no means surprising. The medical profession in the United States in the period immediately preceding the Civil War was not in the highest repute. Some very eminent scientists did have medical degrees, such as Asa Gray and F. V. Hayden. One can speculate that men interested in science acquired medical degrees because that was a good way to earn a living prior to the full development of scientific occupations. Perhaps an even stronger reason was the fact that acquiring the M.D. was one of the better ways of receiving scientific training in the period before the development of science teaching in the liberal arts colleges, in the specialized scientific engineering schools, and in graduate schools. After about 1840 the number of men who became scientists by way of the medical school dropped markedly.[63] William Harkness, a post–Civil War astronomer of repute at the Naval Observatory, had a medical degree, but he is an exceptional case.

Most medical men, now and in the past century, are conscious of belonging to a very distinct tradition. They might refer to themselves or be referred to as scientists, but very few of them did any research, medical or otherwise. In the latter decades of the last century a relatively small number of medical men with scientific tastes would turn their talents and interests to clinical subjects and to related areas of basic science. Some of these men can certainly be called scientists and counted in whatever enumeration one makes of the scientific community, but the medical community in the United States has always been a quite distinct body of men with a strong sense of belonging to a specific tradition. Relatively few were part of the emerging scientific community in the mid-nineteenth century.[64]

As to law, my overall impression is that scientists tend to find the law quite alien to their way of thinking. It was thought "unscientific" to take one side or the other for money without considering on which side truth may lie. Law simply did not appeal to many people in the scientific community as a model for their disciplines.[65]

The clergy were certainly held in fairly high repute by many scientists during the early decades of the nineteenth century. Henry was for many years a trustee of the Princeton Theological Seminary and had

clerical friends. Nevertheless, I find in the record no particular awareness by scientists that the clergy constituted a learned group whose
practices should be emulated or shunned. There was a growing feeling
on the part of some clergy and some scientists that their interests were
antithetical, but this feeling did not become serious until much after
Darwin. By that time the scientists were on their way to establishing a
professional existence and the clergy was involved in a complicated
series of adjustments to a post-Darwin industrialized America.[66]

During the antebellum period, engineering did not constitute an organized profession. It was emerging, just as the sciences were. If anything, it was the scientists to whom the engineers pointed, rather than
the other way around. The more advanced and well trained were inclined to describe themselves as representing a scientific and learned
tradition. Even the more practical men in the group would stress the
role of knowledge, although their definition of science in an empirical
sense was not at all in accordance with the tastes of a Bache or a
Henry. The engineers were not a group scientists would emulate in
setting up their own organizations.

What the scientists did in establishing their institutions was to rely
upon a knowledge and an understanding, almost mythological, of the
history of science, of the development of particular institutions, and of
the role of various great men. They would point to Europe, citing the
Royal Society, Sir Isaac Newton, Lavoisier, Laplace, the Paris Academy, and the École Polytechnique. Later in the century, the German
university was singled out.

Being based on a mythology, all the published and unpublished discussions on professionalization—all the literature, the essays, the letters exchanged—assume that the entire goal in developing science in
the United States was to yield American research on a scale comparable to that of the leading European countries. There was a real desire
to make a contribution commensurate with what many felt was America's greatness; to play a larger role in Western civilization; and, on the
part of some, to repay America's debt to the culture of the Old World.
The proponents of professionalization quite sincerely confused accomplishment with certification. If scientific posts went to proper scientists, accomplishment would follow. But most of the people involved
and most of the posts were not of the kind that would necessarily provide much opportunity or incentive for notable research. There would
be a few scientists of considerable talent, a few we can even describe as
great. Usually, most would be practitioners, just like today. Advanced
degrees and professional societies made no difference whatsoever. Most
of the practitioners were necessarily concerned with teaching, administration, applied developments, and other essential, grubby tasks.

Looked at from the standpoint of the practitioners—aided and abetted
by the researchers, who were their friends and teachers—the plethora
of posts was a consequence not of specialization and the expansion of

knowledge but of increased social participation. The growth of specialized and graduate schools was as much a means of certification as a means of increasing research accomplishment. Diminishing the general liberal arts classical course was exchanging the old learned culture for a practitioners' culture justified by a research ideology. The number of publications increased by 1920 because of pressures from both the ideology and careerism.

The Researchers

Despite differences with the cultivators and the practitioners, the researchers—the leaders of the emerging scientific community—maintained their hegemony. They managed to overcome the strains of the differing goals of accomplishment and certification, as well as the problems occupational egalitarianism presents to an emerging status-conscious group. Most important, although failing in their specific aims, the researchers were largely successful in their grand goal of making the United States a great scientific power. Complaints were also blunted by the real openness of the community. Despite the few incidents, what strikes me in the pre-1860 period is the surprising ease of entry. The researchers were most conscious of being a very small group, of being surrounded by other groups, who were perceived as indifferent and perhaps hostile. They were anxious to expand research. If anyone seemed genuinely interested and had some ability, they were usually eager to promote entry into their community. The letters of Henry, Bache, J. D. Dana, Asa Gray, and James Hall, as well as others, show how often young men came to these scientists for aid in entering the field. In many cases the researchers did their best with all too little resources. In these instances the leading scientists were trying to further the achievement of more significant scientific output.

Further blunting possible hostilities was the fact that the elite researchers were not at all unified on every issue. Henry had reservations about aspects of the Lazzaroni activities; others were indifferent or hostile to the group around Bache. Although an elite intellectually produces a disproportionate share of research in both qualitative and quantitative terms,[67] one can doubt their influence in every sphere and not assume endless occasions for clashes of interests. Elites are, after all, largely self-annointed. Accounts of their power usually stem from themselves and from bitter opponents, hardly objective sources. An elite is best described as a group forever surprised by the course of events; we need add, perhaps unkindly, that as the course shifts some rush frantically to the new front line.

The continued dominance of the researchers is simply explained. All three groups—cultivators, practitioners, and researchers—had a high degree of agreement on essentials. This agreement was rooted in both the sciences performed and in the ideas of a scientific lifestyle. The content of science was received from scientific fathers, the leading scientists of the previous generation who had brought the state of the art to the point perceived by all three groups. But the lifestyles of the scientists came from their scientific grandfathers. I do not mean this in a literal sense, but there was at least a generation jump in perceptions of scientific roles. Not the fathers but rather the semimythological accounts of the great tradition in science dictated how they should act. Henry's peers took their science from, perhaps, Sir Humphrey Davy, but views of the ideal life of science came from, let us say, Sir Isaac Newton. The people in the learned culture who were not cultivators might know the grandfathers but had only an imperfect knowledge of the fathers; this is really what differentiated them from cultivators. A common core of beliefs from fathers and grandfathers explains why the great disparity between certification and accomplishment did not give rise to any major conflicts. The principal researchers were viewed not only as being on the leading edge of knowledge but also as illustrating the ideal behavior for creative scientists.

This relationship of the major scientists to practitioners and to cultivators is greatly obscured by some of the catchwords that are used to describe the scientific situation in the past and the present. Some assume the existence of so-called Baconian science in the United States in the past, a kind of routine data-gathering operation. It is also assumed that there was, in the past, something called "little science," as opposed to something called "big science," which we have today. Big science is that of teams—of faceless investigators operating under plans and producing copious quantities of articles. Yet I find it hard to believe in this picture of big science that is so readily drawn by many. Twentieth-century science is not dominated by faceless robot investigators operating under more or less rigid plans; rather, it is marked by animating intellectual or administrative leaders, just as was nineteenth-century science. There is a real distinction between bigness and changes in kind.

Scientific investigations in the past were routinized and even done by teams. Consider the *Manual of Scientific Inquiry*[68] issued by the British Admiralty for the benefit of scientific travelers overseas—and not the only example of literature of that kind. During the Civil War, Charles Henry Davis, who was one of the Lazzroni, planned a similar document for Americans. The assumption behind issuing some kind of a manual is that one can describe basic scientific operations in various fields in simple terms and that anyone with reasonable intelligence and education can then perform these operations. What we have here is neither Baconian science nor big science nor little science, but simply the problem of whether a scientific operation can be described and routinized so that it need not be done by people of exceptional ability and training or under unusual circumstances. The idea of having large numbers of cultivators conduct surveys under carefully prepared instructions was one of the early notions of the British Association for the Advancement of Science.[69]

Manualizing was not confined to natural history. Larger portions of meteorology and astronomy were so routinized that men, very skilled it is true, could grind out data in a relatively routine, undeviating manner. In other words, they could be practitioners of science. This leads to one of the real definitions of a practitioner that transcends the occupational category. A mere practitioner is a man who receives from his scientific fathers or his scientific peers a body of scientific method and doctrine which he repeats more or less in an undeviating manner as part of his work. In this sense most of the cultivators were also practitioners, that is, they took a certain kind of scientific activity and performed it, usually in an undeviating manner. Some of the cultivators and some of the practitioners deviated from the set procedures with enough success to become creative researchers.

Two further explanatory points need to be made. First of all, I am not speaking about normal science in the Kuhnian sense, although some normal science does fall into this category. If a particular kind of problem solving can be done by a standard routine method, even if it requires a person of considerable ability, I would describe this person as a practitioner. T. S. Kuhn, of course, thinks of normal science problem solving as being a very high-skilled scientific activity that, when done properly, is the most certain kind of knowledge obtainable from the sciences. What I am concerned about here is whether or not scientific operations have been so described in the literature and are so followed in practice that they can be applied in a straightforward manner, given a reasonable amount of intelligence and training. This is simply a stage in the natural history of scientific concepts. Yesterday's brilliant novelties become today's routine practices; tomorrow they become operations turned over to machines. The second point is that quite a number of very eminent scientists in various stages of their career carry out routine tasks, either as training or as incidental to their investigations. That is, they are practitioners for at least that moment in their careers. Of course, this does not affect their overall life style as research scientists. Nor does this deny that in the course of routine investigations a perceptive mind can notice a deviation and perceive its meaning. Think of Roentgen's work with photographic plates that other scientists threw away or returned to the dealer. Nor does this imply that a perceptive mind cannot see the routine in a novel manner. Practitioners exist because some scientific activities were susceptible to routinization. Successful routinization is the basis for occupations.

But occupational egalitarianism posed a great problem for the leading researchers in their efforts to develop an American scientific community over the century 1820–1920. What they did not see was the absence of a literal egalitarianism. Lacking in America was the sense of a hierarchy, a social structure, and a system of deference associated in the researchers' minds with the societies of western Europe. The scientists had a problem with a feeling that pervaded American society, as demonstrated by the way the Census Bureau classified occupations in 1860.[70] Actors and showmen were also professionals, as were dancing masters and riding masters. How could the researchers differentiate themselves from the practitioners and the cultivators if they were classed with dancing masters? By 1900 there was even a problem of distinguishing the heirs of Newton and Darwin from technicians.

The researchers' reaction was threefold. The most obvious was to establish honorific organizations or awards to distinguish the best scientists from mere practitioners; the National Academy of Sciences is an obvious example. A second reaction continuing to this day, was to set up research enclaves, clearly what Joseph Henry had in mind with the Smithsonian Institution, what Alexander Dallas Bache wanted with the Coast Survey, and what Louis Agassiz thought he was doing with his museum.

The third reaction is more difficult to define: the researchers in the mid-nineteenth century did not want to change the organizational system as it existed. I do not believe that most of them wanted to negate the network of organizations that existed in the form of local societies, academies, and institutes of all sorts. I do not think they really conceived of the people in them as being vulnerable, a dying breed; there were far too many of them. I think the astute leaders of the scientific community saw such people as a potential resource. Some might become scientists, others might become friends of science to be counted on when lobbying was needed in Congress and in state legislatures. Because of that, most of the leading scientists of that day usually did not think in terms of the graduate degree. Somehow people would get specialized training beyond what could be had in the liberal arts schools. The researchers certainly did not conceive of what has happened since: the universities, by awarding a degree, actually provide an essential element in the certification of scientists.

At the same time, they did not envision the post-1860 growth of the many specialized scientific societies. The National Academy deliberately brought together various disciplines, just as most of the local societies did. Even the most specialized of these local societies were broad natural history groupings. What strikes my eye when comparing the American case with the British is the fact that the Americans failed to develop the kind of study associations that began appearing in Britain well before midcentury. The Geological Society of London and the Royal Astronomical Society are good examples; there were very few equivalents in the United States. In Britain some of these societies would remain purely study societies, with a mixture of cultivators and professionals, while others would become specialized societies. In the United States, when these societies appeared after the Civil War, they were almost invariably professional societies formed to facilitate and accomplish development of a specialized field by sponsoring annual meetings and publishing a journal. A few of these societies became concerned with aspects of the certification process, that is, with improving the supply of people. They were involved with the standards of education, and in some cases, particularly through the refereeing process, with proper professional conduct.

The result was that while the mid-nineteenth-century researchers would invariably opt for an older institutional pattern implying accomplishment, reality called for certification, for providing a way of securing a large number of proper practitioners. These leaders of the scientific community preferred national or local groups, rather than disciplinary groups. Bodies like the original kernel from which the AAAS came tended to disappear as many scientists succumbed to pressures to operate on a national level, despite all of the risks, in order to awaken the nation to the need for research. Inevitably this took precedence over strong feelings for the mere development of narrow subject specialities. Since the researchers perceived the sciences as functioning best in nonspecialized groupings, it was an understandable position. But, as the century progressed, disciplinary groups became more common.

By the time George Ellery Hale began revitalizing the National Academy of Sciences, such groups were ascendant. Hale recognized that in order to advance astrophysics there had to be cooperation on an international basis and by various disciplines. His scientific father was Sir William Huggins. Institutionally, Hale noted that the work of the National Academy of Sciences, a multidisciplinary body, was in decay while the specialized societies were rising. It was a violation of the way of the grandfathers—Sir John Herschel, let us say. In other words, Hale's particular research problems required cooperation across disciplinary lines, and he wanted to restore the unity of the scientific community that he perceived in the generation of his scientific grandfathers. Being an elitist, he did not quite succeed in this attempt because the trends of the time were quite different. The Academy acquired real power only after World War II, under circumstances far different from those Hale envisioned.

But Sir John was only one of the grandfathers. A surprising number of twentieth-century scientists confused him with the whole class of cultivators, now transformed by historical mythology into gentlemen amateurs. In 1892 the British physicist Arthur Schuster, extolling the dying tradition of amateurism, pointed to Faraday.[71] Faraday was gainfully employed as a scientist that would make him a mere practitioner—but Faraday was an amateur by lack of formal training. Schuster thought this a virtue, being unconstrained by received doctrines in the university, even though he admitted many untrained amateurs did have strange ideas. What Schuster was extolling was a mythical grandfather, the scientist as romantic discoverer, certainly a more appealing candidate for grandfatherhood than the plodding practitioner and the casual hobbyist.

Eager to differentiate themselves from mere practitioners of science, not to mention the myriads of engineers and physicians, many researchers after 1900 implicitly contrasted the purity of research by romantic discoverers with being in trade. It was the purity of social climbing. For the high-minded seeking of the plan of Divine Providence, for the high-minded certainty of science ameliorating man's lot, some researchers substituted the purity of the play element. Did not Rutherfurd, after all, stand on the deck of the *America?*

Notes

1. A recent example is W. G. Holt, "Social Aspects in the Emergence of Chemistry as an Exact Science: The British Chemical Profession," *The British Journal of Sociology* 31 (1970):181–198.

2. On these various incidents, see A. Hunter Dupree, *Science in the Federal Government* (Cambridge, Mass.: Harvard University Press, Belknap Press, 1957), chap. 4; Sally Kohlstedt, "A Step Toward Scientific Self-Identity in the United States: The Failure of the National Institute, 1844," *Isis* 62 (Fall 1971):339–362; idem, "The Formation of the American Scientific Community: The American Association for the Advancement of Science, 1848–1860" (Ph.D. diss., University of Illinois, 1972); Warner Papers, Library of the American Philosophical Society, Philadelphia; John D. Holmfeld, "From Amateurs to Professionals in American Science: The Controversy over the Proceedings of an 1853 Scientific Meeting," *Proceedings of the American Philosophical Society* 114 (1970):23–36; Edward Lurie, *Louis Agassiz: A Life in Science* (Chicago: University of Chicago Press, 1960), p. 180; Nathan Reingold "Two Views of Maury . . . and a Third," *Isis* 55 (1964):370–372.

3. The social science literature on professionalization is vast enough without even considering the many writings of members of particular professions. The following selected list is a starting point for historians unfamiliar with this literature, sticking very close to pieces specifically on professions and professionalization, not obviously related topics like role theory:

> Ben-David, Joseph. "The Profession of Science and its Powers." *Minerva* 10 (1972):362–383.
> Carr-Saunder, A. M. and Wilson, P. A. *The Professions.* London and New York: Oxford University Press, 1933. The classic work, highly influential and still usable.

Cogan, Morris L. "Toward a Definition of Profession." *Harvard Educational Review* 23 (1953):33–50. Has useful bibliography of older works.

Holt, B. W. G. "Social Aspects in the Emergence of Chemistry as an Exact Science: The British Chemical Profession." *The British Journal of Sociology* 21 (1970):181–198. Interesting, but not wholly convincing.

Jackson, John Archer, ed. *Professions and Professionalization.* Sociological Studies, vol. 3. London and New York: Cambridge University Press, 1970. T. Legatt on teaching has interesting comments.

Lynn, K. S., et al., eds. *The Professions in America.* Boston: Houghton Mifflin, 1965. Barber, Beer, and Lewis are most relevant.

Millerson, Geoffrey. *The Qualifying Associations: A Study in Professionalization.* London: Routledge & Kegan Paul; New York: Humanities Press, 1964. The best single work by far. Very British, but with a few pointed comments on the U.S. situation.

Moore, W. E., and Rosenblum, G. W. *The Professions: Roles and Rules.* New York: Russell Sage Foundation, 1970. A fine monograph, but not particularly useful to a historian.

Parsons, Talcott. "Professions." *International Encyclopedia of the Social Sciences.* An unsuccessful attempt at redefinition.

Reader, W. J. *Professional Men: The Rise of the Professional Classes in Nineteenth-Century England.* London: Weidenfeld & Nicolson, 1966. Splendid blend of history and sociology, far superior to Calhoun's *Professional Lives in America.*

Shils, Edward. "The Profession of Science." *The Advancement of Science* 24 (1968):469–480. A good example of a conventional view.

Strauss, George. "Professionalism and Occupational Associations." *Industrial Relations* 2 (May 1967):7–13. Good for comparison with engineers.

Vollmer, H. M., and Mills, Donald L., eds. *Professionalization.* Englewood Cliffs, N.J.: Prentice-Hall, 1966. A well-done anthology.

Weber, Max. "Science as a Vocation." *From Max Weber: Essays in Sociology,* edited by H. H. Gerth and C. Wright Mills. London and New York: Oxford University Press, 1946.

Wilensky, H. L. "The Professionalization of Everyone?" *American Journal of Sociology* 70 (September 1964):137–158. An important article in the American context.

4.   Millerson, *Qualifying Associations,* pp. 1–2.

5.   Bernard Barber, "Some Problems in the Sociology of the Professions," in Lynn, *Professions in America,* p. 18.

6.   John J. Beer and W. David Lewis, "Aspects of the Professionalization of Science," in ibid, pp. 110–130.

7.   Shils, "Profession of Science," and Ben-David, "Profession of Science and Its Powers."

8.   Parsons, "Professions." This presentation is more a reflection of hopes than a description of reality.

9.   Shils, "Profession of Science," p. 473.

10.   Henry W. Menard, *Science: Growth and Change* (Cambridge, Mass.: Harvard University Press, 1971), pp. 202–204.

11. Millerson, *Qualifying Associations,* pp. 10–11.

12. I am greatly indebted to Dr. Cannon for many stimulating insights on this matter of professionalization. He does not merit any blame, however, for what follows.

13. Ben-David, "Profession of Science and Its Powers," pp. 371–73, passim.

14. Only 37 percent of holders of the doctorate are in research and development. National Science Foundation, *American Science Manpower 1970* (Washington, D.C., 1971), p. 14.

15. Counts of this are very hard to come by in the contemporary literature, since attention is understandably on the prolific producers and the authors of great impact. But see J. R. Cole, "Patterns of Intellectual Influence in Scientific Research," *Sociology of Education* 43 (1970):377–403.

16. Menard, *Science: Growth and Change,* p. 99, estimates that 78 percent of all geologists are never cited.

17. E. C. Hughes, *Men and Their Work* (New York: Free Press, 1958), pp. 137–138.

18. For example, C. A. Elliott, "The American Scientist, 1800–1863: His Origins, Career, and Interests" (Ph.D. diss., Case Western Reserve University, 1970). By adjusting his figures, I estimate about 2,200 American authors of scientific articles. More than two-thirds published only one piece.

19. A. D. Orange, "The British Association for the Advancement of Science: The Provincial Background," *Science Studies* 1 (1971):315–329, called my attention to Brewster's usage. Since writing this, several others have called my attention to this usage in antebellum America, including sources familiar to me.

20. Nathan Reingold, ed., *Science in Nineteenth-Century America: A Documentary History* (New York: Hill & Wang, 1964), pp. 17–20.

21. Joel R. Poinsett, *Discourse on the Objects and Importance of the National Institution for the Promotion of Science* (Washington, D.C., 1841). John Kazar of the University of Massachusetts at Amherst called this to my attention.

22. Nathan Reingold et al., eds. *The Papers of Joseph Henry* (Washington, D.C.: Smithsonian Institution Press, 1972), vol. 1, pp. 66, 75, passim.

23. John C. Spencer to Francis Wayland, 2 February 1844, Wayland Collection, Brown University Library, Providence, R.I.

24. U.S. Bureau of the Census, *Occupations at the Twelfth Census* (Washington, D.C.: U.S. Government Printing Office, 1904), p. xxiii.

25. *North American Review* 20 (1825):309–66.

26. The periodicals examined were *American Quarterly, New York Review, New York Review and Atheneum Magazine, Belles Lettres Repository and Monthly Magazine, American Monthly Magazine and Critical Review, The Literary and Scientific Repository and Critical Review, Boston Quarterly Review.* Further browsings have not altered the picture.

27. Linda Kerber, "Science in the Early Republic: The Society for the Study of Natural Philosophy," *William and Mary Quarterly* 39 (1972):261–280, is a fine account of this tradition.

28. "For practically the whole century science and many specialized subjects could serve as hobbies, which did not entail very deep, intricate pre-instruction." Millerson, *Qualifying Associations,* p. 22.

29. Millerson, *Qualifying Associations.*

30. Quentin Skinner, "Thomas Hobbes and the Nature of the Early Royal Society," *Historical Journal* 12 (1969):217–239.

31. Reader, *Professional Men,* pp. 6–7.

32. A fine example of this activity is provided by the Charles Gill Papers, Department of Manuscripts and University Archives, Olin Library, Cornell University, Ithaca, N.Y., with its letters from problem solvers. These include college teachers of mathematics (i.e., practitioners) like Theodore Strong, but also others devoted to this kind of mathematics.

33. Reingold et al., eds., *Papers of Joseph Henry,* 1:212.

34. *Dictionary of American Biography,* s.v. "Root, Erastus."

35. Reingold, *Science in Nineteenth-Century America,* p. 88.

36. Francis Fry Wayland, *Andrew Stevenson: Democrat and Diplomat* (Philadelphia: University of Pennsylvania Press, 1949).

37. For example, the draft of Henry's address of November 18, 1871, to the Philosophical Society of Washington has strong passages excised from the printed version in *Scientific Writings of Joseph Henry* (Washington, D.C., 1886), 2:468–275. The draft is in box 26 of the Henry Papers, Smithsonian Archives, Washington, D.C.

38. National Institute, Journal of Proceedings, entries of 1 January 1849, 6 January and 7 January 1851, 5 January 1852, 15 January 1855, 14 January 1856, National Institute Papers, Smithsonian Archives, Washington, D.C.

39. See note 37, above.

40. Joseph Henry, "On the Organization of Local Scientific Societies," *Scientific Writings,* pp. 511–513.

41. Hugh McCulloch, *Men and Measures of Half a Century* (New York, 1888), pp. 261–269, passim.

42. *Proceedings of the American Philosophical Society* 11:85–91. Delivered March 19, 1869.

43. *Administrators Sale. Extensive Rare and Beautiful Cabinet of Science . . . Catalogue of Valuable Philosophical Apparatus being a collection made by the late Charles N. Bancker* (Philadelphia, 1869).

44. *A Collection of Rare and Valuable Books . . . Catalogue of the entire Private Library of the late Charles N. Bancker* (Philadelphia, 1869).

45. Anna Eliot Ticknor, ed., *Life of Joseph G. Cogswell as Sketched in his Letters* (Cambridge, Mass., 1874), pp. 43–107, passim.

46. Ibid., p. 352.

47. All this is very distressing to me, since I implied in the Bowditch article in the *Dictionary of Scientific Biography* that Peirce is the only one qualifying as a student of Bowditch. Sam Ward was below my level of perception.

48. Robert V. Steele (Lately Thomas), *Sam Ward: King of the Lobby* (Boston: Houghton Mifflin—Riverside Press, 1965), pp. 20–21, 30–32; Louise Hall Tharp, *Three Saints and a Sinner* (Boston: Little, Brown, 1956), pp. 38–39, 45–46, 49–53; Maud Howe Elliott, *Uncle Sam Ward and His Circle* (New York: Macmillan, 1938), p. 47.

49. R. C. Archbald, "Unpublished Letters of James Joseph Sylvester and other new Information concerning his Life and Work," *Osiris* 1 (1936):118.

50. See my Rutherfurd article in the *Dictionary of Scientific Biography,* in press.

51. Benjamin Apthorp Gould, "Memoir of Lewis Morris Rutherfurd," *Biographical Memoirs of the National Academy of Sciences* 3 (1895):417–441.

52. G. Brown Goode, "The Beginnings of American Science," *A Memorial of George Brown Goode* (Washington, D.C., 1901), p. 463. Originally delivered in 1888.

53. W. Stephen Thomas, ed., *The Layman Scientists in Philadelphia* (Philadelphia: American Philosophical Society, 1940).

54. Elliott, "American Scientist."

55. George Daniels, *American Science in the Age of Jackson* (New York: Columbia University Press, 1968); Donald Beaver, "The American Scientific Community, 1800–1860: A Statistical Historical Study" (Ph.D. diss., Yale University, 1966).

56. Reingold et al., eds., *Papers of Joseph Henry* (1975), 2:432–238.

57. Daniels, *American Science*, pp. 10–11.

58. *American Science Manpower* 1970, pp. 14, 19.

59. Pauling: "I decided then to be a chemist, and to study chemical engineering, which was, I thought, the profession that chemists followed." Gerald Holton, ed., *The Twentieth-Century Sciences: Studies in the Biography of Ideas* (New York: Norton, 1972), p. 281; Warren Weaver, *Scene of Change: A Lifetime in Science* (New York: Scribner, 1970), pp. 23–25.

60. Reingold et al., eds., *Papers of Joseph Henry*, 1:245–246.

61. This is by no means a position original with me. As an example, I can cite the work of Edward Pessen, which has a strange relationship to the writing of historians studying science in America: it is avowedly anti–de Tocqueville. See his "Did Fortunes Rise and Fall Mercurially in Antebellum America?" *Journal of Social History* 4 (1971):339–357, especially fn. 3.

62. Of course I haven't read everything, and my memory is far from infallible. I would welcome any contrary instances.

63. Elliott, "American Scientist," pp. 107–110.

64. These few were important in their own right. The 1860 census reported 54,543 physicians, however, so that the number doing science was a very small fraction of that total. *Eighth Census of the United States Population* (Washington, D.C., 1864), 671.

65. See Anton-Hermann Chroust, *Rise of the Legal Profession in America*, 2 vols. (Norman: University of Oklahoma Press, 1965).

66. There is surprisingly little specifically on professionalization and the ministry even in such an extensive source as N. R. Burr, *Critical Bibliography of Religion in America*, 2 vols. (Princeton, N.J.: Princeton University Press, 1961), Vol. 4 of *Religion in American Life*, edited by J. W. Smith and A. L. Jamison.

67. Beaver, "American Scientific Community," pp. 15–16, indicates that the leading scientists produced at least 50 percent of the literature in this period.

68. Three editions appeared: 1849, 1851, 1859.

69. Orange, "British Association for the Advancement of Science."

70. See note 64, above.

71. British Association for the Advancement of Science, *Report of the Sixty-Second Meeting* (Edinburgh, 1892), pp. 628–630.

3□ American Indifference to Basic Research: A Reappraisal

The belief in American indifference to basic research rested on very thin, actual knowledge of the history involved, even on the part of such eminent historians as Richard H. Shryock and I. Bernard Cohen. At least that was my impression as a young Turk prowling in the archives. Influenced by Shryock's writings on medicine, I was particularly struck by the widespread existence over time of situations with complex mixtures of pure and applied research. Dividing basic research from applied research was often difficult, if not impossible. And sometimes sterile. I had dedicated my documentary history of science in nineteenth-century American to Shryock and, of course, sent him a copy of this paper. Genuinely surprised, he mildly congratulated me while affirming his belief in the contrary view.

At one time, just before and after World War II, on many campuses there was something called social and cultural history, particularly of the United States. (The historians of Europe played other games.) The specialty was amorphous and open ended, tolerant of both the signs of the emerging of "real" social history and of the great ideas tradition. In the former intellectual economy of the discipline of history, "social and cultural" had considerable implications for many developments in the next generation. Today, history strikes me as far more fractionated and beset by insulated sectarianisms.

Shryock and others like him each gave a distinctive thrust to social and cultural history. In his case, the emphasis was on the history of pure and applied sciences in particular societies. To those fortunate enough to have him as a teacher, it was impossible ever to look simply at a scientific discipline as the be-all-and-end-all of historical analysis. Nor was his teaching conducive to glorification of the scientist, the physician, or the engineer. (For more on Shryock see my "Eloge" in the March 1973 *Isis*.)

Like so many issues in history, the indifference theme implicitly told us something about its supporters, whatever its value for understanding the past. Like those eminent historians I criticized by name or cavalierly as a group, I favored the support of research and wanted more in the United States. But when I started out as a historian, the role of "experts" in a democratic society was a topic that concerned many of us. Shryock talked about that issue, as did other leading historians. The laments about indifference to basic research (not those from Shryock) almost asked for a blank check, a privileged position in American society. Accompanying the laments was an implicit over-idealizing of the Cartesian mandarinate of France, the Herr Doktor Professor of the German university, and the Oxbridge don. Nor was it uncommon to encounter American intellectuals with a great animus against what I would describe as the vernacular culture of the United States. No doubt they relished folk costumes and dances in the Old World. I threw in a reference to the comics near the end of the paper as a polite way of distancing myself from the world view of some of my colleagues. My roots are in the vernacular culture, and I have no qualms on that point.

Although the indifference theme had long interested me, I doubt in retrospect whether I would have written without the stimulus of an invitation to give a paper at a conference at Northwestern University in 1971 on science in nineteenth-century America. What surprised me when I delivered this paper and afterward when the proceedings appeared in print was the reaction of a fair number of individuals. Despite my words, they persisted in believing I was arguing for the opposite of the well-accepted cliché. Clearly, some believed I thought Americans in the past were avidly for basic research. What I attempted was a more realistic definition of terms, hopefully enabling historians to show changes over time. My underlying motivation was to remove a roadblock in the way of research on the historical past of the sciences in this country.

All of this came back to me in 1986—the defensiveness about national attitudes, the smugness about western European virtues, the surprisingly large attendance at Evanston eager to learn and to discuss—while I shared the national horror of the *Challenger* on television. Senator John Glenn, the former astronaut, was asked if this was the start of the end of Americans in space. Not at all, he replied. The United States, the senator averred, had supported basic research since the start of the Republic. What a change! Did it presage the day when a future historian would write an essay showing that Americans had not avidly supported basic research starting with the first inauguration of Washington? Would I rate a footnote?

American indifference to basic research, decried both in academic circles and among the educated public, is an old theme in the literature of national self-appraisal dating back at least to de Tocqueville. We Americans abjectly trot it out when bested by Europeans. When Soviet Sputniks orbited the earth, American indifference to basic research was blamed for this technological defeat. When we landed on the moon, quite properly no praise was given to basic research. In recent years, American science—the product of this supposed indifference—was castigated for neglecting the problems of our society. Almost forgotten are the enormous successes achieved by a research and development community highly responsive to national needs.[1]

The American research and development community is the envy of the world and the object of fascinated, avid study by science policy experts in international and foreign national bodies. Obviously, American practical achievements first hit the eye. Practical results are not necessarily incompatible with indifference to basic research—the Japanese have done very well economically with relatively little basic research. But ironically enough, of all Western nations the indifferent Americans devote a greater percentage of their gross national product to research and development in general and probably to basic research in particular.[2] What does this mean historically? The purpose of this question is neither to deny nor to confirm American indifference. I do not know if Americans were indifferent to basic research, and neither does anyone else. This widely discussed topic has become a sterile historiographic concept lacking adequate evidence for either refutation or confirmation.

Beliefs affect perceptions of reality. As these perceptions were applied to the evidences of the past, scientists and historians concerned with the sciences in America inevitably produced writings reflecting their attitudes toward the indifference theme. Two related tasks are proposed here: first, a survey of the characteristic ideas or criteria in the indifference literature to suggest some of the underlying connections to the real world that these writings purport to describe; second, suggestions of how we might find out more about the relationships of Americans and research in order to predict what these relationships might be.

The extensive literature on American attitudes to basic research goes back to the early years of the Republic and is absorbing reading, even to educated laymen interested primarily in our current dilemmas.[3] Besides de Tocqueville, there is a host of foreign commentators who failed to find their image of a cultural center in the United States. Their laments and similar American ones are clearly related to the assumption that Americans were not interested in literature, music, and the arts. Someone might profitably compare the artistic and the scientific laments. Although overtaken by events, these assumptions still persist on both sides of the Atlantic. Around 1876, the centennial brought forth a number of article on America's inferior position in the scientific world. The end of the century was another time of national stocktaking in which our scientific deficiencies were publicly decried. In both 1876 and 1900, the critics were usually scientists. American scientists have always decried American callousness toward the search for truth. Foreign experts have sided with the scientists.

Historians have accepted the scientists' criticism as fact. Unconcerned about the sciences, most have probably been convinced that the country was not really interested in scholarship and art. When historians of the sciences appeared in this generation, they found the writings of Richard H. Shryock and I. Bernard Cohen affirming the scientists' complaints.[4]

Richard H. Shryock started out in American political history but soon became interested in medicine, and public health in particular, as well as in more general questions of American social and intellectual history. His work in medical history has a very strong concern with social context. Aspects of medical science and practice in the Shryock canon are often examined in terms of social factors which affect application. Shryock has investigated how scientific advances impinge on medicine and how nonintellectual forces determine the acceptance of scientific developments. His work has had considerable influence not only because of its quality but because he has always tried to relate his findings to broader questions in American history and in the history of science.[5]

In 1948, Shryock published an article entitled "American Indifference to Basic Research During the Nineteenth Century."[6] It is the classic article; when Richard Hofstadter wrote on American anti-intellectualism, Shryock was cited to make the point about basic research.[7] The structure of Shryock's piece is simple. After asserting the fact of general indifference, Shryock tests and rejects various alternative explanations of the phenomenon: availability of European science; preoccupation with conquering a continent; and clerical control of colleges and universities. Accepting a fourth thesis—extreme emphasis on utility—Shryock asserts that preoccupation with utility was not evident in the colonial period and became increasingly apparent only as the nation industrialized. The indifference ultimately stemmed from the industrial and mercantile leadership. When Americans recognized the implications of science for technology in about 1900, we have, in Shryock's words, "the rather sudden emergence of basic science in the United States."

Shryock's explanation does not seem tenable. Take the two most spectacular philanthropists of the early twentieth century, Carnegie and Rockefeller. The former did support technical education but backed pure research in the Carnegie Institution of Washington. The Institution has rarely deviated from pure research. There is ample evidence that this is what Carnegie and his trustees wanted. The principal Rockefeller aid to research was not in physics, chemistry, and geology, which might ultimately have redounded to the benefit of petroleum industry holdings, but in medical research and public health. Even when the various Rockefeller endowments broadened their scopes, one could argue that concern for cultural life was at least as strong a motivation as any desire for ultimate applied benefits. The point here is that there is, indeed, very little evidence which indicates that Carnegie and Rockefeller realized that implications of science for technology and pumped money into science. The sources of support for science prior to the massive infusion of federal funds starting in World War II are so varied as to cast doubts on any explanation stressing a single factor. The industrial and mercantile elite was as likely or more likely to put funds into stadiums and art museums as into laboratories.

I. Bernard Cohen was originally a mathematician. A student of George Sarton, who was the pioneer historian of science in America, he succeeded his teacher as editor of *Isis,* the most eminent journal in the field. Cohen has published extensively on many topics and has trained a significant number of the historians of science in America. Although discussed here in the Americanist context, Cohen is essentially a historian of astronomy and physics with emphasis on the seventeenth and eighteenth centuries. He is not primarily concerned with social context, that is, the external history of science. On the contrary, Cohen is a leading practitioner and proponent of the internalist history of science—the study of the development of scientific concepts and of the scientific knowledge ordered by these concepts. Social and other contexts are decidedly secondary in intellectual interest to these internal developments.[8]

The internalist position is—to oversimplify—a reaction against Marxian determinism as proposed in the 1930s and the somewhat naive sociological approach of Robert Merton's "Science, Technology, and Society in Seventeenth Century England,"[9] in favor of the intellectual history of the context of science, particularly its conceptual aspects. Most of the dispute in the 1930s centered around the origins and natures of the scientific revolution of the seventeenth century.[10] American events played no role in this dispute. With the general victory of the intellectualist historiographical approach, American events were largely neglected by internalists since the indifference literature already affirmed the relative insignificance of American scientific achievements from a conceptual standpoint. What historical work did occur largely stemmed from men like Shryock, the elder Arthur Schlesinger, and Merle Curti, who were, broadly speaking, American social and intellectual historians.

The differences between the two schools are describable in terms of three foci of concern: the contents of science, the scientists themselves, and the society at large. The orthodox internalist historian of science is primarily concerned with the contents of science and only with those parts of the scientists' lives that influenced their professional work (i.e., religion, general philosophic beliefs, institutional affiliations). The conventional historian is concerned with aspects of the society at large. Shryock and others studying the sciences in America are in the uncomfortable position of being primarily concerned with the scientists as such, and, moreover, wanting to understand the scientists in terms of both the scientific content and the social environment.[11] Cohen, nevertheless, has produced an appreciable body of writings concerned with the sciences in America. Unlike Shryock, he does not venture far into social or cultural history. Like Shryock, he stresses the debilitating effects of pressures for practical results.[12]

In addition to their stress on the pernicious effects of practical pressures, the two men agree on other aspects of the indifference theme. Both pay considerable attention to the absence of great men and great contributions to major advances in the sciences. As is typical in this genre, they usually cite Benjamin Franklin, and then note the exceptional cases of Joseph Henry and J. Willard Gibbs. Both men also stress the small number of American basic scientists in the past.

Shryock and Cohen both emphasize colonial science. Both, viewing the colonial period almost as a golden age, attribute is scientific fecundity to an aristocratic, or at least patrician, sympathy for basic science. From this position is derived a view of the history of science in America as being like a Bactrian camel, a nineteenth-century trough between colonial and twentieth-century humps. Shryock and Cohen have both written of the colonial period and are understandably enthusiastic about it. However, Franklin, John Winthrop, John Logan, David Rittenhouse, and Cadwallader Colden hardly constituted a scientific community. With all due allowance for Franklin, a great man in many respects, the overall colonial scientific output was not very large or significant.[13]

Enthusiasm has often overpowered critical faculties. Consider, for example, the question of the number of Americans who were made fellows of the Royal Society. That there were many more in the colonial period than later is offered as proof of the existence of a colonial hump. Forgotten is the fact that colonials were British subjects; after 1783, Americans had to become foreign fellows, a severely restricted honorific category. Further, the standard source of information on colonial fellows lists West Indians, a Canadian, and even Elihu Yale, who was born in Boston but lived his adult life in the Eastern Hemisphere. Eighteen bona fide American colonial fellows from 1660 to 1788 seem a very small hump, indeed, considering that many of them made few scientific contributions at all.[14]

To return to the more general characteristics of the indifference literature, periodization is a problem to many authors. What date does one use to mark the point when indifferent Americans became favorably disposed to basic research? Is it Johns Hopkins University's effective launching of graduate education in the United States in 1876? Alternatively, 1900 is a possible candidate, probably because it is easy to remember. World War I is another alternative, because American strengths became more apparent in contrast to Europe's distress. The arrival of the refugees fleeing Hitler is a conceivable turning point. The refugees, after all, were notable as mathematicians and physical theorists, specialists in areas in which Americans were weak in comparison to Germany. If drama and social significance are the criteria, 1945 and the first atom bomb at Alamogordo could signal the end of indifference. Obviously, the date selected is arbitrary. Not so obvious is the frequent assumption of a historical discontinuity—that two distinct periods exist and that little or nothing prior to the turning point requires consideration in explanation of the present situation. Evidences of this continuity in the history of science in America are quite plain; a good number of native scientific institutions antedate the Civil War, and our university pattern was in existence before 1900.

A more important characteristic of the indifference literature is the loose use of words like *basic* and *pure* as modifiers of *science.* Many authors, unaware of semantic difficulties, use the two interchangeably. *Basic* refers to intrinsic merit, usually scientific activities involved in formulating and verifying hypotheses and general theories. *Pure,* in contrast, refers to a psychological motivation unsullied by concerns other than the growth of scientific knowledge. Stress on the pernicious effect of the American ideal of practicality on science leads authors to the assumption that purity yields the basic and that the existence of the basic is proof of purity. In practice, the situation is not quite so simple. How does one determine purity of motive? How does one classify Irving Langmuir, a great theoretical physical chemist, who worked at the General Electric Research laboratory and even took out patents? What do we do with a university physicist who consults with the Atomic Energy Commission on the development of nuclear power? And what do we do with the very many pure-minded scientists on countless campuses whose contributions to the development of basic theory will never amount to even a footnote?

Related to this semantic confusion is the assumption that we can always or usually clearly differentiate between some activities designated as science and others designated as technology, invention, or development. No writer in the indifference literature explicitly and seriously raises the point that research situations are quite often mixed. In addition to the problem of differentiating purity of motive and intrinsic merit, there is the problem of differentiating science (theory and its underlying data) from technology (applications meeting human needs). A mixture of scientific and technological concerns typified a significant proportion of research activities in the past; quite clearly, such mixtures exist today.[15] The existence of "pure" or "basic" scientists, and of practical men concerned solely with solving immediate problems, is not being denied. What is being stressed is that the significant presence of mixed situations renders analysis in terms of basic and applied science increasingly sterile. For convenience, we establish arbitrary distinctions, but the results often do violence to the real world under discussion. Writers convinced of the truth of American indifference to basic research will enthusiastically seize on every kernel of practicality to downgrade scientific contributions in the American past. Historians of the opposite persuasion will exalt every nugget of theoretical concern as proof of the scientific glory of the American past.[16]

As George Daniels points out, many indifference-theme authors, discussing motivations that lead to applications, make no distinction at all between the scientists and society at large.[17] Society may sponsor research for presumedly practical reasons, while the researchers themselves are often primarily interested in expanding knowledge. An example is the case of a great paleontologist-geologist of the last century, James Hall.[18] The New York State Survey, which he headed, undoubtedly originated in large measure in the concerns of many for more accurate information about exploitable natural resources. Yet, year after year, Hall turned out a series of abstruse volumes of little apparent applicability. The state legislature grumbled; Hall vented his ire at his critics; but the money continued to come in support of the research. Hall did not convert the Philistines to the true faith of science. He was just tough enough to survive until his work was taken for granted as part of the overhead of state government.

In complaining about the paucity of great men and great contributions to major scientific developments in the United States, the authors of the indifference literature pay particular attention to the physical sciences, especially astronomy and physics. Even authors interested in the biological sciences, such as Shryock, reflect this awe for the Comtian hierarchy. When Shryock and others consider biology, the stress is on the lack of American contributions to the cell theory, to the germ theory of disease, and to physiology, all areas presumably characterized by a respectable degree of experimentation, quantification, and theoretical structure. Anti-intellectualism, they assume, leads to a mistrust of abstract or theoretical work. As a consequence, the argument goes, Americans tended to go in for taxonomic fields—natural history and geology. Although neither experimental nor quantitative, a taxonomy is, after all, theoretical statement of relationships. Nor is the work involved in developing a taxonomic scheme necessarily Baconian data-grubbing.

The Baconian position is ahistorical and distorts the nature of the past research communities. One picks out high points in contemporary science, looks backward in time to establish a "main line" of development, and largely blots out all research not in or near the "main line"—physical theory and experimental biology. Geology, a very important field in nineteenth-century Europe and America, is ignored, except in relation to evolution. A conventional concern of historians of science is to overturn the accepted description of sections of the "main line" by pointing out omissions of additional intellectual factors. Because the indifference theme has a social context, it apparently never occurs to most historians of science to question the oversimplification of the "main line" thesis in studying science in America.

Take the example of mathematics. This field is one of the presumed weaknesses of the national scientific community, as one would expect of people who are pictured as anti-intellectual, not given to abstract, theoretical work. As far back as 1940, L. C. Karpinski published a magnificent bibliography of mathematical works printed in America to 1850. He lists 1,092 appearing in a total of 2,998 editions. (These include many reprints and translations of leading European works.) Karpinski concludes:

No reasonable person should expect that in the period before 1850 America would have produced mathematicians to rank with Newton and the foremost European scientists. It was too much occupied with settlement, exploration, and expansion. The wonder is that at this time the population could furnish a body of readers of so much mathematical work that was really valuable. *Upon the work of these years America built,* and comparatively rapid progress after 1850 is due in no small measure to the broad and solid foundation laid in the earlier period.[19]

The wonder is that historians have not reconsidered glib assumptions about Americans and exact sciences. While Americans prior to 1850 or just afterward did not spawn schools of great mathematicians, it is true that the evidence requires us to rethink our positions on many aspects of the history of science and technology in the United States. Would we have a different historiography if our predecessors had started from the books of Karpinski and the thought of such men as Nathaniel Bowditch; Bowditch's student, the mathematician-astronomer Benjamin Peirce; and Peirce's students, Charles Sanders Peirce and the astronomer Simon Newcomb? Suppose they had been impressed by William Ferrel's work on the mechanics of the earth's atmosphere?

Most indifference-theme authors implicitly compare the United States with all of Europe, or at least with all of western Europe. A comparison of the two as they were in 1800 is invalid. America and Europe were probably not comparable as late as 1900. At issue are not only relatively simple matters of population and gross national product but intangibles such as the nature of educational systems, the quality of scientific institutions, the efficiency of scientific communication, and, indeed, the nature of the entire relationship of the scientific community and the society at large. The implicit comparison with all of Europe is not based on any findings of statistical equivalency; it arises from a pattern of unreasonable expectations. For example, in Denmark, the succession Brahe, Römer, Oersted, Bohr is roughly one significant physical scientist per century,[20] the same average as the United States of the eighteenth and nineteenth centuries with Franklin and Gibbs. I am not aware of any breast-beating or finger-pointing literature about science in Denmark. Neither the Danes nor anyone else expects Denmark to dominate or to match the other nations in scientific fecundity. As a matter of fact, some smaller European nations contribute more to science relative to their population than larger powers. Obviously, a different pattern of expectations is applied to the United States.

The indifference charges often have strongly nationalistic undercurrents. Not only should the United States participate significantly in the great human achievement know as science, but it should lead. One can understand why scientists would want to see an expansion of science in America and even seek the pride of first place. It is unclear how historians of science who stress the autonomous development of the sciences largely, if not wholly, determined by the internal logic of the sciences, can point to numbers and wealth as reasons why America should have or might have contributed more to science. Nor is it clear what is so strange or horrendous about Americans not being among the founders of the cell theory or any other major scientific development. There are sometimes rather prosaic explanations of the improbability of such American contributions. Except for nationalistic pride, is there any reason why Americans or any other nationality *must be* among the founders of any major scientific concept? As the number of original theorists is necessarily small, an honest historiography will disappoint many nationalisms.

Perhaps the strangest characteristic of the professional historians' reactions to scientists' complaints of public indifference is the former's lack of critical scrutiny. Much modern historiography of the sciences originates in attempts to complete or to correct the traditional accounts in scientific literature. Many historians of science now consider these accounts spurious myths that serve to inculcate orthodoxies about the nature of science. The very men who will insist that students of the history of physics must scour the printed sources and rummage through bundles of old manuscripts blandly accept scientists' assertions on the indifference theme. Are the statements factually accurate? Were the authors self-seeking or otherwise motivated? Because past historians were predisposed to agree with the scientists' laments, they often misunderstood the context from which the writings originated and missed key developments. For example, from the studies produced—amid charges of public apathy—near the turn of the century, no one could predict the development of genetics. Yet the quality of much American biological work in 1900 is quite evident in historical perspective.

As to complaints about the lack of support and appreciation, are they confined exclusively to the United States? When haven't scientists complained? Observers from the "other culture" sometimes feel that the scientists are perpetually dissatisfied out of sheer greed and the drive for power. A more reasonable explanation is that scientists have tended to complain because of the growing pains of their disciplines. We can speculate that, since the time of Newton, the sciences have grown rapidly as a consequence of a series of major intellectual advances. Men attracted to the sciences were caught up in the sheer excitement of the advances, which, in turn, attracted more men to science. At any given moment, support for the expanding sciences probably tended to lag behind needs. Strongly committed scientists were very conscious of the gap between needs and actual support. Not surprisingly, they complained. We can even speculate that complaints are a sign of a healthy, lusty scientific community. Their absence may denote dangerously smug self-satisfaction, stagnation, or actual decline. In retrospect, we can see that when the French scientists stopped complaining early in the last century and the Germans stopped complaining toward the end of the century, both were entering periods of relative decline. The British, to their credit, have complained fairly steadily since at least the start of the last century.[21]

If we want to trace accurately the development of the sciences in America and what relations, if any, existed between the sciences and other aspects of American life, we have to avoid the easy generalizations engendered by the indifference literature. Perhaps the best way to start is by making two observations.

The first is that, so far as we can judge by available evidence, in the modern period applied research has always outbulked pure or basic research. This is true today in every major scientific community; it was apparently true from the time of Newton to the immediate past and in every country of any scientific consequence.[22] Providing health, foods, and goods understandably occupies much of the energy of mankind, including researchers. The indifference literature, animated by an almost aristocratic disdain, ascribes American applied work to sordid motives. Such motives do exist; however, many of the problems of improving health, food production, and technology are quite challenging to inquiring minds. These challenges may arise from enlightened concerns, as well as from sordid motives. In any case, realism indicates that very few men are endowed with the ability to perform truly basic research.

The second observation is inspired by a very good popular history of astronomy published in 1885. Citing the rise of astronomical research in the United States, the author notes the existence in 1882 of 143 observatories.[23] These ranged from world-famous institutions to rather minor establishments. Some practical applications of astronomy did exist in 1882—time determinations and observations that related to surveying. These hardly justified 143 observatories. Until the space program, modern astronomy was very nonutilitarian. The large number of observatories certainly does not indicate an absolute indifference shattered only by a freakish Gibbs or Henry.

The underlying problem obscured by the indifference literature can be stated as follows: Given the existence of at least some basic research in the United States and the general preponderance of applied concerns over theoretical concerns, in what qualitative and quantitative respects did the American research experience differ from the European? If the differences are slight, we can characterize the American experience as a variant of the European situation; if the differences are substantial, two further questions arise: What are the origins of the differences? Are we faced merely with a gross variant or a new, unique pattern?

Of the qualitative aspects, three deserve mention: the great man issue, the nature of scientific work performed, and the distribution pattern of this work. Great men, great discoveries, and great traditions do exist and deserve careful investigation, but a historiography which is overwhelmingly based on these obscures the reasons why men, discoveries, and traditions are great and, most important, the processes involved in their genesis. Centering on greats guarantees a present-minded historiography making neat longitudinal sections backward in time—the "main line" method. What we really want are successions of cross sections in time, each showing the interrelations of various activities at a given moment. The sequence of cross sections will enable us to formulate concepts explaining changes in time. Too many chance factors are involved in the appearance or nonappearance of great men for us to derive generalizations with any confidence. Who in 1900 could have predicted a Copenhagen school of physics in the twentieth century?

More promising is the study of the nature of the scientific work done in America. We have barely started on this line of investigation. The internalist historians of science posses skills that are essential, provided that they are supplemented by a critical concern for the total environment in which research occurs. Shryock's orientation, not Cohen's, is what we need.

Rather than judge scientific work in terms of nearness to a "main line" which has been determined in retrospect, it is necessary to understand clearly what occurred in a particular scientific investigation; what the origins of the investigation were; how the work fit into its contemporary scientific scene; and what, if any, its consequences were. Of these, the most important are the relationships to contemporary scientific work. Lonely precursors of later trends and geniuses towering across the ages are all too rare, and their appearance is based on too many chance hazards for fruitful insights into the norm of science at any time. The position of scientists and their work in terms of the standards of their contemporaries is what counts in judging the quality of the past scientific community. If this favorable judgment by peers is maintained and enhanced in time, we have a clear sign of the intrinsic strength of the scientific community and the presence of favorable conditions in the society for research. Changes in judgment are signs of change in scientific values.

Beyond the question of quality is the matter of the pattern of distribution of research effort. The indifference literature clearly postulates the existence of an excessive concern for descriptive, nonquantitative, and nonexperimental scientific endeavors in the American past. This type of research was not uniquely "American."[24] Some recent historical work clearly questions the overwhelming dominance of such scientific research in the American past.[25] Americans, after all, read European scientific journals and studied European monographs and textbooks, and a significant number in the last century went to European universities. More interesting is how some fields rose earlier than others in America; how particular fields became ossified; and how others became favored recipients of society's largesse. The indifference literature suffers from being couched explicitly or implicitly in quantitative terms while lacking numerical foundations. "Less" and "more," "increase" and "decline," and "compared to" abound in the genre. Proving or disproving the assumption of indifference necessitates quantitative evidence.

If a colonial status for early American science is assumed, and if a subsequent sudden emergence of a metropolitan scientific pattern is further discerned, there really is nothing to analyze quantitatively. Colonial status in science can be defined not only as being small and dependent on metropolitan scientific communities for training, intellectual approbation, and basic conceptual tools but also as partially replicating the metropolitan research and institutional patterns.[26] Being colonial in this sense does not mean doing particular and inferior kinds of science; such activities are also found in metropolitan scientific communities. A colonial scientific community is distinguishable from a small but mature nations' science by the fairly complete range of research and supporting institutions of the latter. Although participating in an international endeavor, the small but mature national scientific community will not have an almost complete intellectual dependence on others.

The colonial period in American science was more or less over by

1825. (This date and the following ones are not meant to imply abrupt, complete transitions, only approximations.) From that date to about 1875, science in America entered a crucial middle period in which the foundations of our national scientific community were established. By the terminal date, America was clearly a mature but small scientific nation with many of its basic institutions and attributes in existence or in embryo.[27] Obviously, not all fields were equally strong. From 1876 to 1920 the scientific community developed further into a major world intellectual body. Opinions will naturally differ, but 1900 is a fair approximation of the date the United States became the scientific peer of any nation in Europe. Please note, *peer* does not mean *leader*. Sometime between 1920 and 1940, that role passed to this side of the Atlantic. The situation varied from discipline to discipline.

If these views are correct, we will not find a quantitative curve that shows a pre-nineteenth-century rise, followed by a relative decline and then an upsurge in this century. More likely is an unspectacular curve going from small (1825) to medium (1875) to large (1920) and then to very large (1960).

We need historical statistics for the number of American scientists and engineers, the number of articles and other written contributions, and the amount of financial support given to research. These should be further broken down by kind of scientific and technological research and by institutional location. But even these series are insufficient, giving rise to a fundamental problem, the difficulties of comparison. How can one prove, rather than merely assert, that Americans are more or less indifferent to basic research if there is no European yardstick? Only the preoccupation with greats and the emotional belief in anti-intellectualism as a basic feature of American life can explain this curious oversight in the writings of so many talented men. We need qualitative and quantitative data for *both* Europe and America to know if research patterns differ and in what degrees.

It is likely that future historical studies will show that the American research pattern in the last century was not overwhelmingly different from the as-yet-undetermined European norm. Recent historical work in this area points to the discovery of much more basic or pure research than was predictable from the traditional accounts, in addition to the previously mentioned recognition of more scientific work of an abstract, theoretical kind (i.e., non-Baconian) than that indicated in the indifference literature. Deviations from the European norms are, in all probability, minor, and result from simple time lags or temporary local conditions. The American pattern is not unique, but a variant from the European norm.

A proviso is that the American data will differ most from the European data at the start of the nineteenth century, when the nation was still dependent on Europe for training, intellectual norms, and concepts. By 1860, the qualitative pattern and the gross quantitative data in the United States might be between those of the kingdoms of Prussia and Portugal. One can wave the flag slightly by predicting results somewhat closer to Prussia than Portugal. By the end of the century, the United States' commitment of a proportion of its gross national product to research and development was probably roughly equal to Germany's—perhaps even slightly larger. By 1920, the American commitment certainly justified comparisons with a war-ravaged Europe. Indirect evidence supports these assertions. The basic American university pattern was in flourishing existence by 1900. One can cite the favorable comments of foreign scientific visitors. A rather impressive number of new men and new institutions became prominent in the years 1890–1910.[28]

Why, then, was this not recognized in 1900? The explanation is that the improvements were not at all obvious at the time. The payoff on this national investment in research and development did not come immediately or all at once. A time lag was involved. Wasn't it the case, however, that the physical sciences were still conspicuously lagging at this date? A community of physical scientists in 1900, including A. A. Michelson, Simon Newcomb, Henry A. Rowland, J. Willard Gibbs, S. P. Langley, T. W. Richards, G. E. Hale, R. W. Wood, Ira Remsen, and many others of note, was quite respectable, even if not at the top of the heap. Here, too, the full harvest would come in not too many years.

A second proviso relates to technology. The qualitative and quantitative studies previously called for will undoubtedly disclose a technology quite different from the traditional accounts in which inventors loom so large. The Rube Goldberg gadgeteer struggling against the odds in a garret will prove less important than the applier of science and the engineer of systematic production and distribution. What we will need in the future are studies of technologies existing in a domain between the mere empirics of the inventors' world and the aristocratic dispensers of Newton's mechanics.

If the writers on the indifference theme were impelled by beliefs in the overwhelming effects of practical pressures and the pervasiveness of anti-intellectualism—in spite of rather dubious evidence—the views cited here also arise from an intellectual bias. Even in the absence of the requisite evidence, the scientific analogue of the Turner thesis seems dubious. In F. J. Turner's view, frontier conditions determined a unique American historical development; the indifference theme is an obvious parallel. It is inconceivable that science in America was a unique case because of American conditions. There are too many evidences in print and in manuscript of European-American scientific contacts. If anything, early American science is a splendid example of the opposing historiographical view of the transit of civilization westward—followed in short order by a return flow of research findings.

An early, persistent minor note in the indifference literature and related works on the state of American science is on the paucity of supporting scientific institutions. Complaints of this nature become increasingly rare in this century. The lack of professorial chairs, learned and professional societies, scientific periodicals, university laboratories, and scientific bureaus was once very apparent. America was rich in land, people, and natural resources but poor in institutions. As these were needed to start and to sustain science, the last three-quarters of the nineteenth century in America was a great period of institution building. Quite often the scientists were consciously trying to create American counterparts of European science—or at least what was believed to be European science. The effort was, by and large, a great success. Not surprisingly, however, this scientific community differed in many respects from the original European model.

Rather than a relatively small group of scientists accorded great esteem, official status, and financial support, the American scientific community started out as an amorphous group struggling to differentiate itself from amateurs. Thanks to the institution building, amateurs were largely excluded. But American society continued to rate men of practicality as the equals of men of science. As American scientists passionately believed in the importance of their labors, the equation with inventors and the like was a matter of great concern, if not anguish. The anguish possibly arose from unfulfilled aspirations to the supposed social status and power of European scientists in the face of competition with technologists in a presumably anti-intellectual society concerned with getting ahead.

Was science really valued more for itself in Europe? Quite a number of interesting historical studies remain to be done on the American image of scientific Europe and how that image changed as Americans gained more firsthand knowledge. The image was based largely on a few great names—men with seats in ancient universities, men who became noblemen or knights, men who received subventions from the state. The many who could not find positions or even get education at all were forgotten, as was the class structure, which largely determined who entered the circle of science.[29] Gauss and Faraday were not at all typical. Forgotten also were the abundant signs of European concern with the practical uses of science.[30] Those who praised European support of abstract science often overlooked the degree that support originated in and depended on desires for national prestige and glory for the sovereign. Earning the psychological benefits of prestige and glory are practical goals, just as much as building a better mouse trap or growing two blades of grass where one grew before.

To writers on the indifference theme, it was reprehensible that Americans knew of Thomas Edison but had never heard of his great contemporary, J. Willard Gibbs. Americans *should* know about Gibbs.

We could prepare new history textbooks exalting theory over gadgeteering, in which Gibbs looms large and Edison is simply a name in a list of inventors. But why worry about public acclaim if advancement of knowledge is the real concern? Gibbs did much scientific work; loyal to basic research, he was indifferent to mass adulation.[31] There is no point in substituting a pure science myth for an inventor's myth. The man in the street in London is probably not familiar with James Clerk Maxwell's achievements; neither is the French peasantry conscious of Baron Cuvier; nor are there many in the crowds of the Munich Oktoberfest who have ever heard of Karl Friedrich Gauss.

This point illustrates a confusion in the literature about the differences between general or popular knowledge and professional knowledge. In part the confusion stems from the assumption that the two were once identical, i.e., that the knowledge of the public and the knowledge of the working scientist largely coincided. As science became more complex, it is further assumed, the two knowledges diverged and science became the arcane property of an elite. This thesis is faulty on two grounds: first, in the past most people were not literate and obviously did not partake of whatever science existed; second, even among the educated of the past, it is very doubtful if all could comprehend the full development of theories and facts, even if so inclined. Ptolemy was not exactly easy to comprehend; neither was Copernicus; nor was Newton's mind readily accessible to the public. Obviously, a large portion of the professional scientific knowledge remained professional. What the public should know, and how to reach them, is a major problem, especially in a modern, democratic society.

American scientists looking at Edison in the limelight and Gibbs in solitary splendor at Yale could not overcome their ambivalent feelings about the relations of science to technology. The ties between the two were probably closer than in Europe. The number of scientists was growing, and their institutions were flourishing. Yet the feeling persisted that science was not adequately recognized by society and that the scientists were not honored as such. The scientists did not (and do not) want to be loved for winning the frontier, for advancing technology, for helping to make bombs and missiles, or even for refurbishing ghettos and restoring the purity of the environment, but for advancing knowledge.

The indifference literature is a statement of dismay, maybe despair, at the insensitivity of the American people to the importance of advancing knowledge. The scientists fail to recognize the difference between the American and the European situations. The issue here is not the pressures for practical results or the urge for power and status, although scientists are not immune to these. The main difference is a matter of numbers and the question of how a large, nonelite scientific community can function in a mass culture with an ideology derived from an elitist, European situation.

By 1900, almost all leading American institutions of higher education differed from those in Europe in one basic feature. Instead of one professor for each major field, American institutions had departments with two or more faculty. Instead of a small number of very highly esteemed professors, there were many professors and therefore more opportunities in science. Only after World War II did some European institutions slowly start departments on the American model. In America, unlike Europe, science was an activity carried on by a large number of people—not counted in tens or hundreds, but in thousands, tens of thousands, and hundreds of thousands. All were presumably properly educated and formally equal, although in talent and productivity differences, naturally, did exist. All, however, were supposed to do research. Despite recurring complaints about the glut of scientific literature, the number who performed research and the quality of publications are impressive.

The leaders of the scientific community—perhaps the rank and file also—were not always conscious of the change. They were thinking of Europe and looking for a few great men. Two leaders of the American scientific community proposed ways to overcome what they saw as America's scientific deficiencies, especially in the physical sciences. In 1902, R. S. Woodward, who would later become head of the Carnegie Institution of Washington, proposed to its trustees the erection of a laboratory for a few outstanding physicists.[32] George Ellery Hale, in post–World War I planning for the National Academy of Sciences, similarly proposed laboratories for a handful of great investigators.[33] Woodward's proposal was a bit too extravagant for the trustees, who leaned to other fields. Hale's proposal lost to the argument that these few investigators should be at institutions of higher learning. The leaders of the scientific community—at least up to World War II—never seemed to realize the nature of their problem. What was needed was not support for a few great or near-great physicists, zoologists, chemists, physiologists, or mathematicians, but support for the labors of a mass scientific community.

The problem was peculiarly American. English scientists were utterly indifferent to Cockney ignorance of James Clerk Maxwell. After all, only a small percentage of the British population would go to universities. The class structure and the system of interlocking institutions would require that only a small number need know of Maxwell and appreciate science. Nothing like this existed in the America of the last century, already on the way to a mass culture. Edison outpolling Gibbs was bad enough. How could the scientists compete for popular esteem with politicians, show business types, athletes, and evangelists? How could basic research be sold to a society where everyone reads the comics?

To this day, the scientific community is still somewhat ill at ease in a mass culture. It still tries to sell basic research by aligning it with practical goals of popular appeal.[34] More significantly, starting in the last century, the scientific community has fairly consistently attempted—not always successfully—to form privileged sanctuaries for its way of life. Yet the secure position that science now occupies in America is a great credit to the institution-building talents of the scientists. It also indicates that Americans, who may or may not have been indifferent, were not particularly hostile and were remarkably tolerant of this deviant group in their midst.

Notes

1.   A recent and quite superior example of this concern with science and technology in the United States is the report of the Organization for Economic Cooperation and Development, *Reviews of National Science Policy: United States* (Paris, 1968).

2.   The United States was apparently the first nation to expend over 3 percent of GNP on research and development. See ibid., p. 32. (At the time, Soviet data were not available in a readily comparable form.) Clearly, the 1969 estimated support of basic research, $3.73 billion, is the largest in absolute terms, and perhaps relatively as well. See National Science Foundation, *National Patterns of R & D Resources* . . . (Washington, D.C., 1969).

3.   There is no complete bibliography of such works. The best published starting point is I. Bernard Cohen, *Science and Society in the First Century of the Republic* (Columbus, Ohio, 1961), with its useful bibliography.

4.   For reactions of two leading postwar scholars, see A. Hunter Dupree, "The History American Science—A Field Finds Itself," *American Historical Review* 71 (1966):863–874; Edward Lurie, "An Interpretation of Science in the Nineteenth Century: A Study in History and Historiography," *Journal of World History* 8 (1965):681–706.

5.   See Richard Shryock, *Medicine in America: Historical Essays* (Baltimore, Md., 1966). A bibliography of Shryock's writings is in the *Journal of the History of Medicine and Allied Sciences* 23 (January 1968):8–15.

6.   Originally in *Archives Internationales d'Histoire des Sciences* 28 (1948):50–65.

7.   Richard Hofstadter, *Anti-Intellectualism in American Life* (New York, 1963), p. 26.

8.   See *International Encyclopedia of the Social Sciences,* s.v. "Science, history of."

9.   *Osiris* 4 (1938):414–565.

10.   A good introduction to the controversy is George Basalla, *The Rise of Modern Science: Internal or External Factors?* (Lexington, Mass., 1968).

11.   For an Americanist-internalist confrontation, see Shryock's comments in "Conference on the History, Philosophy, and Sociology of Science," *Proceed-*

ings of the American Philosophical Society 99 (1955):327–354; and Herbert Dingle, "History of Science and the Sociology of Science," *Scientific Monthly* 82 (1956):107–111.

12. Cohen's work includes: *Benjamin Franklin's Experiments* . . . (Cambridge, Mass., 1941); "How Practical Was Benjamin Franklin's Science?" *Pennsylvania Magazine of History and Biography* 69 (1945), 284–293; *Franklin and Newton* . . . (Philadelphia, 1956); *Some Early Tools of American Science: An Account of the Early Scientific Instruments and Mineralogical and Biological Collections in Harvard University* (Cambridge, Mass., 1950); "Some Reflections on the State of Science in America During the Nineteenth Century," *Proceedings of the National Academy of Sciences* 45 (1959), 666–677; "La Vie scientifique aux États-Unis au XIXe siècle," in R. Taton, *Histoire générale des sciences* (Paris, 1957) vol. 3, pp. 633–644; "Science in America: The Nineteenth Century," in *Paths of American Thought,* edited by A. M. Schlesinger, Jr., and Morton White (Boston, 1963), pp. 167–189; *Science, Servant of Man* (Boston, 1948); "American Physicists at War: From the Revolution to the World Wars," *American Journal of Physics* 13 (1942), 233–235, 333–346; "Science and the Revolution," *Technology Review* 47 (1945):6; "Science and the Civil War," *Technology Review* 48 (1946):5.

13. In Philadelphia, especially, and in Boston there were signs of emerging small groups of the scientifically minded toward the end of the colonial period. This could hardly be compared in scale and quality with events in the next century; for example, compare the Harvard and Yale faculties in 1760 and 1860. A recent dissertation estimated that Americans in the years 1800–1860 produced approximately 10,000 scientific papers. Donald Beaver, "The American Scientific Community, 1800–1860: A Statistical-Historical Study" (Ph.D. diss., Yale University, 1966), p. 102.

14. R. P. Stearns, "Colonial Fellows of the Royal Society of London, 1661–1788," *Osiris* 8 (1948):73–121. Stearn's purpose, of course, is not to list merely those from the thirteen original colonies but from all of British North American (including the West Indies), for he is concerned with the total intellectual traffic across the Atlantic.

15. See Nathan Reingold, "Alexander Dallas Bache: Science and Technology in the American Idiom," *Technology and Culture* 11 (April 1970):163–177: idem, "Cleveland Abbe at Pulkowa: Theory and Practice in the Nineteenth Century Physical Sciences," *Archives Internationales d'Histoire des Sciences* 17 (1964):133–147; Charles Davies, *The Logic and Utility of Mathematics* . . . (New York, 1850), pp. 3–4, a nineteenth-century statement of this view.

16. In recent years the problems arising from the terms *pure* and *basic,* not to mention *applied* and *development,* have been met somewhat by the device of postulating "mission-related basic research." Part of the problem arises from different ways of looking at research because of different needs (i.e., where funds originate, where they are spent, who performs the work).

17. George H. Daniels, *American Science in the Age of Jackson* (New York, 1968), p. 21.

18. See John M. Clarke, *James Hall of Albany* (New York, 1922).

19. L. C. Karpinski, *Bibliography of Mathematical Works Printed in America through 1850* (Ann Arbor, Mich., 1940), p. 17.

20. T. Brahe (1546–1601), the astronomer; O. Ròmer (1644–1710), the

astronomer; H. C. Oersted (1777–1851), the natural philosopher; N. Bohr (1885–1962), the physicist.

21. See D. S. L. Cardwell, *The Organisation of Science in England: A Retrospect* (London, 1957). Interestingly, British complaints to this day are fundamentally different from American complaints. Where Americans have complained about the lack of support for the advancement of knowledge, the British have usually complained about the lack of support for the proper application of science.

22. The volume of publications is a good indicator. An examination of distribution of funds would also be helpful, but comparable historical data are lacking. For the current publication situation, see Charles P. Bourne, "The World's Technical Journal Literature: An Estimate of Volume, Origin, Language, Field, Indexing, and Abstracting," *American Documentation* 13 (1962): 159–168. See also David A. Kronick, *A History of Scientific and Technical Periodicals* . . . (New York, 1962), passim; H. C. Bolton, *A Catalogue of Scientific and Technical Periodicals, 1665–1895* . . . (Washington, D.C., 1897); and W. A. Smith et al., *World List of Scientific Periodicals Published in the Years 1900–1950* (New York, 1952).

23. Agnes Clerke, *History of Astronomy During the Nineteenth Century* (London, 1885), p. 8.

24. See Royal Society of London, *Catalogue of Scientific Papers,* 1st ser., 6 vols. (Cambridge, 1863–1872).

25. Daniels, *Age of Jackson;* Beaver, "The Scientific Community"; P. A. Richmond, "A Selected Bibliography of American Fundamental Scientific Research During the Nineteenth Century", M. A. thesis, Western Reserve University, 1956).

26. What follows consciously differs from the explanation of scientific colonialism given in George Basalla, "The Spread of Western Science," *Science* [n.s.] 156 (1967):611–622. Specifically, the relevance of Basalla's analysis is doubtful in two respects. First, the stress on a different kind of science based on geographic newness is not basic. Similar activities occurred in Europe. Second, his analysis, with its strong emphasis on non-Western countries, does not apply to the United States. The settlers were of the West; science was part of cultural baggage brought across the Atlantic. The concept of provincialism is preferable. The early colonists were in the same relationship to European centers of science as were Europeans in rural areas and smaller towns and cities.

27. For the early period, see John C. Greene, "American Science Comes of Age, 1780–1820," *Journal of American History* 55 (1968):22–41; idem, "The Development of Mineralogy in Philadelphia, 1780–1820," *Proceedings of the American Philosophical Society* 113 (1969):283–295; and Leonard Wilson, "The Emergence of Geology as a Science in the United States," *Journal of World History* 10 (1967):416–437. For the middle period, see George H. Daniels, "The Process of Professionalization in American Science: The Emergent Period, 1820–1860," *Isis* 58 (1967):151–166.

28. See the *Yearbooks* of the Carnegie Institution of Washington, 1902–1916; G. W. Corner, *A History of the Rockefeller Institute, 1901–53* (New York, 1964). For American higher education, see Lawrence R. Veysey, *The Emergence of the American University* (Chicago, 1965); Joseph Ben-David, "The Universities and the Growth of Science in Germany and the United States,"

Minerva 7 (1968/69):1–35; S. M. Guralnick, "Science and the American College: 1828–1860" (Ph.D. diss., University of Pennsylvania, 1969). Among the favorable comments of foreign visitors are W. G. Waldeyer, "Relations Between the United States and Germany, Especially in the Field of Science," *Annual Report,* 1905, Smithsonian Institution, pp. 533–547 (translated from the original in the *Sitzungsberichte des Königlichen Preusschen Academieder Wissenschaften,* Jan. 26, 1905); Maurice Caullery, *les Universités et la vie scientifique aux Etats-Unis* . . . (Paris, 1917). For the 1920s, see League of Nations, Committee on International Cooperation, *Enquiry into Conditions of Intellectual Work, Second Series: Intellectual Life in the Various Countries: The United States of America* (Geneva, [1924]), Brochure No. 41. See also James McKean Cattell, *Science* 63 (1926):188, for an assertion of U.S. scientific parity with Germany and Great Britain.

29. Cf. Fritz K. Ringer, *Decline of the German Mandarin: The German Academic Community* (Cambridge, Mass., 1969); and idem, "Higher Education in Germany in the Nineteenth Century," *Journal of Contemporary History* 2 (1967):123–138; and Cardwell, *Organisation of Science.*

30. An example of how historiographic commitments affect the secondary literature is Jacob Schmookler, "Catastrophe and Utilitarianism in the Development of Basic Science," *Economics of Research and Development,* edited by R. A. Tyabout (Columbus, Ohio, 1962), pp. 19–33. Schmookler has no difficulty, as an economic historian, in finding evidence for applied concerns, even in citing A. R. Hall, a confirmed internalist, to make the point for the period of the scientific revolution of the seventeenth century.

31. See Muriel Rukeyser, *Willard Gibbs* (Garden City, N.Y., 1942); and L. P. Wheeler, *Josiah Willard Gibbs: The History of a Great Mind* (New Haven, Conn., 1951).

32. Carnegie Institution of Washington, *Yearbook,* 1902, pp. 13–24.

33. See Ronald C. Tobey, "The New Sciences and Democratic Society: The American Ideology of National Science, 1919–1930" (Ph.D. diss., Cornell University, 1969), pp. 35–94; Daniel Kevles, "George Ellery Hale, the First World War and the Advancement of Science in America," *Isis* 59 (1968):427–437; and Helen Wright, *Explorer of the Universe: A Biography of George Ellery Hale* (New York, 1966)

34. Cf. George Daniels, "The Pure-Science Ideal and Democratic Culture," *Science* 156 (1967):1699–1705.

ESTABLISHING SCIENCE:

THE NINETEENTH CENTURY

4.□ Science in the Civil War: The Permanent Commission of the Navy Department

The article originally appeared in *Isis* in September 1958. I did the research at a time when the memory of the World War II scientific and technological mobilization was very much alive. From these events would come much of the post-1945 setting for research and development in the United States. I was interested in how the Civil War scene superficially resembled the situations of World Wars I and II while retaining a dynamic peculiar to its period. Although I did not formally develop it as such in the text, I was engaging in a comparative study of the three periods. Alexander Dallas Bache serving oysters to his colleagues on the Permanent Commission was not really identical to Vannevar Bush lunching at the Cosmos Club in Washington with Rear Adm. Julius Augustus Furer, the Navy's Coordinator of Research and Development in World War II.

In 1958 I realized that science and warfare was an area filled with great historical research problems, as well as a topic of major public policy debates. What I did not realize is that one day I would be studying Vannevar Bush with the same intensity I then accorded men like Alexander Dallas Bache and Joseph Henry.

War is a widespread, significant activity in human affairs. As in the case of medicine and religion, social and intellectual problems are endemic to the maintenance of armed forces and the conduct of warfare. The services to this day display distinctive tensions between theory and practice, between the creators of knowledge and devices and their eventual users.

The Permanent Commission provided opportunity for a case study of behavior at social margins where two or more groups interact. The civilian scientists and engineers and the commissioned officers were both undergoing professionalization. That had two aspects. As a result of professionalization they would cease simply being cultivated gentlemen bound loosely by ties of class and intellectual affinity. At the same time, each group strived to establish preserved, inviolate domains. In

1863–1865 the process was barely underway; even today, we see that the process never reached a point of complete, triumphantly professional hegemony. Overlapping margins and interpenetrations persist.

Within a period of less than a month during the course of the Civil War two organizations concerned with the provision of scientific aid for the U.S. Government were created. On 11 February 1863, Gideon Welles, the Secretary of the Navy, approved the organization of a Permanent Commission to advise the Department on "questions of science and art." The members of the Commission were Rear Admiral Charles Henry Davis, Chief of the Bureau of Navigation; Alexander Dallas Bache, Superintendent of the Coast Survey; and Joseph Henry, Secretary of the Smithsonian Institution.[1] On 3 March 1863, President Lincoln signed the act incorporating the National Academy of Sciences and permitting it to "investigate, examine, experiment, and to report upon any subject of science or art" referred to it by the Government.[2] Bache was the first president of the National Academy.

The appearance of these two organizations indicates a lively interest in the role of science in the Civil War, at least among the small group involved in the two agencies. Most of what follows is an attempt to explain why the Permanent Commission and the National Academy of Sciences accomplished so little during the Civil War in terms of scientific aid for the Government, that is to say, why the Civil War was not a "scientific" war.

The crucial characteristic of a "scientific" war is that a deliberate attempt is made to have the best scientific talents, usually basic scientists, improve existing or devise new weapons, equipment and processes, and that, as a consequence, new or drastically improved weapons appear in battle in sufficient quantity to alter the tactics and strategy of the armed forces. Three antecedent conditions are necessary for a "scientific" war. First, there must be an opponent capable of waging a "scientific" war. The reason why the Navy was relatively more progressive than the Army during the Civil War was that only in naval warfare did the Confederacy threaten the Union forces with novel devices. Because the United States encountered opponents capable of waging a "scientific" war only in the present century, its armed services generally lagged behind European powers in research and development until recently. Certainly, Indian wars or scuffles with the Mexicans did not call for superweapons. Second, there must be sciences at stages amenable to significant applications and scientists versed in those fields. Although these considerations are outside the scope of this paper, the relative scarcity of physical scientists in Civil War America and the fact that physics and chemistry in the mid-nineteenth century were not obviously pregnant with warlike possibilities are significant bench marks. The third condition is the existence of industries requiring and performing research and capable of translating research data into military hardware. In the absence of such industries the Federal Government was forced to rely on the chance, unreliable labors of inventors and amateurs of science. The position of the private amateur inventor in relation to the professional scientist was the crucial issue in the role of science in the Civil War. This paper will attempt to demonstrate how the role of the inventor largely determined official actions and policies in utilizing science in warfare.

Evaluating innovations in military technology was a familiar peacetime problem to the Army and Navy. Stimulated by a mixture of patriotic fervor and avarice too complex for historical analysis, inventors literally besieged official Washington after the outbreak of war. By the end of 1861 Secretary of the Navy Welles established a Naval Examining Board to appraise inventions. His instructions stressed that recommendations to adopt an invention should state "the advantages and the economy that will result from its use and the total expenditure that it will occasion." For, as the Secretary had previously admonished, "the money appropriated by Congress for the Navy cannot be applied to any experimental purpose but only for objects of undoubted utility." The Board functioned from 2 January to 10 July 1862 and accomplished very little.[3]

In 1863 the Navy Department's need for a means of reviewing inventions coincided with more ambitious stirrings in the American scientific community—notably Davis, Bache, and Henry. Rear Admiral Davis was one of the few naval officers of that period with any substantial scientific competence. His Bureau of Navigation, embracing the Naval Observatory, the Hydrographic Office, and the Nautical Almanac Office, was intended to be the Navy's scientific bureau.[4] Bache, a West Pointer who became a geophysicist, was a first-rate scientist with pronounced abilities as an organizer and administrator. He was the leader of an influential group of scientists and had favored the organization of an American society of savants at least since 1851.[5] Joseph Henry was a very cautious protector of the Smithsonian endowment from proposals conflicting with his organization's primary purpose, "the increase and diffusion of knowledge among men." During the Civil War he readily gave his services and facilities of the Smithsonian Institution to the war effort. Henry was simply seeking a way of accomplishing a disagreeable but necessary task.[6]

The bare outline of the events leading to the formation of the Commission and the Academy is as follows: Some time after Davis returned to Washington in November 1862, the scientific coterie around Bache began to discuss the problem of utilizing science in the war effort. By January two specific proposals had progressed from talk to action. Because of Joseph Henry's initial opposition to the proposed Academy on the grounds that Congress would not pass such an act and that a National Academy would arouse jealousies among scientists, the National Academy was ostensibly dropped in favor of a Permanent Commission, a select standing body to advise the Navy. On 26 January 1863, Bache, Henry, and Davis conferred with Assistant Secretary of the Navy Gustavus V. Fox. Further details were discussed by the three on 5 February. Two days later Henry transmitted to Fox a "programme," which was almost faithfully copied in the 11 February letters sent over Welles's signature to the three original members of the Commission. But without Henry's knowledge, Bache, Davis, Louis Agassiz, and the astronomer Benjamin Apthorp Gould were actively promoting the National Academy. Henry heard of these efforts only after the bill for the incorporation of the Academy was drawn up, and learned of the bill's passage only on 5 March.[7]

While the legislation was pending, the Commission was already in operation. Its first meeting was held on 20 February 1863.[8] Two days later it received 41 proposals from the Navy Department; the earliest was dated 3 July 1862, and the latest 5 February 1863.[9] Some of the early proposals, as well as others subsequently transferred to the Permanent Commission, had originally been referred to the Naval Examining Board. The last meeting of the Commission of which any record survives occurred in April 1865, but the last report, bearing only the signature of Rear Admiral Davis, is dated 21 September 1865. Between 20 February 1863, the date of the first meeting, and 21 September, 1865, the date of the last report, the Commission held approximately 109 meetings and submitted 257 formal reports on proposals evaluated in those sessions.[10] Eighty-two of these meetings occurred in the period up to 24 February 1864, the date of the last meeting for which minutes survive. During that year and four days, 187 of the reports were prepared. After this period of its greatest activity, the Commission began to flag, and by the summer of 1864, it entered a period of marked decline. On the basis of the surviving records a reasonable estimate would be that some 300 inventions were transmitted to the Commission.

The operations of the Permanent Commission were quite informal, as befitted the deliberations of a group of old friends. Secretary Welles's instructions, based on Henry's "programme," simply authorized the use of associates and prohibited compensation for members as well as the associates.[11] The Commission itself adopted only two procedural resolutions during its lifetime. At the 13 April 1863 meeting it accepted a resolution of Bache's that "it is not expedient" for members to receive personal communications from persons having proposals before the Commission. It then approved Davis's resolution that the Commission "confine itself to plans and descriptions presented to it," i.e., those coming through the Office of the Secretary of the Navy.[12]

During its greatest period of activity the Commission held meetings as often as three times a week, usually rotating among the Bureau of Navigation, the Smithsonian Institution, and the Coast Survey offices. On occasions when demonstrations were called for, the Commission met at the Washington Navy Yard or elsewhere in Washington. Inventors or their representatives were encouraged to appear personally and explain their proposals. Those meriting special attention were assigned to one or more members for study before preparation of a report. Others were summarily rejected. In spite of all the serious business transacted at these sessions, they were also rather pleasant social occasions. (On 10 February 1864 the minutes gravely noted, "Mrs. Bache being unwell, the Professor regretfully announced that there were no oysters."[13])

The membership of the Commission was soon expanded to five by the addition of Joseph Saxton and Brigadier General John Gross Barnard. Saxton was Bache's assistant for work on standards of weights and measures. Initially an associate of the Commission in the summer of 1863, he became a full member, at Henry's urging, by the end of the year.[14] Although Henry had written the War Department at the same time as the Navy,[15] Barnard's presence was simply due to the fact that many subjects considered by the Commission were also of interest to the Corps of Engineers (i.e., harbor defences). Prior to July 1863 when he appeared on the scene, the War Department and the Permanent Commission had exchanged inventions. Barnard played a minor role in the Commission's deliberations. According to Joseph Henry, after the general joined the group, "our music is not quite as entirely as harmonious as before."[16]

Apparently the Civil War public was somewhat confused as to the relation of the Permanent Commission to the National Academy of Sciences. Both were founded at about the same time and in part for the same purpose. It was certainly confusing to have Bache, as a member of the Commission, vote to have a question referred to himself as president of the National Academy. The confusion was further compounded by the appointment of Academy committees including the very members of the Permanent Commission who had voted to refer the question to the Academy in the first place! The relationship between the two organizations apparently troubled the Commission members also, and they took pains to explain the situation in the only public announcement issued on the Commission's activities:

The present members of the Commission are also members of the National Academy of Sciences; and the Commission itself would probably never have been created if the Academy had been in existence at that time, since they both have the same objects, and are designed to perform similar duties; it is not impossible that the former may at some time be resolved into a Committee of the latter.[17]

But for the awkward fact that it was founded earlier and within the Navy Department, the Permanent Commission might have served as the operating arm of the Academy. That is to say, the Permanent Commission was an abortive National Research Council.

The questions referred to the National Academy by the Permanent Commission stemmed from the changes which revolutionized naval warfare in the nineteenth century—the introduction of armor and steam. The Navy, for example, encountered a serious difficulty in compass deviations due to the large masses of iron in its new vessels. The problem fell within the jurisdiction of the Bureau of Navigation, which furnished compasses and other navigational devices to the fleet. The same day that Congress incorporated the National Academy, it authorized the Navy Department "to make experiments for the correction of local attraction in vessels built wholly or partly of iron."[18] The matter was referred to the Permanent Commission, which appointed Wolcott Gibbs, Bache, and Henry to witness the correction of compass deviations in the steamer *Circassian* by the application of methods agreed on in conference with "practical" experts.[19] The three met in New York, where the *Circassian* lay, on 19 March 1863. A few days later Welles approved the appointment of the committee to conduct the experiments called for by Congress. This body consisted of the three from the *Circassian* group plus Benjamin Peirce and W. P. Trowbridge. The five met in New York on 21 April 1863, shortly before the first meeting of the National Academy. After the Academy was organized, the Permanent Commission referred the problem of magnetic deviations to it on 8 May 1863.[20] The Academy committee consisted of the five from the Navy group and Fairman Rogers and Charles Henry Davis. The problem was obviously of great interest to Bache and Henry. But although the Academy committee labored diligently, it did not make any signal contributions. The English astronomer Sir George Airy had previously proposed a means of coping with the problem, and several individuals, some of whom were hired by the Navy, were already applying Airy's method.[21] Nevertheless, the report of the Academy committee filled a definite need for expert appraisal of a significant practical problem which impinged on basic issues in physics. The Bureau of Navigation immediately established stations to correct compasses by Airy's method.

The experiments on the expansion of steam had origins similar to the work on magnetic deviations. They also stemmed from an act of Congress authorizing experiments, passed on 3 March 1863,[22] whose origins can be traced to the labors of "practical" experts, in this case Horatio Allen, president of an iron works, and B. F. Isherwood, Chief of the Navy Department's Bureau of Steam Engineering. As in the case of the magnetic deviations, members of the Permanent Commission were also interested in the general subject area. (One of Bache's earliest triumphs was the investigation of the explosion of steam boilers at the Franklin Institute.[23])

The experiments were originally referred to a Commission composed of Allen and Isherwood, but the instructions drafted by them were sent to the Permanent Commission. What happened between March 1863, when the Commission considered the proposed experiments, and February 1864, when the National Academy got into the act, is not clear. On hearing from Isherwood that the plans for the experiments were in the hands of the Permanent Commission, Allen wrote Bache that he would be most happy to cooperate with the Commission.[24] But no action was taken until February of the next year. At Fox's suggestion, the original two-man group was supplanted by a nine-man tripartite body composed of representatives of the Navy, the National Academy, and the Franklin Institute.[25] Davis, along with Allen and Isherwood, represented the Navy. The experiments were conducted at Allen's plant by Navy engineers and were never completed. The project was top-heavy; there was little need for a tripartite commission to supervise a fairly routine testing operation.

The committee on protecting ironclads from corrosion and fouling in salt water offers by far the most interesting example of Navy–National Academy relations during the Civil War. There is a reasonable doubt that the Department even considered the Academy's committee as acting in its behalf. It involved the Commission in the first of its two exasperatingly delicate encounters with Professor Eben N. Horsford, a food chemist who resigned in 1863 from the faculty of the Lawrence Scientific School at Harvard to engage successfully in the manufacture of baking powder and who during the war devised methods of condensing milk and producing field rations. One of the first actions of the Commission at its initial meeting was to make Horsford an associate to report on means of protecting ironclads from corrosion. On 27 April 1863, Horsford dispatched a long, comprehensive report on possible methods of combatting corrosion and fouling. Horsford devoted most of his attention to the possibility of electroplating vessels with copper, but he also discussed possible use of paints. The experiments on electroplating would take four to eight months and would use ships already constructed. On 8 May 1863, the Commission referred Horsford's report to the National Academy of Sciences and so informed Horsford.[26]

After an interval, Professor Horsford penned a letter to Welles on 25 June, protesting against the Commission's action:

I beg respectfully to suggest to the Department that my report contained the results of much thought, considerable labor and some expenditures, which are embodied in proposed methods of protection to which I attach great value, and which I am unwilling should pass to the knowledge of many persons, particularly in their unverified state.

Horsford asked that his report be restricted to members of the Permanent Commission; he would carry on all research at his own expense and forward the results to the Department.[27] The Commission immediately ordered the withdrawal of Horsford's report from the Academy on 9 July, 1863.[28]

The next day Davis wrote Bache:

Please send me the report of Mr. Horsford, "on protecting the bottoms of iron-clads," for reasons which I will explain to you at some very distant future period—say a hundred years hence.

Your very truly,
C. H. Davis

P.S. Lest this should appear too mysterious, I will mention the real reason, viz. that Mr. Horsford has written to the Secretary (under date of June 25th, & admitting the receipt of our letter of the 8th of May informing him of the reference of his Report to the National Academy), to say that he considers it confidential & does not wish it to go beyond the Commission.

Although Davis assured Horsford on the same date that his report had not gone beyond Bache, he subsequently learned that the efficient Bache had already sent out a copy to the committee of the National Academy. Bache returned his copy on 17 July and, after being reminded by Davis, returned the copy furnished the committee.[29]

One consequence of Horsford's action was that the committee of the Academy, formed to consider his report, had nothing to deliberate upon. As the members of the committee were not versed in the problem, their report was inconclusive. Henry, however, offered facilities in the Smithsonian Institution to carry on research if Congress appropriated funds.[30]

The Commission's second encounter with Horsford began on 3 August 1863, when Gideon Welles referred to it Horsford's proposal for a submarine. During the previous winter Horsford had conducted experiments on underwater transportation at the Washington Navy Yard—interestingly enough at the very time the *Alligator*, the first Navy submarine, was tested there[31]—and had discussed his vessel with Fox. On 8 June, 1863, he wrote Fox: "I have no doubt of my ability to open the way for the Monitor fleet to Charleston Harbour." The next day he dispatched a letter to Gideon Welles giving additional details: "I am prepared to contract to remove the obstructions to Charleston Harbour. I understand the offer of a million of dollars will be made for the accomplishment of this object." No such offer was ever made, but the rumor certainly acted as a stimulant to Horsford's imagination. He proposed to build two submarienes (the extra one in case of an accident to the first) in six weeks. The Government was to transport them to Charleston and to provide an ironclad or a monitor to aid the submarine. In the event of an early peace Horsford asked only for reimbursement of costs.[32] In the following month the plans for the submarine, the *Soligo*, arrived in Washington. The cost was $54,000, and Horsford now estimated that construction would require from three to four months.[33] Welles referred the plans to Commodore Joseph Smith, Chief of the Bureau of Yards and Docks, who was unimpressed by their originality, utility, or clarity but thought the price reasonable if Horsford guaranteed success.

The day after his proposal was received by the Commission, Horsford appeared before them to explain his invention. The Commission actually wrote two reports on the Horsford submarine. The first, dated 7 August 1863, gingerly noted that, although ingenious solutions were offered for purely scientific difficulties, the Commission hesitated to rule on the practicability of the vessel for warfare. Ten days later the Commission raised several criticisms. The method of steering in a vertical plane was questioned, as was the proposed use of telescope and compass to aid the navigator. The Commission finally pointed out the ease with which the skin of the submarine could be ruptured or penetrated.[34]

Henry wrote Bache:

He [Horsford] thought the contract would be entered into without sumitting the matter to the Commission, but in this he was mistaken. The plans were referred to the Commission, which, though it wished as far as possible to deal kindly with the Professor, could not indorse the invention. The result was Mr. Fox could not accept the proposition; for said he, we have no special appropriation for this object. . . . I informed the Professor that, although I was sorry that he should be troubled in regard to the matter, I was not sorry that his proposition . . . was not accepted.

Henry felt that Horsford had "considerable suggestive power and abounds in kind feeling." Henry further stated that Horsford was making a fortune in baking power but that submarines were beyond his depth.[35]

Undaunted by the initial rejection, Horsford once more proposed the construction of his submarine in 1864. This time the Commission agreed to test it because Congress had voted an appropriation for the tests of submarine inventions.[36] The Commission was willing to pay part of the construction expenses. But Horsford, who was trying to interest the Governor of Massachusetts and the Mayor of Boston in subscribing $10,000 and $5,000, respectively, was apparently unable to raise funds in time. The *Soligo* remained high and dry in Horsford's imagination.[37]

Contacts with scientist inventors like Horsford illustrate some of the awkward situations confronting the Commission. But it is the less dramatic relations with the Navy Department that provide significant clues for an understanding of the use of science in the Civil War. The most crucial day in the life of the Commission was probably 24 July 1863. On that date it approved a report on William Norris's design for an ironclad, recommending that it be examined by a "professional board" of navy officers. The report stated: "It would take the liberty, in this connection, also to suggest that it would, in general hardly venture to express its views upon purely technical questions, but rather to act upon such as involve principles or applications of science not forming a part of familiar knowledge."[38]

Bache yielded to the views of the majority while doubting the propriety of the Commission's reasons for not considering the plans: "True they involve much study & the aid of a professional board for the special occasion may be needed but there must be principles involved of which in my opinion the board should take cognizance."[39] The Commission had previously turned Norris down in an eleven-page refutation of his claims.[40] Referring Norris to a board of officers was a neat way of "passing the buck" to the Navy.

But the weakness in the position of the Permanent Commission majority was that most of their labors did not involve "principles or applications of science not forming a part of familiar knowledge." Except for the topics referred to the National Academy of Sciences and a few inventions such as submarines and balloons, the proposals evaluated largely fell within the area of "familiar knowledge" of the Navy and Army. Most could be rejected on the basis of experience; few were rejected solely on theoretical grounds. Bache, the West Pointer who became a geophysicist, was alone in sensing the desirability of having a board of officers test specific devices in cooperation with a high-level scientific group.

On 24 July 1863, the Commission also asked the Navy Department's permission to refer plans for experiments on armored vessels to the National Academy.[41] To this proposal the Secretary replied with an unequivocal rejection:

I think the subject somewhat professional, and should be investigated and reported upon by Officers connected with the Department, or a commission instituted by it, rather than the National Academy of Sciences.

A transfer of duties from the Department to that institution does not strike me favorably. I know not what funds the Academy has for extensive experi-

ments, the Department certainly has none which it would feel justified in placing at their disposal.[42]

The Commission had received many proposals on methods of armor-
plating vessels and the design of ironclads. "Practical" experts had proposed solutions that were in dispute. The members of the Permanent
Commission had obvious interests in the problem—Davis, as a naval
officer who had commanded armored vesels and had served on the
board which recommended the construction of the *Monitor*,[43] Henry because of his service on the board evaluating the Stevens Battery, and
Bache and General Barnard as engineers designing the coastal forti-
fications against the dreaded depredations of Confederate raiders. The
most significant difference between the proposed tests on armored vessels and the steam expansion and magnetic deviation tests was the
lack of statutory authority buttressed by an appropriation. What
Welles's reply said, in effect, was that the time for commissions and
boards was past—Davis and Henry had already served on these—and
now the Navy was concerned with the problems of production and combat utilization. The Department had a war to win and could no longer
wait upon the deliberations of scientists.[44]

The difference of opinion among the members of the Commission
persisted. The first overt sign was Report No. 174, 4 February 1864, on
the use of corrugated iron armor, which laconically noted that the
members "are not unanimous" in rejecting the proposal. Shortly afterwards a report rejecting John Ridgway's "vertical revolving battery"
was sent out at Bache's request without his signature. Ridgway's tur-
ret was presumably capable of firing in any direction or elevation, a
most beguiling idea but fraught with practical difficulties. Ridgway
had overcome the objections of the Naval Examining Board, which recommended a trial. When the Permanent Commission came into existence, it referred the device to Benjamin Peirce, the noted Harvard
mathematician. Peirce submitted a report on 23 March 1863, certifying
that the invention was theoretically sound, but stating, "I have made
no allusion to the difficulties of its construction . . . because I regard
the enquiries upon this point as belonging to a board of practical engineers and constructors." This report probably did not impress the Department, which had also received a report from one of its engineers,
A. C. Stimmers, pronouncing the invention impracticable. It must have
been galling to the Commission to have the opinion of Peirce set aside
by the opinion of a man who was regarded as a partisan of John
Ericsson and was subsequently the builder of a little fleet of unfloata-
ble ironclads.

The Commission was handicapped by the uncertainty of the boundary between the labors of its members as "scientific gentlemen" and

the labors of "practical" men. In Ridgway's case there was such a weight of authoritative opinion behind the invention that Bache was moved to open protest. Even the Commission's first report acknowledged the possible merits of the device. On 23 July 1864, the Commission reversed itself and reported that improvements in the design justified tests. The recommendation was disregarded.[45]

By raising his objection in the field of ironclads, Bache was applying pressure to an exposed nerve of the Navy Department. Navy policy had already committed public funds to an extensive construction program of monitors and other ironclads—in the words of Gideon Welles, all "objects of undoubted utility." Within the Department there was considerable doubt about these vessels, which were designed not for duty on the high seas but for harbor assault and defense. Outside the Department there was considerable pressure to use new designs of vessels, other turrets, and even different steam engines. Under these circumstances it is understandable that the Navy Department would be sensitive about letting outsiders do research on ironclads and even more sensitive about having designs or devices certified as superior to the "objects of undoubted utility" on which so much money had been spent. And the Commission members, with the exception of Bache, were not disposed to challenge the status quo.

Since the amount expended on a function is often a touchstone of Executive and Congressional esteem, the status of scientific research and development in the Navy during the Civil War is roughly proportional to the availability of funds. The Navy Department's position was quite clear; unless there were specific appropriations for tests or experiments, it could commit funds only for objects of undoubted utility. Even when the Department decided on the construction of a novel vessel, as in the case of the *Monitor*, all the risk was borne by the contractors, who would receive not a penny if the ship did not meet performance requirements written into the contract.[46] The environment was markedly different from the present situation, when vast sums are expended for research and development as a matter of course.

Bache and Henry were well aware of the situation. In a letter of 21 August 1863, Henry, in discussing the possibility of further referrals to the National Academy, stated, "It will be necessary that the Heads of Departments be not frightened" by proposals involving expenditures. He went on to relate an experience of his. At the request of Dahlgren, who needed them in the assault on Charleston, Henry prepared lighting devices. In the absence of Fox, he had to present his bill for reimbursement to Welles who insisted on personally approving every expenditure of $100:

Goodman, [sic] he has been so much in the habit of dealing with those who have no other aim than that of self advantage that he could not imagine that any one could have any other motive. Disgusted with being placed on so low a level I concluded to do nothing more with light and even not to make a charge for the expenses actually incurred.[47]

A trifling sum and a trifling incident but indicative of the Navy's parsimony toward research. In this environment the Navy Department could conceive of the Commission only as a negative device for rejecting inventions, not as a positive instrument for improving the fleet's capabilities.

Uppermost in the minds of the Secretary and his able Assistant Secretary, Gustavus V. Fox, was the embarrassment occasioned by the flood of inventions, not the desirability of coordinating the proposals with the research activities of the Department's bureaus (which were quite limited). Nor is there any evidence suggesting the conception of "mobilizing the nation's scientists" (the most significant professional groups being botanists and geologists, not physicists and chemists). What the Navy was violently criticized for during the Civil War was its lack of support of private inventors.[48] But the Army and the Navy knew from experience that most inventions submitted were not usable.[49] Inventions presented opportunities for financial bonanzas; and the resulting pressure on the military to adopt inventions raised the specter of misuse of public funds.[50] To winnow the few meritorious proposals from the dubious chaff was a serious problem, if only from the standpoint of placating public opinion impatient of the war's progress and eager to adopt promising devices.

Of course, the problem was not to be unique in American history. In World War I the Naval Consulting Board, headed by Thomas Edison, was set up to examine inventions. The National Inventors Council during World War II and up to the present has acted in a similar capacity. This remarkable solicitude for the individual inventor is the result of three American attitudes present today as well as in the Civil War. The first is pragmatic; in times of national emergency it would be folly to overlook any possible source of talent. The Wright brothers were, after all, only bicycle manufacturers when they became interested in aeronautics. The second attitude is the American distrust of experts, possibly because reliance on a few individuals violates egalitarian notions. Every major American war has been accompanied by criticism of the weapons used. That some of the criticisms were valid should not obscure the fact that most advances in military technology are the result of the labors of professional scientists and engineers, not of inspired amateurs. But encouraging private inventors seems an almost unconscious compensation for an irksome dependence on experts. The third attitude is a result of the prominent place occupied by inventors in the national folklore. Nurtured on tales of inventors whose discoveries transform the nation and reap wealth for themselves, the average

American was, and still is, quite receptive to proposals to facilitate the realization of this small part of the "American dream."

Notes

1.   Gideon Welles to Charles Henry Davis, 11 Feb. 1863, Letters Received, Permanent Commission, Naval Records Collection of the Office of Naval Records and Library (RG 45), National Archives. Unless otherwise cited, all records are in the National Archives. "RG" is the symbol for record group.
2.   12 Stat. 808.
3.   See Minutes of the Naval Examining Board and Inventions Referred to the Naval Examining Board (RG 45). Welles's instructions of 27 Dec. 1861 are attached to the front of the Minutes.
4.   For Davis, see C. H. Davis, Jr., *Life of Charles Henry Davis, Rear Admiral, 1807–1877* (Boston, 1899).
5.   Frederick W. True, ed., *A History of the First Half-Century of the National Academy of Sciences* (Washington, 1913), pp. 7–8. The possibility of splitting the American scientific community, many scientists being antagonistic to Bache's group, the "Lazzaroni," was recognized by the founders of the Academy. Louis Agassiz wrote Bache on 6 March 1863 (Rhees Collection, Huntington Library), "How shall the first meeting be called. [*sic*] I wish it were not done by you, that no one can say this is going to be a branch of the Coast Survey and/or the like." See also Merle M. Odgers, *Alexander Dallas Bache, Scientist and Educator, 1806–1867* (Philadelphia, 1947).
6.   An example of this attitude: on 15 July 1863 he wrote Assistant Secretary of the Navy Gustavus V. Fox, advising tolerance of an inventor whose letter "is more that of a visionary enthusiast than that of a modest benefactor of his race." He went on to state, "I never seek to be employed on public commissions since they are generally attended with more disagreeable than pleasant consequences. On the other hand I never refuse to lend my services gratuitously to the Government in any way which they may think of importance." In Fox Papers, New York Historical Society.
7.   Based on the following: Henry to Bache, 26 Jan. and 5 Feb. 1863, Bache to Henry, 14 Feb. 1863, and "Locked Book" extract of 28 Oct. 1863, Henry Papers, Smithsonian Institution; No. 103, Miscellaneous Letters Received of the Secretary of the Navy, Feb 1863, vol. I (RG 45), is Henry's letter transmitting the "programme," followed by the enclosure and Fox's penciled draft of the 11 Feb. 1863 reply of Welles; Davis, *Life*, pp. 289–292; Andrew Denny Rogers III, *John Torrey* (Princeton, 1942), p. 274.
For information concerning the founding of the Permanent Commission and the National Academy, I am indebted to Dr. A. Hunter Dupree (University of California, Berkeley), who has examined the Asa Gray and Benjamin Peirce Papers at Harvard University. Dr. Dupree has specifically called my attention to a letter of Henry's to Louis Agassiz, 13 August 1864, Peirce Papers, which outlines Henry's connection with the founding of the Academy.

8. Minutes, Permanent Commission, 2 (RG 45). The records of the Permanent Commission in the National Archives upon which this paper is largely based consist of the following: 3 vols., Jan. 1861–Dec. 1865, of Proposals Referred to the Commission; Minutes of the Permanent Commission, Feb. 1863–Feb. 1864, 1 vol.; Letters Received Feb.–June 1863, 1 vol.; Letters Sent, Mar. 1863–Sept. 1865, 2 vols. of press copies. The originals of the Reports of the Permanent Commission to the Secretary are also in RG 45 among the records of the Office of the Secretary of the Navy. The three volumes of proposals do not contain all the proposals considered by the Commission, as many were returned to their authors or referred to other agencies. On the other hand, many proposals in the volumes were never considered or were still pending when the Commission's activities ceased. One minor mystery connected with these records is the absence of any Minutes after Feb. 1864, especially since the surviving volume of Minutes contains several hundred blank pages.

9. Minutes, Permanent Commission, 6–7 (RG 45).

10. See the volume of Reports of the Permanent Commission (RG 45). The last record of a meeting is in Letters Sent, vol. 2, which has a letter, 12 Apr. 1865, calling for a meeting the following Friday. As Report No. 256, 10 July 1865, is signed by the full Commission, it is probable that a meeting or meetings occurred after April. The number of reports is from the Reports volume. The number of meetings is based on a count from correspondence plus the 82 meetings in the Minutes.

11. See note 1, above. The absence of any profit motive in the legislation authorizing the National Academy was probably one of the reasons that the *Scientific American* opposed the Academy. See the issue of 8 (23 May 1863):329. The magazine favored private, rather than public, support, especially in "practical" matters.

12. Minutes, 26 (RG 45).

13. Minutes, 151 (RG 45).

14. Henry to Bache, 13 Aug. 1863, Henry Papers, Smithsonian Institution; Minutes, 141.

15. Locked Book Extract, 28 Oct. 1863, Henry Papers, Smithsonian Institution. These extracts were apparently copied by Henry's daughter after his death from a volume containing a diary and/or copies of outgoing correspondence. The original volume was presumably destroyed afterwards.

16. Minutes 81, 94; AGO Special Order 275, 22 June 1863, in Letters Received, Permanant Commission; Henry to Bache, 13 Aug. 1863, Henry Papers.

17. *Scientific American* 10 (12 Mar. 1864):165. The Commission's statement was elicited by a previous item in this magazine [10 (27 Feb. 1864):135] which informed readers that inventions were to be sent directly to the Commission. The Commission was forced to return inventions so received and asked that they be addressed to the Secretary of the Navy. The 12 March announcement was mainly intended to correct this and other inaccuracies in the earlier item. See also Henry to Mrs. Alexander Dallas Bache, 16 July 1864, and Henry to Joseph B. Varnum, 8 April 1865, Henry Papers, Smithsonian Institution.

18. See True, *History*, pp. 215–217.

19. Minutes, 9–10 (RG 45)

20. Minutes, 19, 36. The Permanent Commission proposed the appointment of the five on 25 March. The Secretary's consent was received two days later (RG 45); True, *History*, p. 216.

21. See B. F. Greene, ed., *The Magnetism of Ships and the Deviations of the Compass*, a collection of papers on the subject by Sir George Airy, S. D. Poisson, and others issued by the Bureau of Navigation in 1867.

22. 11 Stat. 751; $20,000 was appropriated.

23. See *Report of Franklin Institute of Philadelphia, in relation to the explosion of Steam Boilers* (Washington, 1836). House Document No. 162, 24 Cong., 1 Sess. (Serial 289).

24. Minutes, 22–23 (RG 45). Allen to Bache, 3 Apr. 1863, Rhees Collection, Huntington Library.

25. Henry to Bache, 6 Mar. 1864, Henry Papers, Smithsonian Institution; Davis to Bache, 24 Feb. 1864, Letters Sent, Permanent Commission (RG 45).

26. Minutes, 2, 35; Horsford to Permanent Commission, 27 Apr. 1863, Letters Received; Permanent Commission to Horsford, 8 May 1863, Letters Sent (RG 45). See "Eben Norton Horsford," in *Dictionary of American Biography*, vol. 9, pp. 236–237. For other examples of his war activities, see Davis to Fox, 30 Apr. 1863, Fox Papers, New York Historical Society, and *Annual Report of the Smithsonian Institution*, 1862, p. 45.

27. Horsford to Welles, 25 June 1863, Letters Received, Permanent Commission (RG 45).

28. Minutes, 93 (RG 45).

29. Davis to Bache, 10 July and 28 July 1863, Letters Sent; Bache to Davis, 13 July and 31 July 1863, Letters Received, Permanent Commission (RG 45).

30. *Annual Report of the National Academy of Sciences*, 1863, pp. 4–5, 21–23. House Miscellaneous Document No. 79, 38 Cong., 1 Sess. (Serial 1200).

31. Louis H. Bolander, "The *Alligator*, First Federal Submarine of the Civil War," *U.S. Naval Institute Proceedings* 64 (June 1938): 845–854.

32. Horsford to Fox, 8 June 1863 and Horsford to Welles, 9 June 1863, Miscellaneous Letters Received of the Secretary of the Navy, June 1863, vol. I (RG 45).

33. Horsford to Welles, 11 July 1863, in Proposals Referred volume marked, "Unfinished Business, 1862–1864," Permanent Commission (RG 45).

34. Reports, Nos. 156 and 157. Smith's report is enclosed in Report 157. Permanent Commission (RG 45).

35. Henry to Bache, 19 Aug. 1863, Henry Papers, Smithsonian Institution; Bache to Henry, 25 Aug. 1863, Rhees Collection, Huntington Library.

36. By this was meant any method or device for producing an explosion below the water line of ironclads. Civil War usage lumped mines, torpedoes, and submarines into one category.

37. Horsford to Welles, 7 Mar. 1864, Miscellaneous Letters Received of the Secretary of the Navy, Mar. 1864. Horsford was referred to Chief Engineer W. W. W. Wood of the Bureau of Construction and Repair. Davis to Horsford, 27 July 1864, Letters Sent, Permanent Commission, announced the decision to construct the submarine; Horsford to Davis, 21 Aug. 1864, with the details of Horsford's attempts to have the state of Massachusetts support his project. He also enclosed Wood's letter of 1 Aug. 1864, praising Horsford's proposals—in Proposals Referred volume marked, "Unfinished Business, 1864–1865," Permanent Commission. All in RG 45. 13 Stat. 329 (4 July 1864).

38. Report No. 131, Permanent Commission (RG 45).

39. Bache to Davis, 1 Aug. 1863, Letters Received, Permanent Commission (RG 45).

40. Report No. 22, 16 April 1863, Permanent Commission (RG 45).

41. See Minutes for those dates. The letter to Welles is Report No. 136, dated 25 July 1863 but actually signed and sent out on 29 July.

42. Welles to Permanent Commission, 29 July 1863, Letters Received (RG 45).

43. Baxter, *The Introduction of the Ironclad Warship* (Cambridge, Mass., 1933), pp. 245–250.

44. Bache to Davis, 1 Aug. 1863, Letters Received and Davis to Bache, 7 Aug. 1863, Letters Sent, Permanent Commission (RG 45).

The author has located only one other Navy proposal for using the National Academy during the Civil War. The Bureau of Ordnance asked for and received the Department's permission to enlist the Academy's aid in determining whether fulminate of mercury in contact with metal, as in cartridges, changes into the fulminate of copper; if so, the Bureau desired to know under what circumstance, and at what speeds, and what precautions should be taken. The records of the Bureau of Ordnance in the National Archives contain no indication that any formal approach to the National Academy was ever made, nor do they explain why the Bureau dropped the subject. If and when the National Academy of Sciences removes its ban on research in its records, this minor mystery may be dispelled. Bureau of Ordnance to Secretary of the Navy, 9 Jan. 1864, Bureau Letters, Jan.–Apr. 1864 (RG 45); Welles to Bureau of Ordnance, 12 Jan. 1864, Department Letters No. 4, records of the Bureau of Ordnance (RG 74).

45. Minutes 156, Reports Nos. 182 and 214, Permanent Commission to Ridgway, 9, 15 Apr. and 4 May 1864, Letters Sent, Ridgway to Permanent Commission, 20 May 1864, in Proposals Referred volume labeled "Supplemental Volume, 1862–1864," Ridgway note of 27 May 1863 attached to which is a letter of "G. S." of 27 March 1863, on Stimmers and Ericsson, in Letters Received—all of the Permanent Commission; Reports, Examining Board, 9 May 1862 and 13 June 1862; Ridgway to Fox, 25 July 1863, Miscellaneous Letters of the Secretary of the Navy, July 1863, vol. III. All in RG 45. For Stimmers, see F. M. Bennett, *The Steam Navy of the United States* (Pittsburgh, 1896), pp. 484–493.

46. Baxter, *Ironclad Warship*, pp. 261–262.

47. Henry to Bache, 21 Aug. 1863, Henry Papers, Smithsonian Institution.

48. For example, see "Imbecility in the Navy Department," *Scientific American* 6 (22 Mar. 1862):185. In 1863 the magazine proposed formation of a "Board of Admiralty," composed of prominent inventors, to evaluate inventions [9 (28 Feb. 1863): 137–138].

49. See Minutes, Naval Examining Board, 3 and No. 7 of Inventions Referred to Naval Examining Board for an extreme case.

50. Corruption of public officials was a real threat. For an example of a bribe offered for a favorable opinion of an invention, see Major General Joseph G. Totten to President Lincoln, 13 Dec. 1861, Totten Letterbooks, vol. IX, pp. 48–49, Records of the Office of the Chief of Engineers (RG 77).

5. Cleveland Abbe at Pulkovo:
Theory and Practice in the Nineteenth-Century Physical Sciences

The American meteorologist Cleveland Abbe is a minor figure in the history of science in the United States. He lacked the qualities enabling men like Alexander Dallas Bache, Joseph Henry, Louis Agassiz, George Ellery Hale, and Vannevar Bush to marshal resources and to command institutions. Nor was he a great innovator in his field. Abbe, nevertheless, had much influence in his day. Most important, he embodied ideas we can designate as ideology of science made visible in his career and in words both published and unpublished.

By the time I brought Cleveland Abbe's personal papers into the Library of Congress, his life was somewhat familiar to me, the nearsighted boy who wanted to look at the heavenly bodies. I first encountered him as the very first entry in the *Dictionary of American Biography*. (I even toyed with the idea of doing a popular article entitled "First in the DAB.") By then I was very interested in the geophysical tradition from my previous work with the Coast Survey records in the National Archives. In *Science in Nineteenth-Century America* (1964), I specifically dealt with the tradition. Abbe and his papers admirably fitted my research inclinations.

What I did not expect were the documents in the collection explicitly on the ideology of science. Other manuscripts had fascinating clues connecting Abbe's Russian experiences with his ideology. My first task in the essay was biographical, to give facts about a little-known figure, particularly the very rare American–Russian scientific relation of the last century. Adding to our knowledge of the geophysical tradition was my second purpose. By the time of this composition of the paper, I had already surveyed the late 1950s records of the U.S. National Committee for the International Geophysical Year, now in the archives of the U.S. National Academy of Sciences. I decided Abbe was somewhere in the middle of a long sequence of personages still being added to. Firsts are not always that important.

The ideology presented me with a historiographic opportunity I could not resist. In this article (and in the two on Bache and Henry which follow) I used my work on the geophysical tradition to propose a markedly new interpretation of the relationship of theory and practice. Conventional historiography argued for a relationship based on the relatively recent experiences of manufacturing industry with the fruits of physics and chemistry and did not look at the earlier use of astronomy and mathematics for navigation and land mensuration. The polemical stance is typical of much of my work, an argument against the homogenization of the past to the service of the glib certainties of today.

This paper was delivered before the History of Science Society, Bloomington, Indiana, April 5, 1963, and was originally published in the April–June 1964 number of *Archives Internationales d'Histoire des Sciences* as "Cleveland Abbe at Pulkowa . . ." (the transliteration has now been modernized to *Pulkovo* throughout).

When Cleveland Abbe, the dean of American meteorologists and a respected figure both in the American scientific community and in his profession abroad, was composing his autobiography in 1888 or 1890 he remembered with emotion his first day in St. Petersburg, Sunday, January 1, 1864:

I have never heard a Russian church bell, even of the higher pitch, that had not a rich soft tone, and the effect on me of the medley floating away from the church towers was as if I were suddenly transplanted among Nature's own sweet fields where the grass is interspersed with daisies, where butterflies flit over the sunny spots, and the quiet shades are filled with violets. There is nothing of the solemn grandeur of Gothic Cathedral and pealing organ—nothing of the agressive activity of Protestant worship amid stern, bare walls; but a beautiful landscape swims before my eyes, everything harsh is transformed into beauty and harmony; the liquid notes seem not so much as raise one to heaven, as rather to serve to beautify the earth. One's eyes rise to a brilliant gilded dome, but no higher. A quarter century has not and three quarters will not dim my memory of the sweetness of the Russian bell.

But I did not go to Russia to study poetry, music and art—or rather, the poetry and art I was to study was of the more precise kind known as "mathematics and astronomy." So on Monday I moved down to the Observatory which was situated on the hill Poulkowa about twelve miles directly south of the city. A broad straight avenue leads from the heart of the city south-ward, under a triumphal arch, straight to the door of our astronomical palace—for palace it is compared to the other observatories of the world, and to live there is an astronomer's inspiration.[1]

For two years Abbe remained at Pulkovo, a stay which greatly influenced his subsequent career. It is not the intention of this paper to detail the incidents of the stay in Russia nor to chronicle the events of a lifetime devoted to science. Although a justly honored poineer in his field in America, Abbe, in all honesty, cannot be described as a great discoverer or innovator. He has the historical virtue of being one of those minor figures who are almost replicas of an archetype or whose careers approach the status of a paradigm. It is my purpose, therefore, to suggest how some apparent singularities of this one career are in reality aspects of general conditions.

Cleveland Abbe (1838–1916) was a New Yorker of New England stock. Raised in an environment of religious orthodoxy (the family was Baptist), he long maintained an interest in religion and theological questions, even after he had chosen science as his vocation instead of the ministry. The conversion to science was apparently the result of reading and the influence of teachers. For an American of his period, Abbe had a rather good formal training in science. At the Free Academy of New York (now the City College of New York), he was a student of Wolcott Gibbs. In 1859–60 he studied astronomy with Brünnow at Michigan.[2] For the next four years he worked with Benjamin Apthorp Gould at Cambridge, Massachusetts, on the telegraphic determinations of the longitudes of points in the United States for the U.S. Coast Survey. During the stay at Cambridge Abbe was part of the lively group of astronomers, mathematicians, and other physical scientists from Harvard, the Nautical Almanac, and the Coast Survey's Cambridge office, and he participated in courses and what we would now call seminars, mostly of an informal nature. Although not the holder of any graduate degree, Abbe was quite well prepared by the time he left for Pulkovo.

At first Abbe had tried to go to an observatory in the Southern Hemisphere, specifically Quito, under the influence of Piazzi Smyth's *Tenneriffe, an Astronomer's Experiment.* But in 1863 the same author's *Three Cities in Russia* turned his attention northward.[3] When Abbe applied to Otto von Struve (1819–1905) in 1864 for a place at Pulkovo, he was overjoyed at the acceptance.

Struve had asked for and received a recommendation from Brünnow. In writing to Abbe about Struve's inquiry, Brünnow gave the young American some advice. Russia was a "land of nobles," be careful to say "Excellency," even at first to Struve. "You will receive," Brünnow wrote, "a first rate instruction and I hope, you will be very industrious and thus give Struve a good opinion of American students. Post yourself well up before you go so that he sees, you have studied something before."[4] When Abbe passed on this letter to his father, it received a mixed reaction:

Dr. Brunnow (*sic*) has recommended you to Mr. Struve as a student and advises you as you are going to a land of "Nobles" to be very obsequious, they had better understand that we are all Nobles here, but it may be a good place to study astronomy for all that.[5]

The father's tolerance of aristocratic protocol was important. Even before the words quoted above were written, Otto von Struve had penned a letter to Cleveland Abbe offering the post of a supernumerary astronomer and "all facilities for becoming intimately acquainted with the different branches of practical astronomy." Further, the Observatory would supply free lodging, furniture, fire, and light. Cleveland Abbe would have to provide all additional living expenses, estimated by Struve to average about $300 a year.[6] Although his parents had no particular knowledge or interest in science (the father was a merchant) and they admittedly did not quite understand their son's motivation, Cleveland Abbe was sent off to Russia with $500 a year. Looking back at this event, Cleveland Abbe recounted his resolve to give his father no cause for regret.

Professionally, the two years at Pulkovo lived up to Abbe's expectations. Among other projects, he made observations to determine the effect of atmospheric refraction on light from stars and attempted to compute the constant of nutation. In other ways the Russian stay was educational but not wholly successful. Abbe came and left an American Puritan. Forever incensed by the ritual of the Orthodox Church, he was always seeking out Protestant services. Although he unbent slightly, Abbe was shocked by the prevelance of drinking and carousing. The conditions of the peasantry appalled Abbe. St. Petersburg impressed him as a city and as a center of learning. His colleagues won his respect and admiration, as did other Russian scientists he encountered. Evidences of autocracy he detested, and the signs of liberalization under Alexander II Abbe noted optimistically. And he unsuccessfully sought the hand of a sister of Otto Struve, even offering to abandon his native land.

Two years after his return Abbe became the Director of the Cincinnati Observatory. Here his principal activity was the establishment of a network of weather observers in the Midwest and the issuing of forecasts. After the Army's Signal Corps started a national meteorological service in 1870, Abbe became the first regular official forecaster, serving from 1871 to 1873.[7] Abbe's principal contribution to the weather service was the promotion of research. Under his prodding the Signal Corps instituted a "study room" and a laboratory. The former was a unique administrative device in its day, a means of giving qualified individuals the opportunity to think without the distraction of routine duties. Much of the research in the laboratory we would now describe as basic atmospheric physics. These activities of Abbe's reached a peak in the 1880s. In 1893 an economy-minded Secretary of Agriculture (the Weather Bureau was now in that Department), curtailed research and transferred Abbe to editorial duties. In this new capacity and until he left the Weather Bureau shortly before his death in 1916, Abbe continued to promote research and to encourage younger investigators.[8]

Why Russia? Even if we were to accept Abbe's enthusiastic appraisal of Pulkovo, it still would seem more likely that an American would go to any one of the many first-rate institutions in western Europe. Benjamin Apthorp Gould, for example, studied with Gauss at Göttingen, and during the nineteenth century many Americans would make the overseas educational pilgrimage to Germany, France, Britain, and Austria. None besides Abbe, to my knowledge, went to Russia. For that matter how many Russians came to study in American institutions in the last century? Was Abbe's going to Pulkovo one of those historical singularities which generate the charming anecdotes and the conceptual untidiness which irritate some philosophers of history? What links the Central Imperial Observatory at Pulkovo and the meteorological study room of the Signal Corps in Washington, D.C.?

Although separated widely by geography and political systems, nineteenth-century republican America and Imperial Russia had some things in common. Both, for example, were greatly concerned with the exploitation of vast tracts of little-known territory. Both nations were mapping their domains and encouraging public and private scientists to bring back physical and biological data from these new lands. So too were the English in India, and every European nation with colonial aspirations. Only the Russians and the Americans had this problem, as it were, on their back doorsteps and not beyond the seas.

More important, perhaps, was the analogous positions of the Russian and American scientific communities with respect to the scientific communities of western Europe. Both were outposts of the heartland of modern science. Both imported men, ideas, and tools from western Europe. Both copied, aspired for recognition from, and strived to surpass the older scientific communities. Not surprisingly, there was a tendency to invest scientific resources in sound, well-established fields, not always the most exciting advanced research of the day. There was an element of cultural lag. But because this lag also implied the absence of the full authority of scientific tradition, there was another tendency present—the impulse to try the scientifically heterodox. In both countries this impulse occasionally led into scientific culs-de-sac, or over the sometimes hazy border between the domain of science and of pseudoscience. Fortunately, for their reputations, both countries did produce worthy scientific advances.

Although there were parallels between the United States and Imperial Russia, there was not an identity, and the two scientific communities diverged.[9] The most significant divergence, probably, was the growth of a strong mathematical tradition in Russia and its absence in the United States until this century. In spite of the strong parallels, what differences existed to explain why the group at Cambridge, Massachusetts—Abbe's friends—did not become the nucleus of a mathematical tradition in America?

To Abbe's American colleagues the stay at Pulkovo was a wonderful opportunity. Asaph Hall, for example, wrote to Abbe just before the trip to Russia:

I am very glad to learn that you are going to Pulkowa to study astronomy. It is a noble science and you must not give it up. A term of study at such a place will doubtless make you an astronomer for life, and it cannot but be profitable and pleasant. You can do us a good service in this country if you will return from Pulkowa thoroughly informed with regard to their methods of practical astronomy. Among us astronomy is in a low state and I fear it will be for years to come. We import fine and costly instruments but that does not help the matter much. What we need is *good astronomers and a system of practical astronomy which will tend to produce them.*[10]

But Abbe's astronomical contributions were minor, and his ideas went beyond Hall's straightforward formulation. Because he felt obliged to justify the course of career to his parents who were, after all, making it possible, Abbe wrote some rather revealing letters home while at Pulkovo. Upon his return to America he would pen similar sentiments to colleagues and finally publish many of them. To his mother on July 8, 1865, he wrote:

I find every facility given me for my work and therefore it is quite delightful. I believe that I have often tried to explain although I think you as yet hardly understand it that the work which I do here is or ought to be independent scientific researches and is not like the studies of a boy at school. I am making observations of stars near the horizon for the investigation of the laws of refraction of light by our atmosphere. The observations being made with instruments which belong to the observatory and at its expense will belong to it— but I shall have a copy of them and may use them or publish my results as I see fit. Such work is my best means of astronomical culture and is at once useful to astronomy and to me. It is the desire of this observatory to assist and encourage young astronomers in such works just as Agassiz does or professes to do in Zoology at Cambridge. These both [i.e., Pulkovo and Agassiz's museum] are institutions for the increase of knowledge rather than for its diffusion[.] The latter object is aimed at in Universities and schools[.] I do not think it has ever been my highest desire to be a Professor in a College which most of my friends seem to have thought the only practical end of my study. When I have thought of such a position it has only been as a means of livelihood by no means entirely accordant with my tastes. The unvarying experience of teachers is that their work is too exhausting to leave them energy and time for investigations [so] they are rarely able to keep up with the progress made by others. Therefore the Professors in Universities etc. are rarely required to do more than two hours of teaching daily. The astronomical work that I wish to do will however be better done if I have instruments and an observatory entirely independent of any college or university. Institutions of learning have in general wasted large sums of money in buying costly instruments entirely inadequate to the wants of the students and have provided no means by which the professor or director could devote his time and energy to the use of the instruments for the advancement of astronomy—which is one of the sciences most vitally necessary to the progress of the world in civilization.[11]

The dialogue between the young zealot of science and the indulgent but not wholly comprehending parents continued fitfully through the Russian stay. Toward the end of 1865, after a relative had brought up the possibility of merely diffusing knowledge, Abbe tried to differentiate his role from the role of those who spread and utilized science.

When I am satisfied with my own work—that I have done it thoroughly and conscientiously and that no other man can find it possible to disprove my results—then I am indifferent to what may be said. This is you see a simple hearted candid search after the Truth. A investigation based upon careful observations—we study these observations with the help of mathematics and calculations until we come at a result, a fact, a Truth—a real thing—it may perhaps be new perhaps old at any rate it is irrefutable and we are satisfied that we have done our best. We have nothing to do with teaching these results to young boys but are glad to teach young men how to do similar work. We have not even much to do with making practical use for these results that belongs to another class of men. We do not write books and lectures about our discoveries only a plain record of what we observed how we reasoned upon the observations and what results we came to. There we stop. Many a time has somebody asked me—But what is the use of all your work what use is trying to measure the hundredth of a second of time or the distances of the fixed stars. What use! I have nothing to do with that it is not my business. I can point you to many instances where my work is of great practical use if you wish to study that subject but it is not my department. However it is not polite or right or politic for astronomers to thus reply and you may then assure all inquirers that the Royal Observatory at Greenwich is instituted for the purpose of observing the moon and all celestial phenomena that will tend to assist seaman to determine their latitude and longitude when out of sight of land. The Imperial Observatory at Pulkowa is established for the advancement of science as such, but has added thereto all that pertains to the determination of latitude and longitude of places on the fixed ground, and measuring the size and figure of the earth for the making of maps.

In the same letter Abbe declared his intention to found in America an "astronomical institution for the investigation of all astronomical problems that promise to be of valuable practical importance to the world."[12]

Back home in America he immediately started working on the grand scheme. At first Abbe thought that New York City was a promising site.[13] When he went to Cincinnati, Abbe again tried to bring his ideal institution into being; only the meteorological work came into existence and even that on a small scale.[14] What Abbe wanted was an astronomical establishment that was associated with a research program on the Earth, which was, after all,

a very small ball very insignificant in comparison with the heavens and heavenly bodies that he [the astronomer] studies—Yet very important because he and his brother workers are forced to stay here and therefore have to study the movements of the Earth in order to correct their observations for the influence of the same.[15]

Among the research fields of the proposed institution were the tides, meteorology (as distinct from climatology), terrestrial magnetism, the temperature within the surface of the Earth, geodesy, earthquakes (especially in connection with the stability of instruments), gravitation, and so on. Like Pulkovo, it was to be a school of "practical astronomy" to which young men could come for advanced training. Beyond the pure researches Abbe proposed "the application of these sciences to geography and geodesy, to storm predictions, and to the wants of the citizen and the land surveyor."[16]

Some further insight into his thinking is afforded by the article on "Dorpat and Poulkova" (prepared by Abbe upon his return to America) in which he noted that

[F. W. G. von] Struve was pre-eminently an utilitarian, and he found in this opening field of usefulness [i.e., geodesy] the proper opportunity for fully reimbursing the government for the great expenses attendant upon the maintenance of the observatory. But the diversion of the small astronomical force of the observatory from scientific investigations to practical applications, from study to teaching, in the course of time threatened to seriously interfere with the attainment of the first and important aim of the institution.[17]

On first glance Abbe's views may seem to fall into a not unfamiliar pattern. Pure research comes first; the applied results were tendered in partial payment for the opportunity of doing research. Or the applied results were offered wearily to still the clamor of the utilitarian-minded for a tangible return. Or the scientist promised or hinted at real or possible applied results when raising funds for the conduct of pure research, a slightly dubious subterfuge. No doubt all of these were present in Abbe and in other scientists. But in his case, and in similar instances, this is not a wholly correct picture, for we are not dealing with a clear, sharp differentiation of pure from applied research but a mixed situation.

The term "practical astronomy" as used in the 1860s and today literally refers to the making of observations and the computation of the resulting data. But in this particular context it also connotes the existence of an applied tradition in astronomy. The role of the science in time mensuration is too well known for any further comment. Sometimes overlooked, or underestimated, however, is the role of astronomy or astronomical-like concepts and techniques, especially in the modern period, in the development of what we would now designate as geophysics. That is, an important application of astronomy was to the study of the physical characteristics of the interior of the Earth, its surface, and its atmosphere. While this resulted in abstruse theoretical formulations, it also produced humbler, useful products like maps, charts, and various aids to navigation. It is hard for us sometimes to appreciate the value placed not too long ago on contributions to land mensuration and to travel on land and sea. To the modern eye these seem relatively less consequential than the effects of the later applictions of science which so dominate our lives.

Geophysics as applied astronomy partially explains some aspects of the history of science. For example, many observatories also served as the laboratories and research installations, as it were, of geophysics. The interests of a Gauss in geodesy and a Le Verrier in meteorology are thus seen not as aberrations or minor exercises of great minds but understandable and legitimate activities within a particular scientific tradition. To use a figure from biology, this tradition had a complex symbiosis of pure astronomy, of the study of the Earth as a physical object, and of the applications of both pure fields to practical needs.

Because men of the past were conscious of the successful applications of astronomy to geophysics (as I have defined that term here), they were, I believe, further emboldened to talk of science being applied, both at that date, and in the near future. (There were other factors involved, of course.) Many contemporary historians are not so certain that science (in the sense of a developed theory) was actually applied successfully to any significant extent until at least the eighteenth century, but certainly in the nineteenth century. Before the eighteenth and nineteenth centuries and even as late as the present century, many significant applications resulted from causes other than the application of a scientific theory. Often in the past theory was a rationalization after the fact of a useful discovery.[18]

When Abbe put forward his grand scheme, he was impractical not in the sense of proposing something drastically in advance of the time. As a matter of fact, the United States already had an organization with many of the attributes Abbe wanted for his "astronomical institution." The Coast Survey under Alexander Dallas Bache and Benjamin Peirce was engaged in astronomical work and in both pure and applied geophysics. In all innocence Abbe was promoting a concept about to go out of fashion. Of course, he was impractical in the sense of attempting to create a very grandiose institution in disregard of the improbability of obtaining financial support in the America of the late 1860s. But, as his biographer puts it, Abbe almost always disregarded practical obstacles, starting many projects which remained unfinished but accomplishing many things which might not have ever been started otherwise.[19]

The scientific tradition to which Abbe belonged has not received the attention it merits from historians. For approximately 200 years it encompassed a large portion of all the scientific work performed in the world. I am not implying that the activities of the followers of this scientific tradition were in a majority at any given time and place or that the results produced were the most outstanding in their period. All that is here asserted is that a significant percentage of the labors of scientists of the period 1660–1860 fall within this category. The exact percentage, of course, cannot be estimated with any precision. Many events and trends fit in or near this tradition; others are best understood as divergencies or as belonging to entirely different traditions.

These assertions are probably valid if we view Abbe's allegiance to the tradition of applied astronomy as a special case of a more general scientific tradition. Let us assume that not too many years ago there was a focus of scientific activities in a discipline we shall designate "geography." This focus of activities soon became a scientific tradition because it persisted and was not a convenient temporary grouping for the accomplishment of a given research task. "Geography" was concerned with the description and analysis of the objects and phenomena (living and nonliving) on the surface of the Earth, in its interior, and in its atmosphere. There were two principal divisions of "geography": geophysics and natural history. In general both divisions were nonexperimental. Of the two, geophysics was quantitative and used the language of mathematics to some extent. Geophysics had a great advantage over natural history in that astronomy provided a model and tools of greater power than any available to natural history. On the other hand, natural history set itself tasks perhaps more amenable to successful resolution than did geophysics in terms of available intellectual resources. Many of the objects of both natural history and geophysics involved data of time and place. Some of the products of both were of a similar form or general nature. I have often wondered whether the compiler of a flora for a region did not subconsciously regard his product as in the same class as a star catalog. And in both branches of "geography," theory and practice were normally intermixed.

Abbe is not merely an exemplar of a particular vanished tradition. He is a good example of a type not too well represented in the historical literature. Most of the historical literature is understandably concerned with what is judged (and properly so) to be the main line of development of the physical sciences. What about the scientists not in the main line? Or, for that matter, what about the scientists described in the literature in terms of their contribution to the main line but who, in varying degrees, did other work.

Joseph Henry, for example, made his greatest contribution to science in the research on electricity and magnetism. Yet he was concerned throughout his life with "geography," especially meteorology. For polemic purposes one could even argue that part of his work in experimental physics was motivated by a desire to understand terrestrial magnetism. I think Henry is better described as a typical geophysicist of his time and place except for an unconventional talent and interest in experimental physics, rather than as an experimental physicist with some minor activities in meteorology, surveying, and the like. Such a description may provide a clue to his ambivalent attitude when faced with disputes with inventors like Morse. How could he raise the banner of pure research when he really did not see any chasm between theory and practice in his intellectual environment?

Many scientists find themselves, like Cleveland Abbe, working in fields outside of the main line. Some of these fields are outside because they were left behind by the advancing thrust of research after reaching a high state of theoretical completeness—that is, they become classical fields. Had Abbe stayed with his astronomical observations and computations, confinement to a classical field would have been his fate. Meteorology was and is a quite different situation. There are scientific fields which are not obviously amenable at a given time to any major application of experimental methods or analytic techniques. They are outside the main line in part because other fields are obviously so amenable.

What is the strategy for advancing a relatively undeveloped field outside the main line, and what is the role of a well-trained scientist who comes to this field from a more mature scientific discipline? One common strategy is to engage in a massive information-collecting operation—sometimes referred to today with condescension as empirical data gathering. Thinking in terms of laboratory sciences and of fields with a great mathematical apparatus, this strategy seems rather crude. It may, however, be a necessary approach or even the only apparent one. The word "empirical" obscures the considerable degree of operational sophistication required by a successful massive data-collecting project. Abbe and other applied astronomers undoubtedly saw these efforts as analogous to the patient, centuries-old gathering of astronomical data.

An Abbe in meteorology was amassing the raw material for a hoped-for Newton-like synthesis in meteorology. As a trained man from the more mature field, Abbe's role was to introduce rigorous scientific standards into an amorphous discipline with many amateur practitioners. One hazard of a trained scientist in a partially developed field is its stultification by dogmatic insistence on rigid conformance to inapplicable concepts from the investigator's first field. Abbe was not guilty in this regard. Much to his credit is Abbe's success as a purveyor to meteorology of new ideas from physics and astrophysics. One other role evaded Abbe. Unlike John Dalton, he did not discover in meteorology anything of more general applicability. Unexpected finds sometimes turn up in fields not overly neat. The scientist from a mature field practicing in a less developed one also may have a symbolic function. To the others in the less developed field, he stands for what science is and what their discipline may attain. Abbe was most successful in this last role, and yet here he encountered conflicts between theory and practice.

A field may be only modestly developed in terms of its conceptual structure and yet have important applied aspects (c.g., weather forecasting). Not all of the applied practitioners will view the relation of theory and practice as Abbe did. There was an opposition to Abbe in the weather service composed of practical men who were impressed by the results already attained in weather forecasting and wished to curtail or to eliminate research. What needed improvement (in their opinion) was the network of observers and the work of the forecasters. Atmospheric physics seemed of little relevance.[20]

Abbe did not believe that there were sufficient data to produce a Newtonian synthesis. Better than the practitioners he realized how little was known. With the example of Pulkovo always before him, where astronomy had suffered briefly because of concentration on surveying, Abbe repeatedly insisted on the necessity of basic research in meteorology. Not that his stand was unique. It is the proper posture for anyone unconsciously serving as a symbol of the scientific life and as a source of standards in a relatively immature field. Abbe's position reflected what was a wide-spread belief until quite recently that there was no intellectual differentiation between theory and practice. Both were properly part of the same enterprise.

I suspect that most scientists of the past genuinely believed that all knowledge was actually being applied or would be in the future. All investigators, pure and applied, were equally scientists, natural philosophers, or what have you. The differentiation that existed was social and cultural between these gentlemen (many of them near-amateurs) and the mere mechanics. The distinction between theory and practice only becomes significant when separate, self-conscious groups of pure and applied investigators came into existence.

Sometime in the nineteenth century the physical sciences such as physics and chemistry entered in a state of rapid development. The theoreticians became more and more immersed in the exhilarating events transforming their disciplines. Applications seemed remote or not as worthy of attention as theoretical work. At the same time there was an efflorescence of applied research. Technologists usually found the latest theory too abstruse and of little relevance to their problems. They went ahead and developed complex technologies. And if they had to call on science at all they often went back to an earlier formulation, not the latest theory. Unlike the men of the "geography" tradition, the theoreticians and the technologists were not on a single spectrum ranging from abstruse theorizing to routine development work but on several different spectra. This description is, of course, not literally true of all scientists and all technologists; the separation was never absolute. One consequence of World War II was that large, important groups of scientists and technologists are once again on a single spectrum ranging from theory to practice.

After the decay of "geography" as a focus of scientific activity in the last century, astronomy and geophysics went their separate ways, their applied link partly dissolved. The former changed markedly through the influence of astrophysics. Applications were certainly a very limited part of the discipline. Geophysics grew slowly and was a minor field. Except for a few developments, as in prospecting, applications were not particularly important. One consequence of the present pattern of research such as the International Geophysical Year (IGY) and the space program is a great revival of these two fields. Once more they are associated and both now have a decided applied component.

Could it be that the historical norm, at least in the modern period, is the mixed situation, not the pure theory/pure applications division? Is it the fate of pure fields, both theoretical and practical, to give way to a mixture of basic and applied research?

Notes

1. The manuscript of Abbe's unpublished autobiography is in the Abbe collection, Library of Congress. The quotation is from pp. 54–55.

2. Franz F. E. Brünnow (1821–1891) had come to Ann Arbor in 1854 to head the observatory. For a few years the University of Michigan was the site of an ambitious plan, for its time and place, for the development of a true university.

3. *Teneriffe* . . . appeared in 1858. In his manuscript autobiography Abbe incorrectly remembered reading the work in 1857. *Three Cities in Russia* was published in two volumes in 1862. Charles Piazzi Smyth (1819–1900) was the Astronomer Royal of Scotland. If he is remembered at all today, it is not for his real contributions as a scientist but as the discoverer of various mystical meanings in the dimensions of the pyramids. *Dictionary of National Biography*, 22, pp. 1222–1223.

4. Brünnow to Abbe, June 16, 1864, Abbe Collection, Library of Congress.

5. George Waldo Abbe to Cleveland Abbe, July 8, 1864, ibid.

6. Struve to Abbe, June 25, 1864, ibid.

7. The first forecasts were made in 1870 by Increase A. Lapham, whose services were intermittent. See D. R. Whitnah, *A History of the United States Weather Bureau* (Urbana, Ill., 1961), p. 22.

8. This brief biographical account is mainly based upon the manuscript autobiography; Truman Abbe's biography of his father, *Prof. Abbe and the Isobars* (New York, 1955), which, although uncritical, is based upon the Abbe Papers and contains much information; W. G. Humphreys, "Biographical Memoir of Cleveland Abbe, 1828–1916," in vol. VIII of the U.S. National Academy's *Biographical Memoirs,* pp. 469–508; and many references in the Abbe Papers in the Library of Congress.

9. For example, read S. Timoshenko, *Engineering Education in Russia* (New York, 1959), pp. 1–9, which describes an influx of French science of the Ecole Polytechnique variety. A similar influx occurred in America. Similar developments occurred in both countries as a result, but only up to a point.

10. Hall to Abbe, Sept. 6, 1864, Abbe Papers, National Archives. (His italics.) See also Hall's letter of May 25, 1865, in the same collection, for a survey of the state of astronomy in America.

11. Cleveland Abbe to Charlotte C. Abbe, July 8, 1865, Abbe Papers, Library of Congress.

12. Cleveland Abbe to Charlotte C. Abbe, Dec. 29, 1865, Abbe Papers, Library of Congress.

13. Abbe to Elias Loomis, Dec. 12, 1866, Abbe Papers, Library of Congress.

14. Humphreys, "Biographical Memoir," p. 473.

15. Cleveland Abbe to G. W. Abbe, Jan. 19, 1866, Abbe Papers, Library of Congress.

16. Humphreys, "Biographical Memoir," p. 473, and Abbe, manuscript autobiography, p. 10f.

17. Cleveland Abbe, "Dorpat and Poulkova," *Annual Report of the Smithsonian Institution,* 1867, p. 376f.

18. For another expression of Abbe's views on the relation of pure and applied research, see his article, "The Relations of Physics and Astronomy to the Development of the Mechanic Arts," *Journal of the Franklin Institute* 118 (Aug. 1899):81–120.

19. Humphreys, "Biographical Memoir," p. 473.

20. See Whitnah, *Weather Bureau,* pp. 38–40, 68–69, 76–81.

6. Alexander Dallas Bache: Science and Technology in the American Idiom

Reflecting its time of composition and publication (in the April 1970 issue of *Technology and Culture*), my essay on Bache has an explicit attempt to place him in juxtaposition to natural history and to the conservation movement. If written today, such a piece might concentrate more on my familiar attempts at showing the interdependence of science and technology in the United States, particularly in the geophysical tradition. Given the ubiquity of the natural history tradition in this country in the last century, the effect is rather pleasing to me today. The attempt to relate Bache and his Coast Survey to the natural history tradition adds valuable nuances to this attempt at a sketch of a national idiom or style.

Environmental concerns were very much in the air when the article was published. By then public policy issues had taken a turn deviating from the classic stance of the conservation movement in the United States. That movement had two components not always in harmony: the demand to save resources for later use, and the desire to maintain the physical and biological environment in an undisturbed state. The utilitarian motivation of the first starkly contrasted with the largely aesthetic underpinning of the second.

What had changed was the perception in many quarters that the advances of science and technology now placed humans as well as the physical and biological environment in peril. The fallout debate of the 1950s, the writings of Rachel Carson, the thalidomide incident, and disclosures of the degradation of air and water quality engendered legislation and administrative actions. Although scientists, engineers, and physicians played significant roles in the environmental movement, a decided element of hostility to their guilds existed in some quarters.

By the end of the 1960s, the federal government was cutting back on research and development. It is now clear that some segments of the scientific and technical communities overreacted, blaming environmentalism for much of the problems, even for a change in national style. There was no danger of this country replacing its commitment to scientific research and development with a Romantic, Luddite culture as some feared. Or, to put the issue in exaggerated personal terms, that George Perkins Marsh would replace Alexander Dallas Bache as the erstwhile cultural model for attitudes to nature, to the creation and use of knowledge, and to the structures and procedures used to meet national goals. Although largely unknown, even to some scholars working in the period, Bache is a seminal figure in U.S. history.

The bulk of the paper was a continuation of the position briefly sketched in the Cleveland Abbe paper. Arthur Molella and I completed the formulation of the viewpoint in the article on Joseph Henry which follows immediately. Here I attempted to explain Bache's significance beyond his historical position in the succession of individuals who have dominated the formulation of national science policy. My strategy was twofold. First, I attempted to show continuities—changes in scale, not in kind. This suggested that Bache's practices and implicit ideas were exemplary, not simply a successful adaptation to the peculiarities of a geographic location at a point in time. Second, I gave examples of the common presence of mixed motivations, of instances of the linkage of theory and practice.

Given the points made in the Abbe paper (and the Henry paper), the Bache article had a subliminal message. Despite a differing later rhetoric, the actual behavioral pattern in the United States was usually closer to the practices of an earlier period. What was not said was that European practices probably also deviated from the conventional rhetoric about theory and applications, a topic now under intense scholarly scrutiny. If there was an American idiom in science and technology beyond the mere persistence of an older way, the idiom would have to explain the persistence. The most obvious factor was the absence of a rigid social hierarchy, as in Europe, successfully relegating the practical to a markedly lower status. A related phenomenon was the oft-noted national propensity for establishing institutions. We are a nation of joiners, the conventional wisdom asserted. And we act almost as if institutions reflected this belief. Americans were enthusiastic organizers and joiners of organizations. Quite often institution building ambitiously avoided specialization and tried to encompass a wide range of functions. Their proponents tended to behave, let us say, like Bache, not like Abraham Flexner, the founder of the Institute for Advanced Study at Princeton.

A rereading produced an afterthought. The last paragraph reads curiously today in light of our concern over Japanese competition. Some of today's polemics about the problems of using research results for economic advancement remind me of the smug assertions of not so long ago: Europeans doted on theory, while the United States had managerial revolutions and excelled in the organization of practicality. All of us are prone to think that reality and proper method are identical with the comfortably familiar.

In the early summer of 1864, Alexander Dallas Bache, the superintendent of the Coast Survey, became seriously ill. While he lingered at the edge of death, a network of personal relations, institutional connections, and assumptions about scientists and engineers eroded away. After his death in 1867, the memory of Bache became dimmer. By the end of the century he was usually absent from those quaint collections of sketches of investigators that once passed for history of science and technology. But as serious scholars—the Luries, Duprees, and Goetzmanns—began to poke into Bache's area looking for the antecedents of our present community of scientists and engineers, they encountered him, a massive but obscure historical reality. Since over a century has elapsed since his death, it is perhaps fitting that a would-be biographer should try to explain why Alexander Dallas Bache is not simply a massive obscurity but a significant landmark in the history of science and technology in the United States.

When Death's finger first touched Bache in 1864, his prospects had seemed very good. He was president of the National Academy of Sciences, a creation of his and his cronies, the Lazzaroni.[1] Opposition existed, of course, but, as long as Bache was active, his old friend Joseph Henry might not join forces with Asa Gray, William Barton Rogers, and James Dwight Dana[2] to overthrow the Lazzaroni "program." Besides, Bache was wonderfully adroit at effecting compromises and arrangements. There now existed a mechanism, the National Academy, for channeling federal funds to worthy investigators. And this same Academy, by setting standards and judging all by these criteria, would determine who was worthy. Bache's friends were entrenched in Harvard, Pennsylvania, and Columbia; open and tacit supporters were at many other institutions.

Bache's pre-Washington career had prepared him for his role as the self-anointed leader of the scientific establishment. As a graduate of West Point, Bache was well educated in an American version of the tradition of the Ecole Polytechnique. And he had started his professional career as an army engineer. Although he left the army very soon to teach at the University of Pennsylvania, he was at ease with the military, then as now an important factor in the official world of science and technology.

Obviously talented as an organizer and administrator, Bache was active in local and national scientific bodies and also acquired a deserved reputation as an educator. In terms of later trends, it is significant that Bache played a key role in the Franklin Institute investigation of steam boiler explosions, the first real example of federal contract research.[3] All these experiences, plus the successful administration of the Coast Survey, clearly placed Bache in a strategic position to advance the cause of research.

Except for Joseph Henry,[4] an unwilling half-ally in the Academy affair, Bache had neither peers nor effective rivals in governmental circles. From the lowly chemist of the Department of Agriculture to the lordly superintendent of the Naval Observatory, all were apparently friends, students, or sycophants. Buttressing his reputation as a man of solid worth when results were sought, Bache was also vice-president of the Sanitary Commission, an extraordinary medical welfare organization of the Civil War.

At the start of the Civil War, Bache, with his knowledge of the coasts, had served on the Commission on Conference which planned the blockade of the Confederacy. Besides cementing his relations with the War and Navy Departments, this enabled Bache to evade a wartime cut in funds. Coast Survey parties now accompanied armies and fleets into battle; in peacetime many of the same men had been commanded by army and navy officers learning peaceful arts while earning civilian pay.

The Coast Survey was a remarkable creation of this man. Administered with both imagination and very tight control, it was by far America's largest employer of men from mathematics and the physical sciences. In size, the Coast Survey was not surpassed until decades after the Civil War.[5] In addition to geodetic surveying, the Coast Survey was actively gathering data on terrestrial magnetism—the great intellectual passion of its superintendent. Studies of the tides had produced elaborations of Whewell's and Lubbock's theories. A great series of explorations had brought back much data on the Gulf Stream. Officially and unofficially, Bache was involved in various astronomical projects. The Survey had developed the use of telegraphy in determining longitude, known widely in Europe as the "American method." Standards of weights and measures were in Bache's capable hands, leading to concentration on instrumentation and a small, growing interest in various physical constants. The Survey had an enviable European reputation.

Bache as a Historiographic Problem

Here we have a man who was a founder of the American scientific community; a key figure in developing relations between science and engineering; perhaps the most important single person in the evolution of the government's policy toward science and technology in the past century; a man intimately involved in the interaction of the military with research and engineering; and an important educational pioneer. Given all of these attributes, why has Bache evaded the piercing scrutiny of a historical profession notable for elevating nonentities from footnotes to fame? Of course, the history of science and technology is a relatively new field, and much that is basic to our understanding of Bache's career is not yet well researched. I can also testify to the problem of studying a figure whose manuscript sources are very large, scattered, and often unreadable.[6]

But, above all, Bache is hard to grasp because he does not fit into any of our historiographic pigeonholes. The basic facts about Bache are that he started his career as an army engineer; that most of his work involved applications of mathematics; and that he was a physical scientist who was often concerned with practical problems. The consequences of these basic facts do not easily fit into the norms governing the study of the American past. Bache was intensely involved in the study of the environment. But our historical literature on studies of the environment arises from the conservation movement of John Wesley Powell and his successors. Bache, a scientist-engineer, seemingly has no place in the story nor any meaningful relationship to the movement. American historians also attempt to differentiate the new elements from the old in our past while they engage in the clearly related task of segregating European influences from native contributions. How does one deal with a man simultaneously concerned with organizing a solar eclipse expedition and harbor improvements?

Bache and the Natural History Tradition

Historiographically, Bache's reputation has suffered simply because our national tradition either excludes or downgrades the contributions of physical scientists and engineers studying the environment in favor of men from the tradition of natural history, that is, biologists and geologists. Americans find it rather easy, apparently, to connect natural history with the grand theme of the settlement of the land. We can rhapsodize about the plant hunter and the rock gatherer but blink vacantly at the man with the theodolite or the magnetometer. Reinforcing this bias is the impact of the story of evolution on the minds of both the general historian and the historians of science. Despite some notable scholarly efforts, the origins and impact of Darwinism often becomes a historical cliché, a substitute for thought. The existence of other ways of looking at the environment simply does not occur to many, to the great damage of the historical reputations of physical scientists and engineers.

It seems untypical of antebellum America that the Coast Survey was so little concerned with natural history (i.e., biology and geology). Practically every survey and exploration one can think of offhand had natural history components, sometimes the most consequential parts. Charles Wilkes obviously favored the physical topics, and A. A. Humphrey's own research interests were not in natural history.[7] Both, however, faithfully saw natural history reports through the press. By the time of the King, Hayden, and Wheeler surveys, the natural history tradition was clearly dominant.[8]

For the record, the Coast Survey did make some contributions to natural history. It provided bottom samples for J. W. Bailey's work in micropaleontology. Louis Agassiz went with a Survey field party to study the Florida Keys. Individuals associated with the organization, like Louis François de Pourtalès, did biological work. An examination of the very full *Annual Reports* of the Survey for Bache's period, however, does not disclose any methodical, continuing interest in any aspect of natural history—only a thin scattering of references. The hydrographic sounding volumes in the National Archives have a place for noting bottom specimens. But only the earliest and some odd later volumes bear such annotations. The evidence of print and manuscript clearly indicates that natural history research occurred on odd occasions, or infrequently, and then because of personal rather than official interests.

It can be argued that the absence of marine biology simply reflects the lack of any stimulus on Bache's talent for finding practical excuses for doing pure science. Such stimuli appeared later, giving S. F. Baird the opportunity to launch the Fish Commission.[9] This argument, besides downgrading Bache's interests in pure research, also overlooks another aspect of the Coast Survey's operations—the topographical mapping of the land immediately adjacent to the coastline. While considerable practical incentives existed for work in natural history in this area, there is no evidence of any concerted attempt to seize an opportunity.

Scientific and administrative pressures would also seem to dictate a Coast Survey plunge into biology and geology. Natural history, in general, was booming in nineteenth-century America; geology, in particular, has a strong claim to being *the* American science of that day. When Bache was first appointed, W. C. Redfield assumed that the Coast Survey would engage in natural history.[10] When Benjamin Peirce succeeded Bache, Agassiz urged him to start a zoological and geological program.[11] For decades after Bache's death, the Coast and Geodetic Survey had minor, occasional relations with state geological surveys, in which one can guess was an element of wistful, wishful thinking.

Although a proposed merger of the great surveys placed the Coast and Geodetic Survey at the heart of the new organization, in the absence of biology and geology it could not attain this position in the 1870s.[12]

The presence of Agassiz among the physical scientists comprising the Lazzaroni, the connection with Bailey, and the work in natural history in surveys and explorations of Bache's followers does indicate some kind of interest or at least a lack of hostility. The paucity of Cost Survey efforts is primarily understandable in terms of two factors: The first is specialization. Both Henry and Bache were rather careful about claiming expertise outside of their fields; neither would lightly venture into completely alien subject domains. The second factor is most fundamental to an understanding of Bache's study of the environment. There was a sense in which he and his circle, notably Henry, were interested in biology. Viewing living matter from the standpoint of the physical sciences, they were interested in finding connections between plant and animal life and various physical phenomena. The interest of Bache and Henry in meteorology is partially due to this objective. In contrast to the endless cataloging of the natural historians, they were seeking recurring patterns. Quetelet, who was in contact with both Bache and Henry, set the tone with his proposals to gather data on periodic biological phenomena. The Smithsonian Institution made an attempt to carry out Quetelet's idea.[13] From the data it was hoped, apparently, that science could derive general laws in a mathematical form. As this hope proved vainly elusive, a practical scientist-administrator like Bache would not overreach himself.

The principal difference between what Bache would or would not touch was the presence or absence of mathematical theories providing structure to the operations and to which data were referred. In the pre-Darwin period, the data of natural history existed in little intellectual clusters unlinked in time except in the minds of a few. Physical data of the earth and measures of the heavens were variable in time—periodic if astronomical; secularly changing if geophysical. A datum of natural history might be literally unique; a reading of a telescope, magnetometer, or tide staff had to fit into a theoretically deducible sequence.

The New versus the Old, the American versus the European

Historians are somewhat thrown off stride by Bache's obvious, yet puzzling, resemblances to contemporary figures. It is hard not to see a succession from Alexander Dallas Bache to Vannevar Bush. But this means interpreting Bache as one of the precursors of the twentieth-century research world. It also means implicitly or explicitly interpreting Bache as a man born ahead of his time. The partial failure of his aims and his later relative obscurity are then explicable by labeling Bache a misfit in time.

But was he really a precursor of the twentieth-century research world? Favoring an affirmative answer are a whole host of factors. Bache was a proponent of large-scale, team research. He developed a close relationship to the military services. His career displays a very high degree of specialization, quite comparable to living scientists. When it came to extracting money from the Congress and the Executive Branch, Bache compares favorably with present-day Pooh-Bahs of science, allowing for the difference in scale.

But these aspects of Bache's career are easily overdone. Sir George Airy, in the Royal Greenwich Observatory, and others had "large-scale" teams.[14] Military and naval personnel show up elsewhere in science and technology in the last century. They were simply gentlemen interested in certain intellectual topics. Real polymaths were very rare by the time of Bache's death. Even in America a high degree of professional specialization existed; Bache and Henry were able to squeeze out most nonspecialized amateurs primarily because that breed was already on the way to extinction. As to politics, writers who find the present lobbying of science new, American, and perhaps reprehensible are blind to the facts of life in the past. What real difference was there between Bache trying to influence an Ohio congressman and, for example, Bernard trying to influence Napoleon III?[15] Are we looking for differences between past and present in the wrong places?

Bache presents another historiographic problem, one encountered in the study of other American scientists and engineers. It is very hard not to fall into the trap of trying to decide whether Bache was more American or European in his attitudes. If the conclusion is that Bache was more European, then his fall from historical grace is due to being out of his proper intellectual place. We can further conclude that his "failure" was not to attach himself and his organization to any major trend in the development of the American nation. In reaching such a conclusion, one is obviously applying a nexus of assumptions that define the terms "American" and "European," assumptions the more hallowed because they are largely untested.

Bache's peers and models were in Europe. A bust of Arago was in his office.[16] Gauss provided the theories for the work on geodesy and terrestrial magnetism.[17] The Bureau des Longitudes was the earliest model for the Survey. Struve's operations at the Pulkovo Observatory in establishing the great Russian geodetic arc were similar in spirit and in many details to Bache's work.[18] Most important of all, in my opinion, was the example of Bache's friend, Sir Edward Sabine. He was an ordnance officer; the Ordnance Survey did the geodetic work in the United Kingdom and elsewhere in the Empire. Like Bache, Sabine's grand passion was terrestrial magnetism, leading to the organization of a global research program. Like Bache, he knew the ins and outs of officialdom and how to maneuver from one scientific group to another in order to gain objectives. Like Bache, who had difficulties with M. F. Maury, Sabine had to content with Airy, the Astronomer Royal, at crucial moments. Like Bache, Sabine was interested in promoting standardization of physical constants, geophysical research, and meteorology.[19]

But these European models did change in Bache's hands. Even before he assumed the superintendency of the Survey, a board reviewing the work of his predecessor, F. R. Hassler, the Swiss-American geodesist,[20] had recommended that doing first-order triangulations was no longer to be the sole prerogative of the superintendent—a symbol of arcane knowledge denied to mere American practitioners, who were obviously not Continental men of science.[21] Carrying forth this mandate in his own fashion, Bache soon after 1844 had many parties working simultaneously, each in accordance with an overall plan and in conformity with specified standards and procedures. All field data were subject to careful review by Washington office staff units, especially the Computing Division. Superficially, the multiplication of parties and the division of labor by specialization in the headquarters seemed simply to speed up the progress of the Coast Survey. But these organizational changes also accentuated the three principal differences between Bache's creation and the typical antebellum survey or exploration: first, there was a higher order of precision; second, the Coast Survey was a continuing if not permanent, venture; and third, the operations were explicitly related to theoretical formulations.

Bache's policies and practices were thoroughly in line with trends we tend to think of as American. They presupposed a diffusion of skills beyond an elite—not popularization with the French connotation of vulgarization, but a reduction of high theory to practice in the form, as it were, of manuals and handbooks containing various levels of knowledge from the abstruse to the elementary. Since the scientific and technical community was so small, such a reduction to practice also presupposed and required an openness of entry into the community. A few gentlemen-scientists could not perform the resulting range of theoretical and applied work. At the same time that Bache was trying to develop an agency on European lines and an elitist scientific community, his actions necessarily had different results. Accompanying the openness of entry was a presumably American hospitality to applications and practical results. Not only was the community open both to the pragmatists and the theoreticians, but the absence of institutional alternatives in America worked against any exclusive absorption in either the purely scientific or the narrowly practical. Mixed motives tend to characterize American efforts in research.

Theory and Practice

In a rare moment of public candor, Bache once asserted that the real justification for the Coast Survey was its contribution to scientific knowledge.[22] Usually, Bache properly noted the reefs charted, the harbors surveyed, and the numbers of charts issued. But there is no doubt of Bache's sincerity in promoting a community of research scientists and in fostering the development of what was then called abstract science. The *Annual Reports* of the Coast Survey are replete with instances of work beyond the literal needs of the navigator. Leaving his estate to the young National Academy of Sciences for the support of research was Bache's last gesture of backing for pure science. It was the Bache Fund which provided money for the Michelson–Morley experiment.

But Bache was not exclusively concerned with pure science. Before coming to Washington he had participated in the notable Franklin Institute investigation of steam boiler explosions, a clear precedent for the "Federalism by contract" which is so common today. More typical was Bache's involvement after 1844 in a series of boards and commissions concerned with improvements of harbors or other civil engineering works. Frequently, his fellow board members were the Chief of the Corps of Engineers, Major General Joseph G. Totten, and Captain, later Rear Admiral, Charles Henry Davis of the Navy. Bache had started his brief army career by working with Totten on construction projects. A translator of Gauss, Davis was an astronomer who played a major role in the founding of the Nautical Almanac and the National Academy.[23]

The composition of these boards is quite enlightening: the Army Engineers; civilian geophysics in the person of a man trained at West Point in the tradition of the Ecole Polytechnique; and a navy astronomer—all worked in smooth harmony. They tried to gather the best data available for each problem presented for adjudication. Most important of all, as their reviews disclosed gaps in the evidence, they attempted to remedy these deficiencies by new observations, calculations, and experiments. Quite similar in spirit was the work of the Light House Board. Bache and Joseph Henry, representing civilian science, served here with two army and two navy officers. The problems of the Board stimulated Henry to do experiments in optics and acoustics.

One last purely civilian example is pertinent to the history of American use of natural resources. Bache accepted with alacrity the rare and minor opportunity to do the cadastral survey of the Florida Keys because he was anxious to place the General Land Office on a sound scientific basis. Just what this meant we can deduce from his friend

Joseph Henry's earlier proposal. Noting that the General Land Office already gathered meteorological data, Henry suggested that the General Land Office surveyors systematically take down geomagnetic and other information to produce eventually a true physiographic map of the United States. A Bache or a Henry in control of the General Land Office prior to the Civil War was beyond the bounds of possibility, but an intriguing topic for historical speculation.[24]

The Bache–Totten–Davis committees and the Light House Board illustrate the existence in the antebellum scene of a rather harmonious interplay between theory and practice, between scientists and engineers. I do not mean to infer that differences of viewpoint and emphasis were totally absent but, rather, to stress the existence of the recognition of a continuous spectrum going from the intensely abstruse to the narrowly practical. In this environment scientists and engineers could easily shift from one portion of the spectrum to another as circumstances and personal inclinations dictated, and their organizations displayed mixed motives and programs.

When Bache needed funds for his terrestrial magnetism observatory at Girard College, he turned to the Topographical Engineers of the army. The observatory was part of Sabine's network which was testing Gauss's theory, among other things. The money was given as a contribution to geographic science; practical consequences in mapping and surveying were taken for granted. Since the use of astronomy to locate precisely points on the land and sea was self-evident, the desirability of the expansion of astronomical theory and data was also self-evident. When the Topographical Engineers Humphreys and Abbot studied the Mississippi Basin, it was not simply to recommend flood-control measures and improvements in navigation. They formulated a mathematical theory of the physics and hydraulics of the Basin.[25]

Elsewhere I have argued that the men in the intellectual tradition Bache belonged to did not see any chasm separating theory and practice.[26] While well aware of the difference between pure research and applications, they assumed that knowledge and action were linked. Since they were thinking primarily of examples of aids to travel on land and sea, Bache, Henry, and their contemporaries could agree that research preceded applications. This accorded well with their experiences. A host of further assumptions derived from this position and the resulting practices of a Bache. First, the employment of science and of scientists was required for any significant use of the environment. Second, routine technical work had to have an accompanying research component. Third, both the routine work and the research were ideally related to a theory in mathematical form—the theory giving direction to the research and the routine work, while the resulting data confirmed, extended, or refuted the theory. Mapping the lands and oceans, applications to civil works, and the various uses of astronomy and terrestrial magnetism as aids to travelers conformed to this position on the relations of theory and practice. Naturally, Bache's experiences had few connections with industrial technology. Very little of the industrial technology of 1860 derived from scientific research.

But different traditions existed in Bache's day; in time these became dominant and influenced historians' perceptions of the past. Most important was the increasing prominence of academic traditions of pure research in the physical and biological sciences assuming a real chasm between theory and practice, between scientists and practitioners. Important also was the growth and increasing self-confidence of an engineering tradition stressing the rationalization of practice and empirical research largely divorced from the advanced pure science of the day.

The academic pure scientists were and are insistent that their labors preceded applications in both a chronological and a logical sense—an insistence not strictly justified in most instances until quite recently. The engineers and technologists insisted on the importance of reduction to practice—the inventors' Holy Grail—and, therefore, of the design process and the kinds of rather narrow testing and research producing design data. Both pure scientists and engineers tend now to agree on the same metaphor to explain the relations of theory and practice. Pure science produces capital assets of knowledge; as applications use up older knowledge, withdrawals of this intellectual capital occur, necessitating new research to replenish the balance.

This metaphor is at variance with Bache's experiences. It also does not correspond to the actualities of recent research and development efforts. In the world of the Coast Survey, applications were preceded, accompanied, and followed by basic research. Rather than a chasm clearly differentiating theory and practice, there was a large gray area in which neither the motivations of the investigators nor the intrinsic nature of the work could unambiguously separate the two—except by applying the most arbitrary definitions. It is both frustrating and meaningless to characterize, for example, Bache's sponsorship of solar eclipse expeditions as "pure," and work on the telegraphic determination of longitude as "applied." Equally pointless would be attempts to calculate the "purity percentage" in the work of Humphreys and Abbot. This is not to deny the existence of scientists who engage in "pure" research, nor that there are individuals who are inventors or applied scientists. What is significant is that research situations in the gray area are increasingly common, and this condition is not historically a change in kind but in degree.

The pattern disclosed in Bache's Coast Survey is quite similar to notable research efforts of this century. For example, Willis R. Whitney of the General Electric Research Laboratory and Vannevar Bush in the Office of Scientific Research and Development found it both necessary and expedient to broaden the spectrum of their operations. Whitney went into pilot plant operations and production engineering; Bush had to send men to the war zone to introduce new devices to the combat troops. For these research systems to function best, there simply had to be a range of activities from the abstruse to the routine.[27]

All of these research and development systems in which pure research preceded, accompanied, and followed applications had to attain some practical success, however modest, to keep the process going. Continuation of the research and development process then depended on the respectful faith of laymen in the abilities of investigators and in the seriousness of their efforts, as well as a belief in usefulness of even the most esoteric investigation. Once this pattern was established, expectations of both scientists and engineers matched the faith of the conditioned laity that applications and pure science would suffer if subsequently separated into discrete systems. Attitudes of this sort explain, in part, why an astrophysicist like George Ellery Hale was so anxious to have engineering and industry involved in the fledgling National Research Council after World War I.[28]

In the last century there was no general indifference or hostility to basic research we can characterize as typically or uniquely American.[29] Sabine's friends complained about the difficulty of getting support of research not connected with trade.[30] Similar laments undoubtedly exists for other countries. The difference between America and the leading European scientific powers in the matter of support of pure research was twofold:

First, the Old World had institutions affording support to pure science, as well as traditions justifying this support. These were largely nonexistent or rudimentary in the United States until well into the century. Institution building, in the broadest sense, was a principal occupation of American scientists in the last century. In contrast to Europe with its academies, universities with professorships, botanical gardens, observatories, government bureaus, and other research institutes, America was underdeveloped. Existing or newly established organizations would get activities that might have been functionally or hierarchically separate if alternate organizations had existed.

Second, American conditions multiplied the numbers of men in applied endeavors while giving them a social and intellectual status often denied their European brethren.[31] Believing passionately in the worth of their calling, scientists were irritated, if not outraged, when they were classed with inventors, engineers, and other practitioners. Added to this was a chronic shortage of funds as scientific needs inevitably outstripped available financial resources. Simply building American equivalents of European scientific institutions would not change the egalitarian values in American society, nor would it convert engineers and inventors into members of the scientific community. Practitioners want to practice, not to explore the margin of knowledge. While the institutional framework was under construction, Bache and those who followed him were trying to educate the practitioners in the usefulness of pure science. And the best way to convince the practitioners was to pass beyond exhortations to example—to make real research ubiquitous in all fields, not just the prerogative of a few in a small number of academic disciplines.

The attitudes I find in Bache help explain why in contemporary America there is usually concern over basic research and relative complacency about technology, while the reverse is true in Europe. We take applied work for granted and worry about basic research because we increasingly assume that basic research is vulnerable and important for its own sake *and* that the way to solve practical problems is by applying science and scientists.

Notes

1. To the best of my knowledge, Edward Lurie was the first serious historian to investigate the Lazzaroni. His *Louis Agassiz, a Life in Science* (Chicago, 1960) is the best introduction to the subject. Further useful information is in A. Hunter Dupree, *Asa Gray* (Cambridge, Mass., 1959) (see also, Dupree, *Science in the Federal Government* [Cambridge, Mass., 1957], pp. 115–148); M. M. Odgers, *Alexander Dallas Bache, Scientist and Educator, 1806–1867* (Philadelphia, 1947) is quite innocent of these topics. Although I shall talk of the Lazzaroni and of their "program," readers should understand that I am not talking of a group with a fixed membership and an explicitly formulated program. When the historian gets further away from a small group of socially congenial men and starts talking implicity or explicitly of invisible colleges and conspiratorial elites, the facts become less and less relevant. All that I mean to state is that there was a small group around Bache and a larger group with less close ties to Bache, both of which shared some attitudes toward science and its role in the nation. The leading members, besides Bache, were the mathematician Benjamin Peirce, the zoologist Louis Agassiz, the astronomer B. A. Gould, and (with reservations) the physicist Joseph Henry. Other significant members of the group were O. W. Gibbs, the chemist, and the Navy astronomer Charles Henry Davis.

2. The botanist Gray, the geologist Dana, and the geologist Rogers (founder of M.I.T.) really represented the group of professional natural scientists; Agassiz did not. To their formidable opposition must be added many in the physical sciences who were not particularly close to Bache and his circle.

3. John C. Burke, "Bursting Boilers and the Federal Power," *Technology and Culture* 7 (Winter 1966):1–23; Bruce Sinclair, *Early Research at the*

124   Establishing Science

Franklin Institute: The Investigation into the Causes of Steam Boiler Explosions, 1830–1837 (Philadelphia, 1966).

4. Nathan Reingold, *Science in Nineteenth-Century America: A Documentary History* (New York, 1964), pp. 200–225. Henry and Bache were old friends with a record of close, harmonious relations in developing the sciences in America. Bache was much more of an activist; by the time of the Civil War he was much closer to the views of Agassiz, Peirce, and Gould, who differed from Henry primarily in their belief that the time was ripe for a decisive aggression against those who did not share their views of professional standards.

5. Bache's Coast Survey had reached the $0.5 million mark in appropriations before the Civil War. The U.S. Geological Survey attained that level by 1884.

6. For an example of these problems, see Nathan Reingold, "The Anatomy of a Collection," *American Archivist* 27 (April 1964):251–259.

7. Wilkes, a navy officer, commanded the great exploring expedition sent to Antarctica and the South Seas in 1838. Humphreys was a army engineer who served as Bache's deputy for a few years. He is best known for his administration of the Pacific Railroad Surveys. After the Civil War, Humphreys became head of the Corps of Engineers. Their range of intellectual interests were quite close to Bache's.

8. H. N. Smith, "Clarence King, John Wesley Powell, and the Establishment of the United States Geological Survey," *Mississippi Valley Historical Review* 35 (June 1946):37–58. For further details see Wallace Stegner, *Beyond the Hundredth Meridian* (Boston, 1953); and William H. Goetzmann, *Exploration and Empire* (New York, 1966), pp. 375–589.

9. Baird, the Assistant Secretary of the Smithsonian, was a bête noire of the great Agassiz. But I would suggest that his basic attitudes toward the relations of theory and practice were similar to those of Bache and Henry.

10. Redfield to Henry, December 21, 1843, Redfield Papers, Sterling Library, Yale University.

11. U.S. Coast Survey, *Annual Report* (Washington, D.C., 1867), pp. 10–11, 183–186.

12. See the Records of the Assistant in Charge of State Surveys, Record Group 23, National Archives, Washington, D.C. In 1871, the Coast Survey was authorized to cooperate with the state geological and topographical surveys. One of the proposals for the merger of the surveys placed all the surveys of mensuration in the Coast and Geodetic Survey, limiting the proposed Geological Survey to geology (and, by implication, to various parts of natural history).

13. A. Quetelet, *Instructions pour l'observation des phénomènes périodiques.* This first appeared in vol. 9, no. 1, of the *Bullétin, Académie Royale de Bruxelles,* 1842; translated in Smithsonian Institution, *Annual Report,* (Washington, D.C., 1855), pp. 259–263. It is significant that the Smithsonian attempt was subsumed by Henry under the rubric of "Meteorology." The results were published in U.S. Patent Office, *Meteorological Observations* (Washington, 1864), vol. 2 (House Executive Document 55, 36th Cong. 1st sess. [Congressional serial no. 1053]).

14. For example, in 1856, the professional staff of the Royal Greenwich Observatory had reached seventeen, of whom eight were supernumerary computers. (Information provided from the Navy Estimates for that year through the courtesy of Mr. P. S. Laurie of the Royal Greenwich Observatory.)

15. J. M. D. Olmsted and E. H. Olmsted, *Claude Bernard and the Experimental Method in Medicine* (New York, 1952), pp. 126–127. This example was chosen by sheer chance. My eye happened to fall on this book while composing this paragraph. I do not mean to imply that Bernard was a scientific promoter, nor that there was anything improper in talking to Napoleon III—or to an Ohio congressman. As long as a researcher lacked independent resources, it was necessary to go to those who had them.

16. D. F. J. Arago (1786–1853), whose interests in mensuration, astronomy, and other topics were so like Bache's, was widely known in America because of his official duties in the Institut de France. When Bache delivered an oration on the death of Alexander von Humboldt, Arago was brought in and praised almost as a subtle reproof of Humboldt (see "Tribute to the Memory of Humboldt," *Pulpit and Rostrum* (June 15, 1859):127–140). The two were actually friends and colaborators, but, in my opinion, Bache was downgrading Humboldt because of the philosophical trends evident in *Kosmos*. Powell and the conservation movement could comfortably stem from *Kosmos,* not Arago. In this address Bache is carefully trying to justify his particular mixture of science and technology by specifying which European tradition was being followed.

17. Karl Friedrich Gauss (1777–1855) is, of course, one of the very great mathematicians. He was not, however, a pure mathematician in the contemporary sense; like Arago, his range of interests were quite similar to Bache's, even including the concern with surveying.

18. F. G. W. von Struve (1793–1864).

19. These remarks are largely based on an examination of the Sabine Papers in the Archives of the Meteorological Office, Bracknell, Berkshire, and the Sabine paper in the Royal Society. Comments on Sabine are in Nathan Reingold, "Babbage and Moll on the State of Science in Great Britain," *British Journal for the History of Science* 4 (June 1968):58–64.

20. For Hassler, see F. Cajori, *The Chequered Career of Ferdinand Rudolph Hassler* (Boston, 1929).

21. Minutes of the Board of the Organization of the Coast Survey, 1843, Record Group 23, National Archives, Washington, D.C.

22. U.S. Coast Survey, *Annual Report* (Washington, D.C., 1849), p. 2.

23. C. H. Davis, Jr., *Life of Charles Henry Davis, Rear Admiral, 1807–1877* (Boston, 1899). There is no adequate account of Major General Totten's career. The best sources for the work of the Army Engineers and the Topographical Engineers are the writings of William Goetzmann, *Exploration and Empire* (New York, 1966) and *Army Exploration in the American West* (New Haven, Conn., 1959).

24. For the Florida Keys Survey, see the two volumes of copies of correspondence in the Record Copies of Correspondence Relating to Land Surveys in Record Group 23, National Archives, Washington, D.C. For Henry's views, see Reingold, *Science in Nineteenth-Century America,* p. 156.

25. A. A. Humphreys and H. L. Abbot, *Report upon the Physics and Hydraulics of the Mississippi River* (Washington, D.C., 1861).

26. Nathan Reingold, "Cleveland Abbe at Pulkowa: Theory and Practice in the Nineteenth Century Physical Sciences," *Archives Internationales d'Histoire des Sciences* 17 (April–June 1964):133–47 [Paper No. 5 in the present volume]; Reingold, *Science in Nineteenth-Century America,* pp. 59–62.

27. For Whitney, see K. Birr, *Pioneering in Industrial Research* (Washington, D.C., 1957), pp. 55–56; for Bush, see J. P. Baxter III, *Scientists Against Time* (Boston, 1946), pp. 404–447.

28. Hale's motives were very complex. I am impressed, however, by the fact that he was an M.I.T. graduate. In the Hale Papers at the California Institute of Technology, there is much on education at M.I.T. and on the role that engineers and industry should play in the scheme of things Hale was promoting. Briefly, Hale was afraid that the engineers would go off on their own, to the detriment of both science and practical affairs. It was not simply a case of wanting to head off the establishment of a rival center of power. A detailed account of Hale's career is in Helen Wright, *Explorer of the Universe* (New York, 1966). For a comment illustrating the attitude of scientists toward their engineer colleagues at M.I.T. quite consonant with Hale's views, see Edwin Bidwell Wilson to Ross G. Harrison, May 4, 1922, Harrison Papers, Sterling Library, Yale University.

29. The classic statement of this viewpoint is in Richard H. Shryock's "American Indifferences to Basic Research during the Nineteenth Century," *Archives internationales d'histoire des sciences* 28 (1948):50–65, and recently reprinted in his *Medicine in America: Historical Essays* (Baltimore, 1966). This opinion has a long lineage, from de Tocqueville to laments published at the time of the 1876 centennial celebration, pieces written at or near the turn of the century, and more recent writings by I. B. Cohen. In an article I am preparing on this topic, I shall argue that so far as the evidence exists, in all countries of the Western world in the modern period, pure research is rare relative to applied work; that no assumptions of the uniqueness in degree or kind in American attitudes toward either basic research or applied research are justified until we can speak from a firm base of comparative data; and that, in general, the perception of indifference by scientists arises from a kind of sociological anguish, for which see below.

30. For example, see R. I. Murchison to Sabine, February 9, 1839, in 66(2), Sabine Papers, Meteorological Office Archives.

31. I can perhaps best illustrate this by a personal anecdote. In S. Timoshenko's *History of the Strength of Materials* (New York, 1953), p. 390, is a reference to Felix Klein's visit to America which so impressed him that he organized research institutions for applied mathematics and physics at Göttingen. Although the text was very clear as to his motivation, I was puzzled. Where in America was there a model for these actions? When I visited Timoshenko in 1961, he patiently explained to me that Klein was impressed by the fact that engineers were within the university community and took the same courses in mathematics as students of mathematics. As a American, I had forgotten what I knew from books: European engineers, even a Timoshenko or a Klein, who were applied mathematicians, would not have the status of real university men.

7.□ Theorists and Ingenious Mechanics: Joseph Henry Defines Science

"Theorists and Ingenious Mechanics," written with Arthur P. Molella, is another look at the science–technology, theory–practice relationship. Although a by-product of the preparation of the first volume of the *Papers of Joseph Henry,* the article is not about the American scene. The example from early-nineteenth-century Albany provided a striking occasion for considering two sets of historiographic assertions—the views expressed by scientists in the past, and the positions taken by historians in the present. An implicit assumption in the article is that written history is the product of the interactions of the ideas of the past subjects and of the later scholar. In this case the problem was both curious and important. Why did so many historians want to discount so many utterances from the past?

The two great, classic problems of historiography are explaining what was meant by certain words in the past and relating those past meanings to events and concepts contemporary to us. What struck me then and now is how facile we are in finding justifications in history, philosophy, and the social sciences for whatever positions we really want to believe in. Perhaps I am not given to clear, simple pictures of reality. What I usually find in the past is not in black and white but in complex grays. Complex grays simply cannot match today's need for certainties.

Put rather bluntly, my position is that historical amnesia has blotted out past perceptions of knowledge being applied in navigation, surveying, and other activities loosely connected with the geophysical tradition. Past utterances make sense if we realize what past speakers were referring to. I think this is a special case of the way recent experiences of the history of the physical sciences and of modern industry have distorted the historiographic framework.

An unexpected effect of this particular historiographic exercise and similar ones like the poking about the indifference theme was to soften the conventional distinctions between the United States and Europe. At the very time that we were looking at nineteenth-century America in general and the provincial culture of Albany, New York, in particular, Western European historians were disclosing an Old World past at variance with the orthodox views discussed in this article. The newer findings were more social in their orientations, including greater emphasis on utilitarian motives. The historians involved, I suspect, were reacting against an overly intellectual history of science reifying grand theory.

I was not then, nor am I now, prepared simply to relegate the United States scene—even Albany—to some minor variation of provincial British or Continental culture. Despite similarities, a Henry, a Bache, and countless other U.S. scientists did not easily fit any European mold known to me. In my oral discourse (but not strongly in my writings), I more and more stressed that the United States represented a con-sequential North American variant of Western civilization. If the theory–practice relationship was not *the* defining difference, what was? Inevitably, I was drawn to the conclusion that social structure counted. Henry had to pay attention to the ingenious mechanics. He had to worry about a populace unclear about the distinction between abstract knowledge and empirical practice. His society did not give him a se-cure niche shielded from the impact of hostile and indifferent gazes.

This paper was read at a joint session of the American Historical Association and the History of Science Society, New York, December 1971. It was origi-nally published in 1973 in *Science Studies*. Support of the National Science Foundation during its preparation is gratefully acknowledged.

A familiar historical exercise is to compare the words of an individual or a group with what was, "in fact," the true state of affairs. Secondary texts are placed in juxtaposition with historical evidence; comparisons are made; and conclusions are drawn. Literature exists in which the assertions of many past scientists are confronted by a counterassertion of many present historians. The former affirm that "science" did indeed make many past contributions to the amelioration of man's lot; the latter conclude that, "in fact," the statements of the scientists are not strictly speaking true until the latter decades of the last century or 1850, always granting, perhaps, a few minor exceptions. Rather than crediting science or scientists, many historians of science, of technol-ogy, and of economics ascribe most past technological advances (i.e., pre–1850 or pre–1880) to the labors of practical men–artisans and en-trepreneurs–often unknown to history.[1] A counterchorus of other histo-rians, technologists, and social scientists minimizes the impact of basic science on economic growth today as well as in the past, also citing historical truth.[2] Somewhere in between are the pioneering efforts of A. E. Musson and Eric Robinson purporting to prove that science did indeed play a notable role in the Industrial Revolution.[3]

Our interest in this area arose out of our efforts to comprehend texts we edited for the first volume of *The Papers of Joseph Henry*. On at least two public occasions, in 1826 and 1832, Joseph Henry, the young Professor of Mathematics and Natural Philosophy at the Albany Academy, carefully defined science (or, rather, natural philosophy) and analyzed its relations with practical endeavors. The opinions are of interest because of their banality, not their originality. Contemporaries said the same thing—and so did scientists before and afterward. We could shrug off these opinions as interest group self-seeking or self-deceiving. We are inclined to doubt such explanations, preferring to assume that so many able men over so long a time period were speaking and acting sincerely, even if in error on particular points. The persistence of the belief and the vigor of its repudiation imply the presence of a real issue, and one of importance

Our method of tackling the issue is, first, to give Henry's position, attempting to trace it historically while analyzing its components. Next, we will discuss Henry's views on the ingenious mechanics, the often anonymous improvers of technology. After these particulars we will consider a number of historiographic issues present in the literature. As the volume of pertinent printed matter is enormous, our explicit coverage is limited to a small number of examples, both typical and outstanding. Nor do we pretend to have the competence to settle all matters under dispute. Our purpose is more modest—to use a specific case to suggest future research possibilities.

On 11 September 1826, the Trustees of the Albany Academy gathered to induct Joseph Henry, a former student of the Academy, as Professor of Mathematics and Natural Philosophy. On this public occasion Henry was singularly honored by an invitation to deliver an inaugural lecture. The ceremonies were obviously important to both the new faculty member and those gathered to honor him. For Henry this was the first formal event in his career as teacher and physicist. For the founders and patrons of the Academy, the ceremonies were not only a tribute to Henry but to the educational program they had instituted only a decade earlier. They saw Henry as an outstanding son of the Academy returning to carry on a carefully nurtured pedagogic tradition.

The nature of that tradition[4] was brought out explicitly in the brief introductory remarks by Gideon Hawley, Academy Trustee and Secretary of the New York State Board of Regents.[5] His hopeful words about what the community leaders expected of the Academy's mathematical program reflected a viewpoint commonly expressed at meetings of the Board of Regents and the Academy Trustees—the utility of mathematical and scientific knowledge.[6] Practical benefits were deemed a direct consequence of the Academy's English Program which included the courses in mathematics and the sciences. In the English department, Hawley stated,

the mathematical sciences must of necessity hold a conspicuous place: and without intending to derogate from the importance and value of the classical department, the trustees are free to acknowledge that they cherish with peculiar favour and regard that department of study which is best fitted to prepare the greatest number of pupils for the useful pursuits of active life.

In concluding, Hawley expressed the confidence of his fellow Trustees that, as a former student of the Academy, Henry shared these educational goals and would discharge his teaching duties with proper attention to society's practical needs. The attitudes expressed by Hawley on behalf of the Academy Trustees were widely shared by certain segments of the Albany community who were also convinced of the practical benefit of theoretical knowledge.[7]

These beliefs were elaborated in Henry's inaugural address.[8] The core of Henry's lecture was devoted to pure mathematics, its history, and its applications to current work in mathematical physics, not to technology. The history opened with a popular, rather trite chronology of mathematical development from its prehistoric origins through the calculus. After this display of erudition, obviously designed to impress, Henry demonstrated his awareness of current work in the physical sciences, chiefly the applications of the calculus to astronomy and celestial mechanics.[9] Only the latter part of his speech responded to Hawley's utilitarian theme. Henry assured his audience that the Academy's mathematical program would equip the student "to pursue all subjects of a practical nature to which the rules of the science are applied," The mechanic arts and engineering disciplines were all instances of the application of mathematical and scientific principles. "Navigation, [he wrote] surveying, guaging and mensuration constitute only different branches of mathematical learning," and the steam engine was "that legitimate and gigantic ofspring [sic] of mathematical and chemical science." Henry was especially struck by America's recent program of internal improvements as an emphatic demonstration of the practical power of theoretical knowledge:

The works of internal improvement, at this time so zealously prosecut[ed] in all parts of the union, emphatically teach us how mathematical and mechanical knowledge may be applied not only to the necessities of ordinary life, but to increase the power and promote the wealth of a nation.

These past triumphs, Henry believed, were a mandate for continuing efforts to teach engineering on the basis of mathematical and scientific principles.[10] Ultimately, he foresaw the merger in America of mathematics, science, and technology.

Statements like Henry's typically provoke two kinds of reactions from historians of science: either they assume a peculiarly American bias toward a utilitarian orientation or, on the contrary, that Henry was a secret advocate of pure scientific research, pandering for obvious economic reasons to the practical-minded attitudes of his American audience. A careful reading of the implications of his inaugural lecture, however, indicates a position fitting neither of these historical stereotypes. First of all, Henry was more than respectful of high theory and science for its own sake. His inaugural speech, much of it devoted to mathematical "monuments of the human intellect," and, above all, his subsequent scientific career, amply testify to a commitment to disinterested theoretical research. That commitment did not make Henry's expressions of concern for technology and utility insincere. Henry was genuinely impressed by recent American technological achievements. In the mid-1820s, American engineering and technology were undergoing a crucial transition, spearheaded by the internal improvements program and the beginnings of railway development. Only a year before Henry's inauguration, the Erie Canal had been opened—perhaps the most consequential technological event of the decade.

Henry's linking of scientific and mathematical knowledge with these technological advances emanated from a core of belief nurtured by extensive reading in popular and scientific literature, and by contact with like-minded men in Albany. The beliefs were based partly on interpretation of the historical record: a glance at the past, for example, convinced Henry that the steam engine was the child of theory. And the historical vision was reinforced and partially determined by strong, widely shared preconceptions about the nature of human knowledge.

The opening paragraphs of the inaugural lecture stated the crux of the relationship between mathematics, science, and the applied arts. Involved were fundamental definitions of mathematics, science, and technology. According to these definitions, there were no incompatibilities between theoretical and applied knowledge; in fact, all sound useful knowledge had to have a theoretical basis. Utilitarian goals, if pursued correctly, in no way debased the theoretical value of mathematics and science. Henry wrote that:

Mathematics has been divided into two kinds: *pure* and *mixed. Pure* mathematics studies the relations of quantities abstractly, or considers them independantly [sic] of any substance actually existing. But *mixed,* or physico-mathematics, investigates these relations in connection with some known law of nature or with reference to the transactions of the affairs of society. Thus, in surveying its principles are applied to the measurement of land, in optics to the properties of light, in astronomy to the motion of the heavenly bodies and in political economy to the doctrine of annuites, the amount of interest, etc.

In Henry's day the division between pure and mixed mathematics was thoroughly conventional. Of European origin, it was found in many standard reference works on science and mathematics which Henry, as a newcomer to science, undoubtedly read with care.[11] The distinction belonged to a continuing tradition. Francis Bacon discussed ideas of pure and mixed knowledge in the *Advancement of Learning*[12] and the mathematician Isaac Barrow, Isaac Newton's mentor, analyzed the distinction in his *Mathematical Lectures.*[13] As Barrow pointed out in his second lecture,[14] the pure–mixed conception can be traced to the beginning of Western science. In *The Metaphysics,* Aristotle referred to the Pythagorean division of mathematics into pure or primary and mixed or secondary. Similar ideas were represented in the well-known Greek distinction between *arithmetica* and *logistica,* where the former related to number in the abstract and the latter, roughly equivalent to the modern logistics, applied number to practical matters—to commercial calculations, for instance, or to the deployment of soldiers.

Like most conventional wisdoms, the terms were seldom applied with consistency, especially as they evolved historically. Generally, pure and mixed mathematics were subsumed under far broader categories of pure and mixed knowledge. Often included among the pure were other rationalistic disciplines such as theology and logic.[15] Mixed mathematics, sometimes termed compound or physico-mathematics or mixed science, was more ambiguous than pure mathematics. In its broadest definition, it applied mathematical deduction to natural phenomena discovered by observation, experiment, and induction. Typically, this category included theoretical aspects of optics, acoustics, mechanics, astronomy, and other highly abstract subjects. In other words what we now designate as theoretical physics was formally considered by many a branch of mathematics. Although the usages were in no way uniform, less mathematical disciplines such as chemistry, meteorology, and geography were vaguely distinguished as the "experimental sciences" and placed slightly lower on the pure–mixed hierarchy. The position of technology and mechanic arts was problematic. Sometimes, the problem of classifying applied fields prompted another distinction between pure or "speculative" mathematics and practical mathematics. Note that Henry grouped all types of applied subjects—surveying, political economy, etc.—with the mixed sciences.

But the important point is that, according to these definitions of mathematics, science, and technology, every technological field embodied elements of the theoretical. There was, furthermore, a continuous progression upward toward the theoretical. The boundaries of mixed mathematics were neither precise nor fixed: when an experimental or practical discipline became sufficiently mathematical, it became part of mixed mathematics. Even the dichotomy between the pure and mixed was questioned by some. Isaac Barrow, for example, was one of the first to criticize the separation, maintaining that every field has both abstract and applied aspects.[16] Theory and practice were not separate, antagonistic realms but differing modes of a unity.

Part of the difficulty in understanding Henry's concept of science and technology stems from the fact that a variant form of his view persists to the present day. Although the term "mixed mathematics" has long since passed from the scientists' lexicon, it is a twentieth-century truism in a number of circles that technology ultimately rests on basic science.[17] Yet today there are subtle emphases distinguishing the modern view from Henry's—differences which a modern historian might easily miss in drawing inferences about past science. The twentieth-century exaltation of theoretical purity has often relegated technology to a related but subordinate role, unworthy of a theoretician's attention. Coupled with this status is the implication of a real difference in kind between science and technology. Though a modern scientist would certainly agree with Henry's premise that scientific theory generated technical change, he would puzzle over Henry's solicitous attention to the details of technological processes. To Henry and like-minded contemporaries, the application of science to utilitarian ends was both a theoretical challenge and an overriding social responsibility.

Henry's inaugural speech scarcely penetrated into the ways science has been and should be applied to technology. Admitting that all but the most talented minds were prohibited from the higher reaches of mathematical and scientific theory, Henry assured his audience, however, that "all problems valuable for their practical application can be deduced from principles of a much more elementary kind." There were, then, distinctions to be made between the latest theories in science and those applied to technology. Just how the advanced and the elementary principles are related was never spelled out. In any case, a good general scientific education would seem to answer the professional needs of the inventor, the mechanic, and the engineer. A later statement in the inaugural lecture suggested another way of interpreting the relationship between mathematical learning and practical affairs. Emphasizing the traditional educational role of mathematics as a mental discipline, Henry pointed out that, although particular facts and theorems may be forgotten, the student "will be much indebted to these early studies for that general discipline and enlargement of the understanding so necessary to his professional rise and usefulness." In other words, the practical benefits arose not only from applying specific principles but from the application of methodologies and work habits learned in courses in mathematics.

A more confident and specific statement of interactions between scientific theory and technology appeared in a different context. In the winter of 1832, after six years of Academy teaching and after embarking upon his most creative scientific researches, Henry collaborated in a series of public chemistry lectures given at the local scientific society, the Albany Institute. Despite the title of the series, Henry's introductory remarks dealt with a very different topic—the importance of basic science for technology and the mechanical arts.[18] Certain of the themes from his inaugural speech were taken up in more elaborate form. At his inauguration, Henry perhaps sensed that he was saying familiar things to a receptive "enlightened" audience who shared many of his basic attitudes. The chemical lecture, on the other hand, had a defensive tone and a polemical edge.

The science Henry dealt with was not chemistry but the most theoretical and basic science of all, the "mechanical philosophy." The mechanical philosophy, Henry indicated, was linked directly and indirectly with the highest achievements of man, "with all the arts and with the developing of the human mind and progress of civilization." As a purely intellectual accomplishment, the discoveries of theoretical science were unexcelled and deserving of appreciation for their own sake. The highest value of science lay in its contribution to pure knowledge. Here was an ideology of purity of scientific theory; but Henry also believed that the mechanical philosophy was intimately and properly related to other forms of human achievement—to astonishing recent advances in technology and the practical arts.

Henry took it for granted that technical innovation underlay the dramatic economic growth we now associate with the Industrial Revolution. New sources of power and advances in engineering had facilitated transportation and stimulated commerce so that necessities and luxuries were made available to broad segments of mankind. Revolutionary developments in communications and transportation had strengthened social, political, and ethical bonds around the world. Henry's paean to technology culminated in a quotation from an unnamed popular author[19] whose admiration for the miracles of the technological age bordered on worship.

For me [he wrote] the sheep of Saxony and Spain are shorn and the clothiers of Gloster and Yorkshier [sic] ply their shuttles; for me the cotton of Carolina and Bourbon is rolled upon the spindles of Manchester. . . . To warm me steam engines are raising from the depths of the earth its mineral fuel.

Behind beneficent technology was the guiding light of theoretical principles. "Every mechanic art," Henry declared, "is based upon some principle [or] general laws of nature and . . . the more intimately acquainted we are with these laws the more capable we must be to advance and improve [the useful] arts." To make his point about the practical uses of scientific knowledge, Henry briefly explored the history of technology, an exploration occupying the bulk of his introductory lecture, and one which he obviously enjoyed. Every historical instance confirmed Henry's fundamental belief in the omnipotence of scientific method and principle. In his eyes, the development of the steam engine depended upon James Watt's scientific exploitation of Joseph Black's theory of latent heat. The water wheel was constructed on the scientific principles of the British engineer John Smeaton. An important improvement of the windmill "employed" the mathematical calculations of Daniel Bernoulli. Ship builders utilized Leonhard Euler's mathematical discoveries about the curvature of hulls. The invention of the cotton gin depended upon Eli Whitney's "extensive scientific knowledge," as did Robert Fulton's achievements in steam navigation. Humphry Davy's scientific investigation of fire-damp led to his invention of the mine safety lamp. The construction of canals and railways required scientific knowledge of the highest order. And so the list went.[20]

While vague about the actual historical connections between theory and practice, Henry's views clearly evolved from those briefly suggested in the inaugural address. The practical arts were no longer restricted to more elementary mathematical and scientific principles. Rather, basic science on all theoretical levels was intricately interwoven with technology. Certain scientific principles, Henry believed, were immediately applicable to, and indeed directed toward, utilitarian ends. Yet, even science pursued solely for its own sake, without practical regard, ultimately bore upon the "wants or luxuries of society."[21] Moreover, the full body of scientific knowledge, be it old or newly discovered, bore practical implications. In relating the connections between scientific and technological principles, Henry used an ambiguous terminology: practical discovery is "dependent on," "constructed on," or "based upon" theoretical principles, or "employs" them. Despite the ambiguity, we feel Henry was not making the trivial point that every invention or useful art merely implicitly embodied fundamental natural principles. Rather, for a useful art to advance, its fundamental principles must be scientifically formulated and extracted theoretically.

Although glossing over the details of theoretical development, Henry was aware that other, less theoretical minds such as the mechanic, the craftsman, and the engineer were involved in the process of invention. Henry stood by his statement that every significant practical discovery depended upon the efforts of scientists or men with substantial scientific abilities. Somehow the theoretical discoveries are "reduced to practice," making them available to men directly involved in practice. Though he never bothered to explain this reductive procedure, Henry presumably meant that scientific laws and facts are in some manner codified and simplified for the use of theoretical technicians, mechanics, and other appliers of knowledge.

In general, Henry defined a one-way relationship between science and technology: theory gradually worked its way down to application. Despite his belief in the priority of theory, however, Henry asserted that the most successful contributor to technology would embody the best attributes of the theoretical and practical man: "We have practical men in great numbers without theory and theoretic men without practice. Now it is evidently the union of these two in the same individual from whom we must expect the greatest and most successful efforts of art."

Henry's facile rendering of the still unraveled interrelationship between scientific and technological development would not convince any historian seriously concerned with the details of technological change. Because of his belief in received ideas of technology, Henry simply misread or did not bother to analyze the historical evidence. Many of his examples suggested links between science and technology, though none followed Henry's scheme in any clear-cut way. In seeking historical precedent for his views, he surely encountered cases in which invention manifested no obvious dependence upon basic science. But it appears that Henry dismissed most such advances by denying their fundamental novelty or importance. For instance, in a discussion of Watt's epochal improvement of the steam engine, Henry dismissed over 300 subsequent patents as inconsequential gadgets. For him, all truly original and significant technology was scientific. Empirical or trial-and-error development simply did not fit his definition of technology.

Interwoven with Henry's account of technology was a less sanguine message. However obvious the scientific basis of technology seemed to him, there existed in America a significant counterideology. In the opening remarks of his lecture, Henry complained that the mechanical philosophy, "and particularly the higher parts of it, has not received that attention in this city and indeed in this country which its importance demands." Henry was not particularly surprised that speculative science—science for its own sake—received little public recognition; this was perhaps too much to expect.[22] But what distressed and puzzled him was the widespread skepticism he perceived about the practical utility of basic science[23]—a skepticism which to some degree has survived to the present day. While enlightened Albany might have shared his visions of a scientific technology, these views were far from universal among the American populace. Henry traced this resistance to the application of basic science to the self-conceptions and public images of inventors, mechanics, and practical men, who, at the scientists' expense, claimed to be the chief purveyors of technical progress.

Important as the study of Mechanical Philosophy must appear to every well informed mind [Henry wrote] it is a fact that there is a general tendancy [sic] among men of mere practicle [sic] skill, particularly in this country, to under-value it and to consider scientific principles, as mere hypothises [sic] from which no practical benefit can be derived.

Notice that Henry recognized this hostility as a universal phenomenon but considered it especially virulent in the United States.[24]

A conventional explanation for this neglect of basic science lay in public attitudes he knew from hearsay and the popular literature. His own brief forays in to technology afforded firsthand observations of such prejudices. Specifically, occasional encounters with American pat-ent law convinced him that official policy placed the scientifically minded innovator at a particular disadvantage.[25] Henry's growing sci-entific prominence also attracted the attention of practical men, who often approached him for support or advice. Among Henry's papers, for instance, there are letters from inventors and mechanics asking for aid in constructing perpetual motion devices. Though the inventors had indeed solicited the aid of science, Henry was appalled by their appar-ent ignorance of theoretical principles.[26]

Henry did not for a moment doubt the technical skills of America's mechanics and inventors. They were indeed ingenious men. Nor did he necessarily attribute their hostility to an irrevocable prejudice against basic science. Their disservice to technology stemmed rather from pride born of ignorance (though Henry hinted that the ignorance was at least occasionally willful). A recurrent theme of his chemical lecture was that men of "mere practicle [sic] skills . . . do not recollect, or are ignorant of the fact, that almost every art particularly if it be of ex-tensive utility is founded on the accumulated discoveries of scientific men for centuries." Ignorance and pride were mutually reinforcing. To Henry, the Faustian sense of power over brute nature imparted by technology deluded the practical man about his own omnipotence. The ship carpenter, for example, was astonished by his own power to build a vessel capable of withstanding the fury of the ocean. Likewise, the navigator prided himself on his abilities to penetrate and chart the unknown. Their continual contest with nature's forces rendered them insensitive to the intricacies of higher scientific principles. According to Henry, it was inconceivable to the practical man that mental agility, not manual skill and bodily force, was needed to extract nature's law. In fact, "the very names and the history of the Philosophers to whom they are indebted for all that is valuable in their respective arts are interely [sic] unknown to them."

By the time technological innovations reached the men of practical skill, Henry insisted, the battle with nature had already been won. The theorist had hammered nature's forces into manageable form, dis-played her secret facts, and codified the theoretical principles so that they were accessible to all. The ease with which inventors and practitioners could now manage the phenomena gave then a false sense of accomplishment:

From the facility which continued practice gives, the operative mechanic is often inclined to plume himself upon those attainments which consist rather in a habitual dexterity [of] fingers than any acquirement of the mind and to regard with contempt the principles of scientific, what he snearingly [sic] denominates merely book knowledge.

Ironically, Henry implied that the practical man's unwarranted pride and resentment of the scientists' competing claims arose from his remoteness from the actual innovative process. The priority of science was simply beyond his comprehension. A "mere practical engineer," Henry insisted, was incapable of extending the limits of existing technological knowledge.

Yet another source of public delusion, Henry noted, was that instances of chance discovery occasionally crop up in history. Or, occasionally, men of transcendent genius, capable of circumventing ordinary scientific procedures, make significant contributions. As an example, Henry recounted a familiar tale in the history of the steam engine.[27] A persistent and possibly accurate account has it that a young boy, Humphrey Potter, had, by chance or rare cleverness, substantially improved the valve mechanism of the Newcomen steam engine. To dwell on such atypical cases was, in Henry's eyes,

rather productive of injury than good. . . . The Shakesperes [sic] of Science are as few as those of Literature: for one that has fallen by chance or unassisted talent upon a fortunate discovery, thousands have improved the arts by the aid of calculations baised [sic] upon the scientific principles.

Henry saw dangers in exalting the ingenious mechanic. First, there was the obvious danger of depriving theorists of their due social and economic rewards, something which must have been in the back of Henry's mind. But, above all, by ignoring scientific truths, the mechanic's store of knowledge would soon dry up, leaving nothing but the repetition of old practices. Henry feared that the whole technological process would become distorted and trivialized. The public could become blind to the meaning of significant technological change which Henry saw emanating from pure or goal-directed science. Technology might become associated with the efficient execution of routine practices, with the "habitual dexterity [of] fingers," or with the laborious accretion of tiny improvements. The public might see this as the true path to discoveries of highest utility. Everywhere Henry saw evidence that the ingenious mechanic's views were winning public, and indeed official, acceptance. The patent law was a case in point.

Many historians of both science and economics share a common viewpoint with most of Henry's contemporaries—they exalt the past contributions of the little men of technology, the ingenious mechanics, to technical and economic development and downgrade the contributions of scientists and the scientifically inclined. Joseph Henry would have nothing to do with this belief. An alternative viewpoint like Henry's was only considered seriously by a minority. The practical labors of practical men were self-evident causes for technical improvements to most Americans of that day, judging by the evidence so far uncovered.

Contemporary historian of science and economics awkwardly face a different body of evidence in Henry's age—numerous recent clear instances of the successful application of science to technology resulting in new processes and products. The assertions of scientists and nonscientists about past applications of science are then dealt with by a set of counterassertions. First and foremost by far is the statement that science was not really applied to technology until the last half, or the last two decades, of the nineteenth century.[28] A second counterassertion is that the past statements about science being applied have meanings differing from those currently ascribed to these terms.[29] We will discuss these counterassertions in greater detail below.

The most striking feature of the counterassertions is the undocumented insistence not only that technology was independent of science in the past but that all or almost all the improvements in technology to some point in the nineteenth century were the result of the labors of anonymous improvers—shop foremen, mechanics, craftsmen, and enterprising manufacturers. Rarely is evidence offered in support; some writers do cite, for example, the inventors of textile machinery in the Industrial Revolution.[30] Usually no evidence whatsoever is given. Not only are names absent—after all we are dealing with anonymous little men, not the eponymic heroes of science—but only rarely does the literature even specify particular instances of improvements in processes and products. Technology may or may not be historically independent of science; science may or may not have been "really" applied until 1850 or 1880; but as so much hinges on these anonymous little men, bare assertions unsupported by solid evidence create a serious historiographic problem.

Evidence does exist of the technological contributions of practical men. There are some very convincing case studies in *The Sources of Invention*, but significantly these are from this century, not the seventeenth and eighteenth century.[31] A splendid instance is provided by A. Rupert Hall's work on ballistics in the seventeenth and eighteenth centuries.[32] There are the previously mentioned inventors of textile machinery. Like the views expressed by Henry, the faith in the little men of technology is an old one; Musson and Robinson cite George Atwood to that effect in the eighteenth century.[33] But when a David Landes sentimentalizes the anonymous improver and the inspired amateur while elsewhere noting the importance of trained scientists,[34] we can only suspect a desire to have the best of all possible worlds. When Hugo Meier writes about the relationship of technology and democracy in *antebellum* America, there is a complete absence of theoretical science. There is also very little cited from the actual improvers of technology. Meier relies mainly on lawyers like Thomas Green Fessenden and clergymen like Edward Everett.[35] The uneven literature on particular inventors is inconclusive on the point.

The shortcomings of the secondary literature do not necessarily mean that the assertions about the role of the anonymous improvers are false. Where efforts have occurred to ferret out these men, to study their backgrounds, to specify just what was done, the results so far disclose a complex situation, not in accord with either a Henry-type assertion or the conventional viewpoints of historians of economics and the sciences. Musson and Robinson, for example, provide an enormous body of information about certain economic, scientific, and technological activities in Britain during the Industrial Revolution. Unlike a recent economic historian,[36] we are not inclined to shrug off their evidence. While not wholly agreeing with them, Musson and Robinson's work clearly demonstrates to us that the conventional historical wisdom is suspect. Something was going on in the Midlands and Lancashire involving the sciences. Manuscripts and printed sources exist for Britain in the Industrial Revolution, just as they exist for nineteenth-century America. And we have no doubt that analogous research in the United States and on the Continent will also uncover unpredictable complexities from which we may derive a higher synthesis. Hypothetical anonymous little men are not as usable historiographically as real persons.

Historians of science are like Joseph Henry in not really thinking much of the anonymous improvers of technology in a scientific sense. Obviously, they could never concur with Musson and Robinson, who want to exalt them to the status of scientists. But many historians of science have a historiographic bias strongly favoring the retention of the little men as prime movers of technological progress in the past. Before a professional discipline existed, much of what passed as history of science vaguely talked about science influencing or generating technology. In the 1930s a Marxist interpretation of the history of science briefly flourished before being supplanted, at least in the West, by a historiography revolving about scientific ideas. Central to this viewpoint, which largely dominates the field today, is the interpretation of the Scientific Revolution of the seventeenth century as overwhelmingly conceptual in nature.[37] For most historians of science, including ourselves, A. Rupert Hall's essay "The Scholar and the Craftsman"[38] effectively ended any lingering allegiance to the belief in an operative role of the nonscientist in the Scientific Revolution. Like Hall, most of our colleagues are inclined to limit the role of craftsman to providing occasions for scientists to develop the conceptual novelties now known as the Scientific Revolution. Since that event, historians of science of this persuasion usually assume a largely autonomous development of science. Only at initial stages of a science are influences admitted from technology (or society at large).[39]

The exaltation of the craftsman and the other anonymous men as improvers of past technology serves to reinforce the belief in the autonomy and purity of science seen as a concept-generating activity. Dissociating science and scientists from the Industrial Revolution and other technological epochs minimizes not quite, but almost, forbidden possibilities. If scientists are shown busying themselves with practical as well as theoretical problems, then the autonomy of science is imperiled. The possibility then would arise that technology (or practice in a broader sense) had an appreciable impact on the actual course of scientific development. Rather than simply an initial occasion, one might find instances where science fed into practice and was in return enriched or stimulated. Ruling out any significant interaction until the late nineteenth century, whatever the validity of such a belief, serves to protect a particular historiographic viewpoint.

For the very recent period, conventional historians of science tend to assume one-way flow from science to technology—like Joseph Henry— and to minimize the possibility or significance of a return flow from technology. This not only maintains the purity and autonomy of science as a conceptual system today but evades current instances which might suggest analogous events in the past contrary to the historiographic convention. Associated with the idea of the intellectual autonomy of science is a strong belief in the development of a particular kind of scientific community. Out of the diverse groups of the amateurs, literati, and a few genuine scientists, historians of science postulate the emergence of a distinct body of professional scientists, the carriers of the tradition.[40]

Henry in Albany and in his later career figures in the American development of a professional scientific community. He was clearly against downgrading theory and must have felt exasperated at an Albany newspaper story ridiculing the pretensions of the learned when confronted with the solid virtues of a hardheaded mechanic.[41] Unlike the European situation, in America the ingenious mechanic was often presumed to be a "scientist" and given equal status with men of theoretical pursuits. None of this was satisfactory to Henry, but Henry's views were unlike today's historians of science. He would not banish science from technology, nor technologists from science. As long as the ingenious mechanic deferred to theory and to the theoretician, Henry was content. But there was no exalting or sentimentalizing practical experience. Watt could send a son abroad for education, including the sciences, but also urge the importance for the civil engineer of actual experience in carpentry.[42] Although not as pure minded as his twentieth-century analogues, Henry, the son of an Albany cartman, was of a different tradition than Watt. In 1826, he did not urge his townsmen to have their sons study carpentry in order to advance internal improvements. Study mathematics, he insisted, and give the Albany Academy more philosophical apparatus. Implicit in his words is the injunction to defer to theory and to theorists. The language is not egalitarian, nor does it offer much solace to the anonymous, if ingenious, little men of technology. Henry's words were a program for the future, not a compromise with a current ideology. It was a future in which theory, while exalted, necessarily improved man's lot.

Conventional historians of science did not evolve their view of the relationship of science and technology in a vacuum. When historiographic pressures demanded an exaltation of the purity of science, a ready-made ideology already existed in the scientific community. Well before the reaction against the Marxist interpretation, elements of the scientific community in many nations were firmly committed to a view of the primacy of basic or fundamental research, coupled to a belief that such research flourished best in a pure environment—that is, untrammeled by concern for practical problems. It is a curious coincidence that the secondary literature is very sparse on the origins and development of so important an intellectual position. Like the assertions of the great contributions of the anonymous little men of technology, one finds only snippets of supporting evidence. Clearly, by sometime late in the nineteenth century a good many scientists entertained such views—that is, around the time the literature asserts that science was first "really" applied. When, in fact, did scientists make a virtue of not being involved in utilitarian ventures? How did they reach that position?

Important for our argument is the role of time in the current view of the application of theory to technology. Henry is rather silent about how theory is applied, now recognized as a major problem. Past scientists, we gather, saw no particular problem; theory was obviously applied. The current viewpoint assumes that the time between the formulation of a theory and its application is being progressively shortened. What is applied is therefore likely to be the most recent finding or a very recent formulation. Henry's examples, on the contrary, indicate the application of very old science, as well as the very new. When a present historian of science, like Charles Gillispie, looks back to the eighteenth century he notes no connection between the latest scientific and technological advances. Even though he elsewhere indicates that the latest theory is not always the one being applied, Gillispie clearly feels he has scored a point against the persistently expressed views of scientists.[43]

Beyond the desire to maintain the purity of the Scientific Revolution of the seventeenth century is a clear element of Whiggish history. The argument implicitly takes the following form: If Western civilization was drastically changed from the seventeenth century to the present, science had to play a major role in this transformation, as the Scientific Revolution was the major discontinuity in the Western world.[44] Technological change is a principal component in the transformation. Since science did not influence technology significantly in the seventeenth and eighteenth centuries, when did its influence become manifest? The obvious answer is during the so-called Second Industrial Revolution of the late nineteenth century, which roughly coincides with the rise of industrial research in the electrical and chemical industries. If one probes further to find out why this impact of science occurred then, the implicit answer is that science had obviously reached a stage where it could significantly influence industry. Theory was not applied earlier because past theory was not as good as present theory. Here, in the absence of the restraint imposed by the desire to keep *the* Scientific Revolution free of technological taint, the historian of science can assume that theoretical progress necessarily implies technological progress.

Although the chemical and electrical industries are quite important in the Second Industrial Revolution, are we dealing with a coincidence or with unequivocal evidence of a relationship between economic advance and the application of scientific theory? The single most important industry in the Second Industrial Revolution was steel. Scientific research and development in our present terms played a very minor role in the growth of steel manufacturing in the years 1850–1900.[45] Or to go forward in time, the rise of the automobile industry in the United States had an enormous social and economical impact. In the early decades of this century there was very little, if anything, we can characterize as the conscious application of scientific theory involved in the expansion of this industry.[46] From this perspective, the present attitude of both the scientists and historians of science extolling the economic necessity of basic research is subject to the same reservations expressed about past statements on science transforming technology. Too much is being claimed, perhaps at the expense of nonscientific factors, perhaps masking the real role of scientific theory in social and economic change.

Take for example the emphasis on manufacturing or production technologies as the evidence for the application of theory since 1880. An immediate consequence is the downgrading or silent passing-over of the scientific components in other technologies before 1880. Pasteur's contributions to the silk and wine industries are a case in point. While Pasteur undoubtedly viewed himself as a scientist applying science while doing this work, it did not affect manufacturing, and Whiggery is unimpressed by the absence of billowing smoke stacks. Gauss, Weber, Henry, Wheatstone, Kelvin, and other scientists played a role in the development of the telegraph which is rarely cited in general discussions of the application of science to technology. Although not a production technology, the telegraph had considerable economic and social impact.

Nor are admitted exceptions handled with any methodological deftness. Denials of any significant contribution of science to technology prior to the latter decades of the last century are often softened by admitting some possibilities involving chemistry in the late eighteenth century.[47] These possible contributions are not to agriculture, medicine, or communications but to manufacturing. Although a literature exists discussing many of these instances, there is no inclination, apparently, to consider why this science—but supposedly no other one—was applied significantly at such an early date.

Other exceptions well delineated in the work of Robert E. Schofield on the Lunar Society[48] are, unfortunately, overlooked. If we understand his position correctly, Schofield views science as properly an exercise in the generation of ontological conceptions. On the surface, then, he looks much like what we have described as a conventional historian of science. But a close reading indicates differences of great importance. Schofield finds in Birmingham not only the application of scientific concepts to practical affairs but also technological concerns giving impetus to research on a conceptual level. Schofield's implicit points are that this industrial orientation is not the only instance of science being applied prior to the nineteenth century, and that the interaction of science and technology on the theoretical level is a principal characteristic of the research scene in this century. Ironically, two recent publications on the Industrial Revolution, both stressing the role of science, cite the differing view of Charles Gillispie based on the French situation, rather than Schofield's position with its solid grounding in the British social and intellectual scene.[49]

Gillispie introduced two influential themes into his analysis of the theory and practice relationship. As his work was derived from eighteenth-century France, Gillispie counters the Schofield position with one based in part on assumptions of national differences. The second theme, arising neatly in juxtaposition with differing national characteristics, is one stressing the role of scientific method as an explanation of the past assertions of the role of theory in improving technology. What happened was, in his striking words: "Science was only exploited. What was applied was scientists." The exploiters were typically nonscientists using the methods of science, rather than its concepts. The French scientists, in Gillispie's view, constructed a markedly different, even autonomous, social and intellectual group from the improvers of technology. Being theoretically inclined, their exploitation is not intellectual but personal, perhaps an intrusion of the world into the life of the mind. The Enlightenment "application of science" in this view is not improvement but taxonomic description of the arts and crafts. Gillispie flatly denies that any improvements in technology arose from the body of science or, just as important, from the community of scientists. While conceding the possible validity of the British examples cited by Schofield, he is inclined to dismiss them. Far more important is the French example in which a theoretically oriented elite concentrate on the development of scientific concepts, clearly a paradigm of a presently preferred arrangement. Looming large in his mind, apparently, is the social and intellectual dichotomy between the practical artisans and the pure-minded French scientists. Schofield is no less concerned with science as a system of concepts, but he, unlike Gillispie, sees connections and relationships between technology and pure science.

Gillispie's position rests squarely on the belief that science did not make any significant contributions to technology even though a central quotation in his article is from Berthollet, whose role in the discovery and development of chlorine bleaching is well known and admirably given by Musson and Robinson.[50] He also studiously disregards the very evident concern of the Old Regime and its scientists for the application of both theory and scientists to practical problems of interest to the state. Such concern existed in the seventeenth century, as shown by Harcourt Brown.[51] Roger Hahn's recent work gives additional confirmation of the avid interest of the French government in using science and, just as important, the great involvement of French scientists in useful endeavors. Réaumur comes to mind as an obvious example. It is very awkward, in our opinion, to dismiss these instances as merely a concern for the taxonomy of industry or as merely unsuccessful. And we agree with Schofield that the scientists participated willingly.[52] We really do not know enough about these instances; the very persistence of the use of science and scientists for practical purposes in France may indicate that events confirmed expectations. Certainly, the picture of French scientists in Hahn's study of the Old Regime is not that of a group exclusively devoted to theory in a society solicitously shielding them from utility.[53]

Somewhat related to Gillispie are the views of Robert P. Multhauf, like Schofield one of the few historians of science who has seriously investigated instances of intersections between the community of scientists and the community of technologists. Primarily a historian of chemistry, Multhauf finds possible application of theory in the chemical industry even though he is inclined in his best known article on sal ammoniac to assign most credit to little known practical men.[54] While he stresses the role of technologists as "improvers," in contrast to the older tradition of crediting scientists, he puts far less emphasis on the elite social group of theorists than does Gillispie. In an early general paper Multhauf notes that many scientists earned their living in the early modern period in engineering and related fields.[55] In terms of Henry's presentation, they were doing mixed mathematics. In terms of the present literature, this is another way of saying they worked as mathematical practitioners. Multhauf discounts these employments on the grounds that the fields in question were not "improving." Again, it seems strange that "improving" is limited to production technologies, whether or not scientists played any role in engineering developments.

In retrospect Multhauf and others seem impressed, if not uninterested, in this area—the area of mixed mathematics—because if science is being applied, whatever that may be, it certainly is not very recent theory. If anything, the knowledge applied is often very old knowledge indeed, being sometimes bits and pieces from mathematics known to the Greeks.[56] Henry was impressed, as we think others were in the past, simply by geometry and trigonometry being so useful in navigation, in civil engineering, and in the design of machines. After all, even in the early nineteenth century such skills were rare in the Western world whose population was largely illiterate. How much rarer must it have been in the Tudor and Stuart England of E. G. R. Taylor's compendium?[57] In such cultural environments an application of mathematics, however modest in our eyes, must have impressed. Only anachronistic thinking would deny the existence of application, even if of an ancient concept. The recent study TRACES demonstrated, after all, that even today many decades may intervene between scientific discovery and successful application.[58]

Ideologies have a way of breeding counterideologies. Contradicting the viewpoint that basic science was essential to and responsible for practical advances are assertions designed to establish the dignity of technology and its practitioners. A recent article by Edwin Layton[59] proposes a scheme of parallel growth of science and technology, each with identical structures but differing sets of values. Linking the two, in his view, are a small number of key individuals who straddle the two intellectual areas. Layton cites Operation Hindsight, which created a flurry a few years ago when its findings were interpreted as demonstrating that basic science inputs were not particularly crucial in achieving technological advances. A different yet related downgrading of the importance of basic science for technology emanates from some economists who flatly contradict assertions of the significance of research in the growth of the gross national product (GNP). It says much about the parochialism of historians of both science and technology that they have never faced up to the implications of such writings as, for example, Edward Denison's belief that improvements in factory layout have contributed more to the growth of GNP than research and development.[60] Carroll Pursell stands for another historiographic strain. While not down grading science as such or even theoretically oriented scientists, he has proposed elevating the little men of technology to the status of scientists.[61] In doing this Pursell is consciously reacting against the views of Henry, Vannevar Bush,[62] and many others who see theory filtering down to enrich practice; Pursell wants practice to grow upward to yield concepts, technological and otherwise. Perhaps this is an exaggeration of his position; yet what we see in Pursell is a desire to change both the definition of science as primarily theory ordering data and the definition of the scientific community as a small body of elitist theoreticians.

While our examples are all from the United States, the counterideology is not confined to this country. If Joseph Henry felt constrained to complain about the ingenious but scientifically illiterate mechanics, French scientists of the eighteenth century had cause to fear the hostility of artisans[63] who extolled practice over theory. The intellectual and social separation of the best of British science from realms of utility made the definition of science, technology, and the scientific community a matter of real debate in that nation. The extent of the split is discernible in a recent article on discussions of engineering education in Britain at about the turn of this century.[64] What is largely at issue are night school classes for apprentices—a far cry from Layton's picture of engineering as the mirror image of science. In her book on the mathematics practitioners of Hanoverian England, E. G. R. Taylor looks back on her work and wishes for the day when theory and practice will again be united.[65] We know nothing about Dr. Taylor's views but suspect a relationship to British politics, just as Jewkes and associates' *The Sources of Invention* seems to express a differing political allegiance by stressing the role of practical men over planning. Is it wholly accidental that Musson and Robinson want to redefine the scientific community on a notably less elite basis by showing that their subjects were empirical-minded scientists concerned with applications? Does demonstrating the role of science in the Industrial Revolution say something about the British policies since World War II? Professor Thackray, when he writes about science in the Industrial Revolution, clearly has great nostalgia for a Britain with vigorous regional cultural centers where the metropolis and Oxbridge have not stifled diversity and creativity.[66] We are not implying that current issues have falsely skewed these writings but, rather, underlining that the issues involved are not limited to a particular time and place. What Henry said in Albany in 1826 and 1832 is still echoed in debates in Washington, London, and elsewhere in the world.

What if the conventional academic wisdom is true that very little of technology was science-generated before 1880? The issues are not merely of interest to historians of science, technology, and economics but also to current policy planners. Knowledge of a long time series, rather than the experience of a few decades, may very well provide fresh outlooks. Prophecy is risky, but we are inclined to believe that Musson and Robinson's work is merely the tip of the iceberg, that future research will disclose other instances of science playing more of a role in the past in practical affairs than suspected;[67] we also think their writings indicate the nature of the findings. Instead of a simple picture of theory yielding practice (or the reverse), we will encounter complex, ambiguous social and intellectual situations. Conventional rubrics will not do. In the future we will require new terms for describing the relations of theory and practice, as well as new definitions of the social factors and institutions influencing these relations.

Even if our prediction proves invalid, we still would maintain that historians should not blithely disregard the obvious. If scientists and others insisted on believing that theory produces technological change and acted on those beliefs, this is obviously important. Nor is the importance diminished by negative results or by less than revolutionary consequences. If beliefs persist over the centuries and produce actions, we are faced with a historical factor of importance.

One way of starting this reassessment is simply to note the varied ways science is supposedly related to applications in the historical literature, leaving aside the rich, complex descriptions of the present situation. For example:

(I) The view that theory generates practice, usually assuming today the existence of a complex research and development system with trained individuals manning the various stations of the system. Historians of science usually minimize the return flow from technology; most current science administrators know better. Whiggish historians of science usually downgrade past instances if the theory applied is fragmentary or less than a general case. For our purposes we will consider these as examples of this kind of application of science.

(II) The view that theory and application (if done properly) are both science. In Henry's view (and Isaac Barrow's) both are differing modes of the same thing. In Layton's scheme, science and technology are identical, parallel but complementary and linked by key scientist-engineers. To Henry, Barrow, and others, the linkage problem does not apparently exist.

(III) The application of the methods of science to technology, usually by nonscientists. As given by Gillispie and Kuhn, an important datum is absent. Did the nonscientific applicators know they were using the methods of science, and did they fancy themselves as scientists? Type III seems very close to—

(IV) The use of pure empiricism, i.e., trial and error and Baconian data-grubbing. Some of Musson and Robinson's cases come down to this. While we are inclined to regard science as a system of conceptually ordered data, their point has a degree of validity. Not only in Britain but in the United States also the avoidance of rationalism in favor of a simple empiricism was considered by some in the past as "scientific." Undoubtedly, some of the ingenious inventive mechanics regarded themselves as sounder scientifically than mere spinners of theory.

(V) The use of scientific data. By this we mean merely that the solution of a practical problem involved the use of information found in the literature. The use of older pieces of scientific knowledge, as cited by Henry, falls into this category. It does not matter if the user is or is not a trained scientist. What counts is the action, the conscious use of scientific sources for reliable information.

What we need to know is what was actually done in the past, by whom, in what way, for what reasons, and with what consequences. Again we venture into prediction. If this research is done untrammeled by the Whiggish preoccupation with recent theory feeding into a research and development system, more instances of theory influencing practice will show up prior to 1880. This prediction is especially safe if investigators look at scientific fields other than physics, mathematics, and astronomy, the sources for the conventional view. The ubiquitous noting of chemistry as an exception is a clue. More than a decade ago Koyré noted that chemistry was simply different from the mathematical sciences in its relations with industry.[68] If we look field by field, perhaps we will discover a multiplicity of differing relations, as well as the obvious similarities. Geology is a possible example. One of its great theorists was at a mining academy. The discoverer of the correlation between fossils and strata was one of Professor Taylor's mathematical practitioners. Geologists have frequently had relations with extractive industries. Only the parochialism of medical historians has inhibited analysis of medicine in terms of the interaction of biological theory and a crude human engineering—perhaps because physicians prefer thinking of their work as science or an art and not akin to engineering. But why blame the historians of medicine when many historians of engineering play down the role of concepts in favor of hardware and structures? Agriculture is still another obvious area for research. In all these investigations we need to avoid slogans and clichés.

In asserting the existence of connections between science and technology, whatever they may be, we are simply reiterating an older view which we hope to translate into modern terms. After all, if allowances are made for changed terminology, it was very well stated early in the Christian Era by Pappus of Alexandria:

Since geometry is, so to speak, the mother of the arts, it is not harmed by aiding in the construction of engines or in the work of the master-builder, or by association with geodesy, horology, mechanics, and scene-painting. On the contrary, geometry obviously promotes these arts and is justly honored and glorified by them.[69]

Notes

1. A few examples are Robert P. Multhauf, "The Scientist and the 'Improver' of Technology," *Technology and Culture* 1 (1959):38–47; Charles C. Gillispie, "The Natural History of Industry," *Isis* 48 (1957):393–407; Thomas S. Kuhn, "The Relations Between History and History of Science," *Daedalus* 100 (1971):271–304, esp. 283–288; David Landes, *The Unbound Prometheus* (Cambridge and New York, 1969), p. 61 (but note the hedging on p. 63); M. M. Postan, *An Economic History of Western Europe, 1945–1964* (London, 1967), p. 134.

2. Later we will specifically discuss two examples: E. Layton, "Mirror Image Twins: The Communities of Science and Technology in Nineteenth-Century America" and C. Pursell, "Science and Industry in Nineteenth-Century America," in G. Daniels (ed.), *Nineteenth-Century American Science: A Reappraisal* (Evanston, Ill, 1972). Two examples of doubts of economists are Edward F. Denison, *The Sources of Economic Growth in the United States and the Alternatives Before Us* (New York, 1962), Supplemental Paper No. 13 of the Committee for Economic Development; and Harry G. Johnson, "Federal Support of Basic Research: Some Economic Issues," *Basic Research and National Goals* (Washington, D.C.: National Academy of Sciences, 1965). Although not originating in the guild of economists but among engineers, the findings of Project Hindsight are of this nature. C. W. Sherwin and R. S. Isenson, *First Interim Report on Project Hindsight (Summary)* (Washington, D.C., 1966).

3. *Science and Technology in the Industrial Revolution* (Manchester, England, 1969).

4. For Henry's education, see Nathan Reingold et al., *The Papers of Joseph Henry,* vol. I (Washington D.C., 1972), p. 118n.

5. Ibid., pp. 162–163.

6. Ibid., pp. 47–48, 243–247.

7. A striking statement to this effect was made by the "Great Patroon" Stephen van Rensselaer before the Albany County Agricultural Society: "Farmers are prejudiced in favor of their own ideals of agricultural subjects, and are unwilling to allow that much improvement can be made in the present mode of pursuing the art. This is a mistaken idea; much may be learned from those who are not practical men, but who make experiments for their amusement; and still more from scientific individuals, who have attended to the nature of soils." Ibid., p. xxvi.

8. Ibid., pp. 163–179.

9. Among other things, Henry was demonstrating his participation in the European scientific tradition. Even his brief survey indicates his awareness of current issues in European science. Henry's knowledge is not surprising. He was an avid reader, and there is ample evidence that European scientific works were readily available to him in Albany. His personal library, presently on deposit with the Joseph Henry Papers, includes a number of treatises on the nature and development of scientific thought, such as John Playfair's *Dissertation Second: Progress of Mathematical and Physical Science* (Edinburgh, 1819).

10.   Precedent for scientific engineering could be found in the technical program of the U.S. Military Academy at West Point, which Henry had visited a few months before. Reingold et al., *Henry Papers,* vol. I, p. 176n.

11.   Some works stressing the distinction were the *Encyclopaedia Metropolitana* (see Samuel Taylor Coleridge's "General Introduction"), vol. I [1818]; Charles Hutton's *Philosophical and Mathematical Dictionary* [London, 1795/96]; and Peter Barlow's *Dictionary of Pure and Mixed Mathematics* [London, 1814]. An annotated copy of Barlow survives in Henry's library. Later parts of Henry's address reveal that he borrowed extensively from Barlow. Notions of pure and mixed mathematics were also well known in eighteenth-century France. See J. E. Montucla's *Histoire des Mathématiques,* 2nd rev. ed. (4 vols., Paris, 1799–1802), especially the introductory pages to vol. I.

12.   In book III, chap. vi. Bacon's usage is mentioned in Montucla's *Histoire des Mathématiques,* vol. I, p. 5.

13.   We have used John Kirby's 1734 London translation of Barrow's *Mathematical Lectures.* See Lecture II, pp. 14–28. We are greatly indebted to Henry Guerlac for pointing out to us Barrow's analysis and for suggesting that similar distinctions can be traced to antiquity.

14.   Ibid. pp. 13, 14.

15.   See, for example, Coleridge's elaborate classificatory scheme in his "General Introduction" to the *Encyclopaedia Metropolitana,* vol. I [1818].

16.   Barrow, *Lectures,* pp. 13, 18–28.

17.   Henry's view has indeed become pervasive in the present century. One of the most confident statements of this belief was made in a highly influential report by Vannevar Bush, Director of the U.S. Office of Scientific Research and Development. Bush's *Science, the Endless Frontier: A Report to the President* (Washington, D.C., 1945), pp. 13–14, postulated a direct relationship between basic research and practical progress: "Basic research leads to a new knowledge. It provides scientific capital. It creates the fund from which the practical applications of knowledge must be drawn. New products and new processes do not appear full-grown. They are founded on new principles and new conceptions, which in turn are painstakingly developed by research in the purest realms of science." Today it is truer than ever that basic research is the pacemaker of technological progress. In the nineteenth century, Yankee mechanical ingenuity, building largely upon the basic discoveries of European scientists, could greatly advance the mechanical arts. Now the situation is different. "A nation which depends upon others for its new basic scientific knowledge will be slow in its industrial progress and weak in its competitive position in world trade, regardless of its mechanical skill." (Printed in the original in boldface.)

18.   The lecture is printed in Reingold et al., *Henry Papers,* vol. I, pp. 380–397.

19.   The author, probably British, is unidentified. Similar exaltations of technology pervade the popular literature of the day.

20.   While many of Henry's examples did indicate some interplay between science and technology, the relationships were in no case as simple as he thought. See, for instance, the discussion of the alleged influence of Black's theory of latent heat on Watt in Henry Guerlac's entry on Joseph Black in the *Dictionary of Scientific Biography.*

21. "But to those who are incapable of appreciating the incentives to daring enterprises like this and who cannot conceive why science should be prosecuted so enthusiastically for its own sake I would say the study of no department of science however remote it may at first sight appear from practical application but what has in some manner an indirect bearing on the wants or luxuries of society. The discovery which now appears as an isolated fact or the principle which at present is regarded as mere object of idle curiosity may hereafter in the progress of science become a new moving power in mechanics or a general law of nature leading to many important practical results." Reingold et al., *Henry Papers,* vol. I, p. 395.

22. Henry recognized that, in vying for public attention, chemistry had an advantage over other sciences—the spectacle of brilliant experiments, something which many public lecturers of the day were already exploiting. One reason Henry chose to devote his introductory lecture to the mechanical philosophy rather than to chemistry was that he felt the former science lacked the benefits of eye-catching demonstrations.

23. With the exception of chemistry, which Henry believed the public readily linked with practical application.

24. Henry had some kind words for technology in Britain, the "queen of the mechanic arts." He wrote that "in no country is there a more intimate combination of scientific knowledge with practical detail." Reingold et al., *Henry Papers,* vol. I, p. 389. Similar admiration for Britain's supposedly scientific technology was expressed a few years earlier in a report made to the New York State Board of Regents. In Britain, the report noted, "there are but few extensive manufacturing establishments which are not directed in their operations by a practical chemist." Board of Regents Minutes, vol. 3 (March 1828), p. 204 (in the Office of the Commissioner of Education of the State of New York, Albany, N.Y.).

25. A letter from Henry, dated 4 November 1831, revealed that Henry had applied his large electromagnets to the problem of iron ore separation. The letter discussed the possibility of patenting a separation device. Henry wrote that he had been advised that the law prohibited virtually all patents involving the application of scientific principles. Henry's views and the principles at issue are considered in the annotations to this letter; Reingold et al., *Henry Papers,* vol. I, pp. 364–372.

26. Consider his exchange of letters with a canal engineer, John R. Henry. On 25 June 1831, J. R. Henry wrote on behalf of an unnamed Louisville mechanic requesting specifications for a Henry electromagnet in order to build a perpetual motion device. Henry's irritated reply is dated 9 July 1831: "But I advise him to save his money if he only wants the magnet for a perpetual motion as all his experiments in regard to this object will certainly fail. . . . It is a fact that does not tell much for the diffusion of knowledge among the mechanics of this country that there are at the present time many ingenius [sic] but illiterate mechanics engaged in attempts to invent self moving machines." Reingold et al., *Henry Papers,* vol. I, pp. 345–346, 348–349. Actually documenting the viewpoints and alleged prejudices of these still unknown mechanics is a difficult and intricate task. Even if they did constitute a coherent group with defined views, American artisans, inventors, and mechanics have left historians little in the way of ideological literature. A reply from the un-

named Louisville mechanic would be welcome evidence. What literature does survive has rarely received historical scrutiny. A major problem in investigating them is the persistent confusion of science and invention in the literature, both popular and scholarly. Inventors are frequently grouped indiscriminately with scientists, with little attempt to distinguish their roles in American science and technology.

27.   The story apparently originated in J. T. Desaguliers's *A Course of Experimental Philosophy,* vol. II (2 vols., London, 1734–1744), p. 533. Confirming Henry's fears, Desaguliers recounted Potter's reputed invention (in 1713) to dramatize that the development of the steam engine was due not to a "thorough knowledge of philosophy" but to chance. Since repeated and embellished, the tale has recently become something of a historiographic issue, dismissed by H. W. Dickinson in *A Short History of the Steam Engine* (Cambridge and New York, 1939), p. 41, but brought to life again by A. E. Musson (in his unpaginated introduction to the London 1963 edition of Dickinson's *Short History*).

28.   See examples cited in note 1.

29.   Kuhn, "The Relations"; Gillispie, "The Natural History."

30.   Musson and Robinson, *Industrial Revolution,* p. 81f.

31.   John Jewkes et al., *The Sources of Invention* (London, 1958).

32.   A. R. Hall, *Ballistics in the Seventeenth Century* (Cambridge and New York, 1952).

33.   Musson and Robinson, *Industrial Revolution,* p. 83f.

34.   Landes, *Unbound Prometheus,* pp. 61, 63, 92, 113, 187, 258, 274, 345.

35.   Hugo A. Meier, "Technology and Democracy, 1800–1860," *Mississippi Valley Historical Review* 43 (1957):618–640.

36.   M. M. Postan, *Economic History,* p. 154.

37.   See A. R. Hall, "Merton Revisited, or Science and Society in the Seventeenth Century," in *History of Science* 2 (1963):1–16; and H. F. Kearney, "Puritanism, Capitalism, and the Scientific Revolution," *Past and Present* (July 1964):81–101. Kearney sees no direct connection between economic and scientific developments but suspects that increased wealth provided more training in abstract thought, especially mathematics. And this training in abstract thought, we presume Kearney believes, somehow had an effect on future economic growth. A more explicit treatment occurs in John U. Nef's "Genesis of Industrialization and Modern Science," in *The Conquest of the Material World* (Chicago, 1964). Although the language is different, the point is quite like Henry's views. Generalizations increase the probability of making advances of great magnitude, certainly more than the limited step-by-step advances likely from routine practice.

38.   M. Clagett (ed.), *Critical Problems in the History of Science* (Madison, Wisc., 1959).

39.   See Kuhn's remarks in his *Daedelus* paper, "The Relations," pp. 285–286.

40.   For the relationship between the group and its research, see Herbert A. Shepherd, "Basic Research and the Social System of Pure Science," *Philosophy of Science,* 3 (1956):48–57.

41.   Letter of "Franklin" in the [Albany] *Age* of 4 March 1828, p. 27.

42.   Musson and Robinson, *Industrial Revolution,* p. 373.

43. Gillispie, "The Natural History."

44. Compare this with R. M. Hartwell's "The Great Discontinuity: Interpretations of the Industrial Revolution," *Historical Journal* 1 (1970):3–16, and the volume he edited, *The Causes of the Industrial Revolution* (London, 1967).

45. Such is the impression we get from, for example, C. S. Smith, in M. Kranzberg and C. W. Pursell, Jr. (eds.), *Technology in Western Civilization*, vol. I (2 vols., New York, 1967), pp. 592–597. This does not deny that science was applied in some instances but that much of the spectacular growth in steelmaking did not ultimately depend on the use of theory and data from science.

46. John B. Rae, *American Automobile Manufacturers* (Philadelphia, 1959).

47. Kuhn's *Daedalus* article, "The Relations," p. 286, is a pertinent example. The most detailed source, but one not too helpful in our context, is Archibald Clow and Nan L. Clow, *The Chemical Revolution: A Contribution to Social Technology* (London, 1952). Two splendidly contrasting articles exist: R. P. Multhauf, "Sal Ammoniac: A Case of History in Industrialization," *Technology and Culture* 6 (1965):569–586; C. C. Gillispie, "The Discovery of the Leblanc Process," *Isis* 48 (1957):152–170.

48. Robert E. Schofield, *The Lunar Society of Birmingham: A Social History of Provincial Science and Industry in Eighteenth-Century England* (Oxford and New York, 1963).

49. Musson and Robinson, *Industrial Revolution,* p. 128; Arnold Thackray, "Science and Technology in the Industrial Revolution," *History of Science* 9 (1970):76–89; for "The Natural History of Industry," see Gillispie (note 2).

50. Musson and Robinson, *Industrial Revolution,* pp. 251–351.

51. "The Utilitarian Motive in the Age of Descartes," *Annals of Science* (1932):186–192.

52. "The Industrial Orientation of Science in the Lunar Society of Birmingham," *Isis* 48 (1957):415.

53. Roger Hahn, *The Anatomy of a Scientific Institution: The Paris Academy of Sciences, 1666–1803* (Berkeley, 1971).

54. Multhauf, "Sal Ammoniac."

55. Multhauf, "The Scientist."

56. A. R. Hall, "Engineering and the Scientific Revolution," *Technology and Culture* 2 (1961):318–332, clearly looks for something like our present-day situation or a clear use of the achievements of the Scientific Revolution. Not finding these, Hall simply overlooks what the mathematics practitioners were often applying—older mathematical knowledge. He also assumes, as do likeminded historians, that an application does not count if it is not of a full-scale theory or if the technologists simply use bits and pieces of knowledge not welded into a coherent theory.

57. E. G. R. Taylor, *The Mathematical Practitioners of Tudor and Stuart England* (Cambridge and New York, 1954).

58. Illinois Institute of Technology, *Technology in Retrospect and Critical Events in Sciences* [TRACES] (Chicago, 1968). This report was prepared to counter the differing results of Operation Hindsight. Both reports are "correct"; the differences arise from the dissimilar time spans and definitions each project used.

59. Layton, "Mirror Image."

60. Denison, *Sources of Economic Growth.* What is involved here is that in GNP determinations it is easier to come up with hard figures for actual production output changes than for changes in GNP due to education, research, and development. This becomes important for the theory vs. practice problem, as the results, if any, of use of theory are not immediately apparent, and these results may have a qualitative nature rather than a readily measurable quantitative aspect.

61. Pursell, "Science and Industry."

62. Bush, *Endless Frontier.*

63. Hahn, *Paris Academy,* is particularly good in sketching how the feelings of the artisans merged with an anti-Academy view of Rousseauistic origin which made the Revolutionary situation more perilous. On 156 he quotes J. H. Bernardin de Saint-Pierre about the role of "little men": "Let the academies accumulate machines, systems, books, and eulogies. . . . The principal credit for them is due to ignorant men who provided them with raw materials."

64. M. D. Stephens and G. W. Roderick, "Education and Training for English Engineers in the Late Nineteenth Century and Early Twentieth Century," *Annals of Science* 27 (1971):143–163. Peter Medawar, *The Art of the Soluble* (London, 1967), 113–128, ascribes the upper-class association with the purity of science to a genteel association of purity with poetry. Applications had, then, a low connotation.

65. E. G. R. Taylor, *The Mathematical Practitioners of Hanoverian England, 1714–1840* (Cambridge and New York, 1966), pp. 105–106.

66. Thackray, "Science and Technology."

67. For example, our colleague, Robert P. Multhauf, is well into a history of chemical technology during the years 1750 to 1850. He reports, to his surprise, finding science playing a considerable role in the eighteenth century. Dr. Multhauf is still dubious about science having any technological impact in the seventeenth century, at least in the case of chemistry.

Corroborating views from the field of economic history have just appeared in the work of Peter Mathias, "Who Unbound Prometheus? Science and Technical Change, 1600–1900," in Peter Mathias (ed.), *Science and Society, 1600–1900* (Cambridge and New York, 1972), pp. 54–80. This excellent article has unfortunately appeared too late for our present consideration. While seeing interactions between science and technology throughout the period under discussion, Mathias warns against assuming simple one-to-one or one-way influences, arguing instead for the profound complexity of the causative relationships involved. Also, see his review of Musson and Robinson's volume in *Minerva* 10 (1972):343–347.

68. A. C. Crombie (ed.), *Scientific Change* (London, 1963), p. 852. Koyré was speaking in opposition to Henry Guerlac's complaint that historians of science are too idealistic and neglect social context.

69. M. R. Cohen and I. E. Drabkin, *A Source Book in Greek Science* (Cambridge, Mass., 1958), p. 185.

8□ Joseph Henry on the Scientific Life: An AAAS Presidential Address of 1850

Joseph Henry's 1850 presidential address to the American Association for the Advancement of Science was never published, nor does a complete text exist. What exists in the Henry Papers in the Smithsonian Archives are four different and partly overlapping fragments. The fragments provided an occasion for historical reconstruction and speculation. Like so many of my papers, this is a detective story but atypical in its intense scrutiny of a single item. The first problem to be solved was textual, piecing together the fragments in a convincing manner. A reconstructed address is given in *A Scientist in American Life: Essays and Lectures of Joseph Henry* (1980).

The normal editorial scrutiny of words did not really clarify the mystery of what Henry meant and why the address was not printed. Literal meanings of words and phrases had to relate to contexts. There was this bright, strongly motivated individual with a distinctive but not always fathomable personality. As he spoke in New Haven, hindsight indicated changes were occurring in his nation: professions were emerging; the colleges began evolving; new bodies like the Smithsonian and the American Association for the Advancement of Science started operating. Subtle and not so subtle frictions existed on matters of values, methods, and social relationships.

Even when sources are voluminous, they are never literally complete. Because gaps always exist, the historian has the pleasure of using ingenuity and imagination to rectify omissions. While I find this account of one event in Joseph Henry's life personally convincing, it is not completed, like a crossword puzzle. This is a mental framework I enjoyed building. A future historian will, I hope, tear it down for a better one.

This paper was originally delivered before the American Association for the Advancement of Science in 1974.

"The topics which should be presented in the annual addresses [to the Association] include all those which have a bearing immediately or remotely on the objects of the Association.

"They should present to the public, from time to time, the character, the doings, the past history and the present condition of the Association.

"They should vindicate the claims of science to public respect and encouragement, and set forth the nature and dignity of the pursuit. [They should] give instruction from past experience in the methods of making new acquisitions from nature; and point out new objects and new paths of research. They should expose the wiles of the pretender and suggest the means of diminishing the impediments to more rapid scientific progress.

"They should call attention to the relation of science to the moral part of our nature and in every condition of the times point out all things which may have a bearing through science on the material and spiritual improvement of man."[1]

This was Joseph Henry speaking in New Haven on August 22, 1850, as an outgoing president of the American Association for the Advancement of Science. His opening words were a sincere, still usable recipe for a platitudinous exhortation, a sermon-like recital of what the audience wanted and expected to hear. As such, they can even serve statesmen of science today. Like Henry, today's scientists can select those themes closest to their hearts.

Henry's presence on the platform recognized his influential position in the small American scientific community. Starting with research in Albany, New York, his native city, Henry had been increasingly recognized as the leading physicist in the United States; his work on electricity and magnetism was known and respected by many of the leading European workers in the field. In 1832 Henry had left Albany for a professorship at Princeton. Not only had his investigations expanded, but Henry had become more visible and respected in the centers of science in America. Intellectual achievements and high personal regard undoubtedly helped him gain the Secretaryship of the newly formed Smithsonian Institution in 1846. When Henry spoke in New Haven, it was both as an acknowledged contributor to science and as the executive head of a body charged with the high hopes of many for "the increase and diffusion of knowledge."

But Henry's confident words did not match his state of mind. A few days after, to his friend Dorothea Dix, Joseph Henry wrote: "My address too was well received and was unanimously called for publication. I was somewhat doubtful as to some of the points I touched upon and thought I might tread upon the toes of some of my friends rather roughly."[2] Doubts must have persisted. Despite the unanimous vote, the address was not printed. I have no specific evidence as to what points Henry regarded as doubtful. So many toes were being treaded on, it is hard to single out any as especially bruised by Henry's stompings. Yet, so impressed was the Association by the address that its standing committee on August 26, 1850, designated Joseph Henry a subcommittee of one to draw up a system of ethics for scientists. The subcommittee never submitted its report.[3]

Although Henry would print similar ideas—even the exact words—in the widely distributed *Annual Reports* of the Smithsonian Institution and elsewhere, one has to assume a reluctance to spread forth in print the full range of his thoughts on the behavior of scientists in light of the nature of their calling. A complete text does not survive to help us explain what happened. What remains in the Smithsonian Archives are four fragments of the address: two are scribe's copies of the introductory passages; one is an incomplete scribe's copy with revisions inserted by Henry; and one is a messy rough draft.[4] There are also a sprinkling of scraps which someday may prove related to the unsubmitted ethics report.

The address has four sections. There is, first, an introduction about the nature of science and of the Association. As much of the opening is a statement of negatives, Henry easily turns to recounting methodological evils. In both this section and in the first section, pointed comments occur on American conditions. What's specifically wrong in his country for the practice of science next occupies Henry's attention. Here we have a problem because of the incompleteness of the surviving texts. Henry has introduced a discussion of religion here as an answer to charges of atheism, particularly from the South. Henry's treatment, although necessarily limited to a few pages, becomes very broad, not so much on the nature of religion as on the nature of knowledge. I think the passage on religion is part of the discussion of national evils; it may have served as a separate unit bridging to the next portion. Yet it has all the earmarks of a text leading to a fusillade of rhetoric in one of the glorious perorations Henry and his era were so fond of. Instead, in our surviving manuscripts we have a sober last section on the philosophy of science which ends abruptly.

The introduction defines the aims of the Association as "the promotion of all branches of knowledge which may be stated in general propositions," a rather broad statement. Although no generalizations of any nature are discarded, the Association has particular interest "in promoting discoveries in the operations of the physical world." "Physical" here includes biology (in our terminology). The contrast is with the immaterial world of spirit.

Having qualified the range of general truths, Henry next conveys a stream of negatives about the functions of the Association: not the direct promotion of religion; "not the application of science to the arts of life"; not a preference of any one branch of "material science"; not the diffusion of knowledge. But in each case there is an additional bit of reinforcing rhetoric. Note a contrast between the treatments of religion and applications:

"Yet we cherish the connection of science and revelation. . . .

"We leave to others with lower aims and different objects to apply our discoveries to what are called useful purposes."

A positive note intrudes in the introduction's affirmation that the advancement of science by the Association does not imply discarding the good and the beautiful. Henry ends the passage with these words: "There is poetry in science and the cultivation of the imagination is an essential prerequisite to the successful investigation of Nature."

The positive tone of the introduction continues with the theme of specialization, citing geology as a specific case of a specialized field. And men of different fields learn from each other. New contributions, new thoughts, new facts are exchanged in the "collision of mind with mind." These meetings of the Association, like other past gatherings of learned societies, are essential because of an attribute of science, the small numbers of the specialized minds. To Henry, the sparseness of the research population was a basic attribute setting off science from other intellectual and artistic domains:

"The historian, or the critic, or the poet finds men everywhere who can enter into his pursuits and who can appreciate his merits and unite with him in his labors; but the man of science whose abstract researches pertain not immediately to the wants of life, finds few men who can sympathize with his pursuits or who do not look with indifference on the objects of his research."

To protect the worthy seekers of truth, Joseph Henry continues in the next section with negatives, now a recital of the means by which the unscrupulous "invade the domain of science." In so doing, the unscrupulous play into the hands of the narrow-minded who despise and ridicule science. As for the unscrupulous, how Henry viewed them is clear from a passage crossed out in the scribe's text. Their "fitting reward would be a chair in the penitentiary." (I am not claiming any technological priorities for Henry on the basis of this passage.) This invasion of the unscrupulous is by means of deceptive and immoral practices passed off as science. Here Henry turns to a work he read first as a young man in Albany, Charles Babbage's *Reflections on the Decline of Science in England,* using its classification of unethical practices. To this reader, Henry's pages on hoaxing, forging, trimming, and cooking are the least interesting portion of the presidential address. Sincerity does not save the words from a flatness, from a smell of staleness largely unrelieved by striking examples or verbal acerbity. It is the dutiful student glibly reciting a well-learned lesson.

But Henry extended Babbage's classification by two additional "species of deception"—humbugging and quackery. Passion and involvement now surfaces in this portion of the address. Henry defines humbugging in these words:

"This species of deception frequently begins in folly and ends in fraud. The author of it generally imagines at first that he has discovered some very important principle in nature which is directly applicable to useful purposes in the arts.

"To this belief like the Arabian impostor he first converts his friends and neighbors. His discovery is afterwards published in the newspapers and he is soon elevated to an unenviable notoriety from which he has neither the magnanimity nor the courage to let himself down by confessing his error. He therefore seeks to discover methods by which the deception may be extended and continued, and generally ends in defrauding all who may have become his dupes."

Although the definition cites false applications of science, the one example given is not technological, medical, or agricultural. The one example, given without names, is the lawsuit in which George Foster sued James Hall and Louis Agassiz for libel over his geological map. It was an unsuccessful attempt to stop suppression of the use of the map based on Ebenezer Emmons's Taconic system. Henry was a witness in the trial.[5] In the address Henry worried about the hazards facing a scientist daring to expose humbugging.

Quackery is so general it enters into all the classes of deception according to Henry. It comprehends "a great variety of petty artifices by which the vain, the superficial, and the unprincipled endeavour, generally at the expense of the labors of others, to elevate themselves into notice and to impose upon the credulity and ignorance of the public." Henry's words about profiting from the labors of others do not lead to a recital of acts of intellectual plagiarism—that comes later. Instead the theme of the small numbers of qualified specialists reappears:

"It should never be forgotten that true reputation must always be based on the favorable opinion of the few in any country who are capable of properly appreciating the labors of him who would claim to have enlarged the bounds of human knowledge, or to have done any thing worthy of commendation by his fellow men.

"The higher and more abstruse the character of the investigations he professes to have made the smaller is the number of those who are capable of rendering a proper verdict. In this case especially the votes must be weighed, not counted."

That means going to a proper learned society or scientific journal, not seeking notoriety through newspapers. "How many wonderful surgical operations are performed in our country every year . . . if the public prints are to be credited. . . . Be not quick to trust your purse or your life in their hands."

Passion and involvement quicken as Henry turns specifically to American shortcomings. But even earlier glances at his nation had aroused him. In the rough draft, for example, the section on humbugging noted it as "more numerous in this country than perhaps in any other part of the world because there is among us a greater diffusion of elementary science and a greater confidence of our people in their own powers of originality." But this supposed American characteristic runs counter to Henry's belief in the specialized few. His first complaint is of the practice of nonscientists recommending scientists for posts in government expeditions. Even worse are cases of "gentlemen of established reputation in one branch of science vouching for the attainments of a candidate in another branch." Interestingly, Henry's wrath is intensified because he believes the officers of the government are sincerely interested in obtaining the services of the best men.

Henry quickly mutes this theme. A minor theme of a prior passage reappears: the denying of credit to scientists for their work. First, there is indignation over manufacturing elementary texts by "shears and the paste brush," invariably excluding the names of the creators of the filched materials. Even worse is the absence of copyright, making it profitable to pirate European books. "English elementary scientific books are proverbially bad," asserts Henry. Besides, he continues, "as a general rule they give no credit for what they borrow from the American Investigator." Henry wants both American and British authors to have the protection of an international copyright.

The rough draft and the scribe's copy differ significantly on how Henry answers the charge of atheism leveled against science by some of his countrymen. In the former Henry cautions that much of science is provisional and that "the man of faith [should recollect] . . . that he may possibly not have caught the precise meaning of the sacred writer. . . . [N]ature rightly understood can never conflict with Revelation rightly interpreted." From this point the rough draft disjointedly considers the relations of scientific law to God's will, ending with a statement of the boundary of human knowledge: "Our investigations must always be bounded by mystery which we cannot penetrate and if God should see fit to reveal to us the actual plan of nature and providence, he would enlarge our capacities." Parallel arguments appear in the writings of other scientists in that era.[6]

The scribe's draft reaches the conclusion of mysteries impenetrable by a different, more interesting route. After the discussion of scientific ethics, Henry defines science before plunging into the matter of the controversy with religion:

"What is science? Is it a collection of facts? Does it consist in a knowledge of the mere results of experience? Is a man entitled to be called a man of science because he can give the name and may be [*sic*] the properties of every stone or of every plant he may meet with in his path? Then the gossip who knows the name and history of every man, woman, and child of his village is a profound savant.

"Science does not consist in a knowledge of facts but of laws. It essentially relates to change. . . .

"But what is the character of this change. Is it capricious without order of succession? Or if order is observable is it that of perpetual recurrence? Do all things after a given lapse of time return to precisely the same condition in which they were at some previous time? To these enquiries philosophical experience answers, that there is an order of succession but that this order within our experience is not recurring. No moment has its fellow—each instant of recorded time has its separate history. We are not the same beings we were a moment ago and we never shall be again what we now are. Every breath we exhale carries off dead portions of our body and every pulse of our heart tends to supply the loss by the deposition of new matter. We are constantly wasting away and constantly being renewed. Every motion of the body, every manifestation of life is at the expense of the death of a part of our living organism."

But what is constant in this universe of ceaseless, ever-varying change? Back comes Henry's reply, the response of a boy who studied Paley well back in Albany: "The answer is the laws of change. These are as immutable as the purposes of *Him* who knows no change." Joseph Henry is now ready to answer those critics shouting atheism at science.

If matter is inert, as Henry believes, and cannot change its state without external force, this implies divine intervention: "the higher meaning of the term *law of nature is our conception of the mode in which divine wisdom invariably operates in producing the phenomena of nature*" (his emphasis). Divine wisdom produces a universe of law without which science is impossible. It is all in accordance, Henry notes, with the Westminster Confession of Faith—"God hath foreordained every thing which cometh to pass, but in such a way as not to interfere with the freedom of will or moral responsibility of the individual." Why this is so, Henry cannot say. Even were God to explain this mystery, other mysteries would succeed it "in boundless succession." In this version, divine enlargement of human capacities no longer exists as a possibility for penetration of the mysteries.

Abruptly, religion is left behind as Henry notes that "an infusion of German philosophy has taken place among us which . . . at first has had a tendency to perplex and unsettle the Anglo Saxon mind." No specific philosophers are named. The draft elaborates by noting attention recently given in the United States "to the connection of psychology and material science" and to the methods of arriving at truth in both. Although the material and the immaterial are subject to the domain of law and investigated by the same methods, Henry asserts that speculations about the noncorporeal, the spiritual, are not part of "positive knowledge." But such speculations are approved by Henry because they turn attention from physical wants to "the higher aspirations of the soul."

Having made his obeisance to idealist philosophy—the nontheological equivalent of religion—Joseph Henry turns to the important question, "what is truth? and what is the method of obtaining it?" Again we get evidence of the good student's well-learned lesson. Plato's cavern and the shadows of the exterior objects are cited; we know phenomena, not essences. And ontological speculations on shadows in the cave form no part of positive knowledge.

Henry then defines a scientific truth as "that definite conception of the operations of nature which we have called a law and which enables us to explain, to predict and in some cases to control the phenomena of nature. The only proof of a truth of this kind is the exact agreement of all the logical deductions from it with the actual phenomena of nature." Henry illustrates the definition by a brief example from the transits of Venus and Mercury. From this and other writings, we can loosely categorize Henry as a believer in what we now call the hypothetico-deductive method.

All of this is prefatory to Henry's real purpose—succinctly appraising the competing *a priori* and *a posteriori* system of philosophy: "We are informed in works on Logic that induction is that process by which a general law is inferred from a number of particular facts, but if the question were asked what is the nature of this process, what are the rules to be observed in making a discovery, the answer would not be so readily found." With that fairly typical scientist's affirmation of the uselessness of philosophy for research, the text ends.

Were I not, in a modest way, also a follower of the hypothetico-deductive method, I should end this paper now. A prudent Baconian would also pause, knowing that strange fragments in archives might prove to be the missing ending and that a sifting of innumerable letters might disclose prosaic reasons for nonpublication of the text and nonsubmission of the scientific ethics report. But, to paraphrase Henry, the actual phenomena are in agreement with logical deductions from the historical truth—at least to my eyes. Clues to Henry's beliefs and his behavior sprinkle the browning sheets of this address in the Smithsonian Archives. In literature, music, and art are examples of analyses of unfinished works. Perhaps not positive knowledge but speculations, yet I shall also rely on their precedents for my further words.

Despite orotund prose and ritualistic sentiments required by the occasion, the address is a highly personal document. When Henry, for example, concludes his exposition of what happened in the transits, he added: "I took some pains several years ago to compare notes on this subject with some of the most prominent scientists of Europe, and found that the method [described above] was identical with theirs." Such a statement presents a nice problem for volume 3 of the *Henry Papers!* Sure enough, in the notebooks of his 1837 trip to Europe occur statements like this one on a visit to Sir David Brewster at Melrose Abbey: "Sir David's opinion of Bacon's rules for making investigations the same as my own. . . . No working man of science advocates Bacon's method, Whewell of the same opinion."[7]

Or consider the matter of specialization. When given in the introduction as a reason for meetings like those of the AAAS in contrast to the needs of the historian, critic, and poet, Henry adds in the passage given earlier these words: "His [i.e., the scientist's] world consists of a few individuals, in some cases less than ten or twelve in a whole country who can fully appreciate him. . . . With this little world it is highly important that he should have personal intercourse." Naming the ten or twelve is no mean problem for the Editor of the *Henry Papers!* But was not this a problem even greater to his audience, most of whom were being summarily placed beyond the Pale? Or did each one in the audience view himself as one of the apostles of science? Small as the number of illuminati was, according to Henry, it is roughly of the same order of magnitude given in recent studies of so-called invisible colleges hypothesizing *international* (not national) groups of from fifty to one hundred. Whether true or not, the small numbers indicte clearly Henry's sense of isolation in his intellectual life.

The sharp words about omission of credits in texts and articles reflect, I think, a need for recognition to diminish the isolation. Priority means more than personal reward; it is scientific visibility and scientific contact replacing the loneliness of near anonymity. It defines identity. In the address Henry dwells on an unnamed "American Savant" whose findings given in a paper at the British Association for the Advancement of Science are used without credit in a book by a Briton who was there. What is worse, the British author sends the book to the American; it is also reprinted in the United States and adopted in the schools of the American's home state. Henry is referring to himself. Henry's copy of John Frederic Daniell's *An Introduction to the Study of Chemical Philosophy* . . . (2nd ed. 1843) has an indignant note in Henry's hand at the point of noncitation. Three other specific acknowledgments elsewhere by Daniell do not cancel the one omission. Being in competition with a Faraday was cruel enough; losing credit for other hard-won achievements was intolerable. Being known outside the national circle of ten or twelve was imperative.

I suspect that almost every reference in the manuscripts of this address not clearly otherwise attributable is from Henry's experiences. There are the echoes of Albany in the use of Babbage and in the language on applications. In the address, Henry does not attack technology even though the higher truths of science are more important. What outrages him is the lower aim of seeking wealth rather than adding to knowledge.[8] Perhaps the most poignant and telling of all these biographical echoes occurs in the rough draft. Here appear the words given previously about poetry in science. This version continues quite differently from the scribe's copy:

"I remember me a youth . . . age of sixteen [who] had lived in a world of imagination when his attention accidentally turned to science by opening a book which happened to be in his path. After study . . . he went forth into [the] world a new being . . . his mind awaken[ed]." Of course, this was Henry, the young apprentice with an interest in dramatics who reads George Gregory's *Lectures on Experimental Philosophy, Astronomy, and Chemistry* (1808) and resolves to be an investigator of nature.[9]

Were I a psychohistorian I might identify these last words of Joseph Henry with a remembered identity crisis. I prefer to think Henry is describing a process of conversion to a deeply held faith, science, which, not surprisingly, has some of the attributes of Anglo-American Protestantism. Having committed himself to an almost priestly dedication to a faith in which truth, beauty, and goodness are one, Joseph Henry found it hard to tolerate lapses. Even rather small things were deeply felt. In matters of his faith, Henry was often intolerant.

Because the faith is in science, not in the God of Calvinist Presbyterianism, the passages on religion may seem an intrusion to many modern readers who could excise those words without detracting from the meaning of the address. While later scientists essentially did just that, such excisions were far from the mind of a man to whom natural theology was unquestionably natural. After Darwin appeared, Henry's friend, Asa Gray, similarly an upstate New York Presbyterian, immediately introduced teleology into evolution in his favorable reception of *The Origin of Species*.[10] If the laws of change were constant, the direction of change was obviously preordained. It was the start of a great quasi-scholarly industry of reconciliation. Many later scientists, uncomfortable about the loss of religion in their milieus, would easily work into their worldviews reconciliations of freedom of will, ethics, or whatever with the prevailing body of science. It was essential and not too difficult for some whose allegiance to their scientific work arose, in the first place, from a kind of religious commitment.

If Henry felt so strongly, why didn't he publish the speech? Why did he not submit a report on scientific ethics? The talk was a public oration, after all; similar ideas appeared in print over his signature in later years, even the identical words. The phrase about weighing not counting quoted previously was actually used in an address after the Civil War.[11] Perhaps the whole matter was too personal for spelling out in print; perhaps Henry was incapable of excising the biographical content.

The events of Henry's full life provide many occasions before and after 1850 in which his views were applied but often ambiguously. Let us consider again the matter of specialization. In a long career of consultation by men in power, Henry wrote many recommendations for positions far removed from his range of personal expertise. We have no idea, to cite another instance, what Henry's testimony was in the suit against Hall and Agassiz. But there is little doubt he had never examined Emmon's evidence; even so, if he had, his qualifications in this area were dubious. Was this simply hypocrisy or a different definition of specialization from ours? We are back at the problem of how the Editor of the *Henry Papers* can identify the ten or twelve with whom Joseph Henry can communicate. Although we have much correspondence and even lists for the distribution of reprints, the criteria for choice involved are not self-evident.

But even if that problem were solved—the dozen identified and their selection explained—a practical and a moral dilemma would remain. Would the excluded accept the jugment and its implications? More important, would nonscientists always trust the wisdom of the community of scientists fairly and equitably to apportion intellectual esteem and its inevitable fellow traveler, institutional power? The issues stir opposing camps even today to rally around banners bearing words like "down with elitism" and "autonomy of science." While the words were different then, Henry was acutely conscious of this area of contention. He was less sensitive about the problems inherent in the phrase about weighing and not counting. What was being weighed, and how was that actually done? A high degree of naïveté is required to believe in the flawlessness of the refereeing process in science and in other intellectual areas. Even if one believes the flaws are anomalies or statistical rarities, their existence breeds skepticism, if not hostility, both in those victimized and those outside the inner group. Again, Henry was aware of this problem. It was omnipresent in the Smithsonian's extensive publications program. What he was not clear about was that others, even scientists, would differ on whether certain questions were "scientific" or "nonscientific." If the latter, they were not weighable but countable, perhaps in the halls of Congress. That would give Joseph Henry some troublesome moments.

An additional explanation is then possible for nonpublication of the address. If the real specialists were so small in numbers, far less than the numbers of the unscrupulous, the narrow-minded, and the indifferent, there was a hazard in full disclosure of doctrine. Why risk ridicule, attack, evasion? Here and elsewhere Joseph Henry implicitly sounded a theme repeated by succeeding generations of American scientists, almost to this very day. The scientific community is described as a minority, sometimes beleaguered, even oppressed, by a hostile majority. This perception sometimes outweighed any facts indicating relative improvements.[12] Better, then, the Fabian strategy actually followed by Joseph Henry, a strategy suited to his personality.

Even though the canons of modern historical editing prohibit ghost-writing to complete the record, I cannot resist composing an ending to the address giving my sense of the strategy Henry thought he followed:

"Hold stubbornly to the essentials of our faith in science. Avoid clashes but give not an inch, no matter what. Remain alert to seize opportunities, even if only for an additional inch. Ours is the future, provided we seize these inches not for vanity and power but for truth, for morality, and for beauty."

Notes

1. These words are from a manuscript fragment containing part of Henry's opening remarks. Here and in later quotations the text is slightly modernized in spelling and syntax.

2. Henry to Dix, August 29, 1850, Dix Papers, Houghton Library, Harvard University. A few days earlier (August 26, 1850) Henry expressed similar sentiments in a letter to Edward Foreman, a Smithsonian colleague (Henry Papers, Smithsonian Archives).

3. AAAS *Proceedings* 1850, [4]:391. Mention of the subcommittee ceases in print after [6]:viii, the account of the sixth meeting of 1851.

4. Similar ideas and even identical language shows up in these four fragments. The two larger manuscripts are not without their textual problems, even the scribe's copy. A properly edited text with variant readings will appear in a forthcoming volume of the *Henry Papers*. Rather than an elaborate editorial apparatus, on this occasion specific references to a fragment will occur in the text only where significant deviations of meaning occur.

5. Edward Lurie, *Louis Agassiz: A Life in Science* (Chicago, 1960), p. 180.

6. Sir Charles Lyell, for example, in *The Quarterly Review* 36 (1827):474–483, in a review of a work on geology.

7. This quotation is in Henry's 1837 pocket notebook, not his more formal diary of the European visit.

8. See Arthur P. Molella and Nathan Reingold, "Theorists and Ingenious Mechanics: Joseph Henry Defines Science," *Science Studies* 4 (1973):323–351, for a discussion of Henry's views [Paper No. 7 in the present volume].

9. Volume I of the *Henry Papers* (1972) deals with Joseph Henry's life in Albany and its environs until his departure at the end of 1832 to accept a chair at Princeton.

10. A. Hunter Dupree. *Asa Gray, 1810–1888* (Cambridge, Mass., 1959).

11.   See Joseph Henry, *Scientific Writings* (2 vols., Washington, D.C., 1886), vol. 2, p. 471.

12.   This is part of the "indifference" interpretation of American attitudes to science, for which see Nathan Reingold, "American Indifference to Basic Research: A Reappraisal," in George H. Daniels (ed.), *Nineteenth-Century American Science: A Reappraisal* (Evanston, Ill. 1972), pp. 38–62 [Paper No. 3 in the present volume].

UNIVERSITIES AND

LABORATORIES

Graduate School and Doctoral
Degree: European Models and American
Realities

At the suggestion of R. W. Home of the University of Melbourne I
collaborated with him in organizing a conference on scientific colonialism. It took place in 1981 in Melbourne. The proceedings appeared
in 1986 as *Scientific Colonialism: A Cross-Cultural Perspective,* from
which this paper was taken.

Comparing the experiences of Australia and the United States was
our initial organizing theme. Both are former British colonies but attained political independence in different centuries. Both are large
land masses where the original inhabitants were thrust aside as European settlers erected versions of Western society. Both remained under
the cultural hegemony of Europe even after gaining political independence. Quite early in the planning, Home and I agreed on the need for
a broader perspective for comparative purposes. Limited funds, however, forced us to concentrate largely on closely related countries like
Canada. The very important cases where Western science and technology encountered large populations with well-developed sophisticated
cultural systems (as in the Islamic nations, India, China, and Japan)
could not be treated. Nor could the conference consider the histories of
the impact on preliterate peoples.

I had the fun of considering, as a species of cultural colonialism, the
strange and wondrous world of the American university. It harbors
masters of the abstruse and has recruited semiliterates to entertain in
stadiums and arenas. Universities inculcate Jeffersonian principles
while simultaneously fashioning weapons systems and rationalizing
naked greed. American university libraries and museums contain
treasures of thought and art barely noticed by the crowds seeking meal
tickets. While very much of and in its society, the university, even in
those cases barely worthy of the name, somehow manages to give a
glimpse of an Arcadia, a vision of a hazy ideal still remembered by
some many years later. One hillock of that Arcadia loomed large in the
minds of the individual scientists I study, and this is my attempt to

sketch how graduate education came into being. Arcadia turns out to exist only in the mind of believers. The real graduate school has the virtues and flaws of the larger society.

By achieving independence in the eighteenth century and before experiencing the Industrial Revolution, the United States of America necessarily had a different "colonial" relationship to Great Britain and to Western Europe from those of other former colonies. Although political independence provided an opportunity to follow different cultural patterns, the United States had no other models to emulate, to modify, or to reject. Political independence did not entail a divorce from Western civilization. Throughout most of its history, the cultural inferiority and the dependence of the United States was assumed on both sides of the Atlantic. In this case, *colonialism,* in the absence of political hegemony, is related to *provincialism.*

After World War II, the high status of the sciences in the United States was widely recognized, posing a problem of historical causation. One nationalistic interpretation located the source of the new status in the graduate school. Doctorates and universities existed in Europe, particulary in Germany, but the graduate school was distinctive to the United States.[1] So too then was the academic multiprofessorial department, and some observers identified it as the locus of United States achievements.[2] I prefer to discuss the graduate school but without in any way demeaning the importance of the department.

In what follows, I am interested in the period to 1920, because the basic pattern of higher education existed by then, but I will make a few forays into later events. My hypothesis is deceptively simple. Because of an American misunderstanding of the German situation and the way this misunderstanding fitted into purely indigenous factors, from 1900 to 1930 (roughly) the better universities developed a policy that produced a doctoral program not only different from the German but superior in certain respects. Although my research is concentrated on the physical and biological sciences, I suspect that the situation was similar in the humanities and the social sciences.

From my standpoint, *colonialism* simply means being different according to some Western European yardstick. I refuse to accept any necessary ascription of inferiority. To understand what was different in the rise of the graduate school requires a comparative consideration of the relations of "higher" and "lower" education; the institutional and ideological roles of vocationalism and of the applications of knowledge; the way concepts of general knowledge and moral upbringing influenced perceptions of research and its inevitable fellow traveler, specialization; and last—but by no means least—the way these were sometimes related to questions of class, ethnicity, and gender. In a brief essay, I can only suggest some of these relations.

By any measure, the German university of the nineteenth and early twentieth centuries was a significant contributor to Western civilization.[3] Andrew D. White, the first president of Cornell University, was exaggerating a bit at the end of the last century when he declared, "Germany was looked upon in the United States as a kind of second mother-country" because of the German university.[4] White and other university reformers did look to Germany for models and a yardstick to measure their accomplishments. They accepted much of the German educational ideology and sometimes the sentimentality implicit in such musical works as the *Academic Festival Overture* and *The Student Prince,* even as they used the German experiences selectively for their own purposes.

Originating in the trauma of defeat and resultant reforms of the Napoleonic era, especially in Prussia, the academic ideology had a decidedly nationalistic bias. Nearly a century later, Friedrich Paulsen could describe the reforms as "a cause which did not [only] concern the German population but the whole of mankind, for the conviction [was] that the distinctive character of the German people was indispensable to mankind."[5] To the German educational reformers, the aim of education was a process of self-cultivation *(Bildung),* the unfolding of personality by a steeping in general knowledge based on fundamental philosophic principles.[6] The yardstick for this education was provided by classical antiquity, to quote Paulsen: "Lastly, since the end of the eighteenth century, the Germans have yielded themselves to the influence of the Hellenic spirit with greater fervour than any other nation."[7] Despite warm words for the sciences, Paulsen could easily conclude, "Intellectual pride finds a much more congenial soil in the domain of aesthetic criticism, of linguistic and literary studies than of natural science and technology."[8] Interestingly, the attitudes explicit and implicit in *Bildung* were echoed by some German scientists.

Although the programs of the Prussian university reformers were only in effect briefly, they persisted as an ideology, becoming particularly strong during the Hohenzollern Empire. Within the faculties of the university—law, medicine, theology, and philosophy—philosophy was the site of the ideology and the special pride of the Germans. Originally an inferior body that prepared students for the professional faculties, the philosophy faculty and its professors were what most Americans meant when they talked of emulating the German universities. Another product of the reform, the classical gymnasium, was recognized as an essential element in the success of the German university and also entered into U.S. discussions of university reform.[9]

Within each faculty, professors supposedly had an absolute right to teach what they wished *(Lehrfreiheit)*, while students were free to enroll anywhere with the *Abitur* from a gymnasium and to take any course *(Lernfreiheit)*. Paulsen and others were proud that teachers were expected to be researchers: "In Germany the scientific investigators are also the instructors of academic youth. . . . The important thing is not the student's preparation for a practical calling, but his introduction into scientific knowledge and research."[10]

Not all American academics believed that Greeks populated the universities in the Rhine Valley and eastward. In 1852, responding to a German-influenced reform movement, Henry Vethake of the University of Pennsylvania noted that in Germany few were in the faculty of philosophy and that most of those were preparing for teaching, diplomacy, and the civil service.[11] Whatever the actual numbers, Paulsen and others recognized that many more students were at the universities for *Brotstudium* than for *Bildung* in the strict sense. As the lower degrees (bachelor's and master's) were in decay, the universities offered only two degrees, the doctorate and the *venia legendi*. Most students left without taking the former, having prepared themselves, for example, for one of the state examinations for a position in government service. What did *Lernfreiheit* mean for those many in *Brotstudium* who needed preparation for particular examinations? To an American, it is odd that those serious, hard-working young men were looked down on by the professoriat.[12]

The Ph.D. was required of gymnasium teachers and for further advancement in the academic world. Of course, the degree was taken by some for prestige and for various personal reasons. It was awarded after at least three years' standing beyond the gymnasium, an oral examination, and "a smaller dissertation." The larger dissertation—perhaps not the right term—was part of the habilitation, a process by which one proved worthy of becoming a peer of the university professoriate, receiving the *venia legendi*. The two stages exactly paralleled the Prussian civil service procedure. Despite words uttered against vocationalism, the universities clearly served the utilitarian needs of the state. Analogously, in the English system and in parts of the U.S. collegiate world in past eras, preparing for the ministry was not considered vocational in the pejorative sense that some applied to engineering.[13]

In the philosophical faculties, there were full *(ordentliche)* professors in whom power resided. In this period the associate *(ausserordentliche)* professors and the lowest rank *(Privatdozenten)* largely depended on the good will of the *ordentliche* professors. Both were unsalaried throughout most of this period. A peculiar feature was that the Dozents could give private lectures, often on the most advanced topics, while the Ordinarius presented the general course in the field. As the professors' salaries depended on enrollment, giving a course required for a state examination had obvious economic benefits. As late as the 1920s and 1930s, some U.S. academicians praised the arrangement by which young Ph.D.'s could give advanced courses rather than a heavy schedue of elementary classes.

Forgotten in such praise was the absence of salaries during most of the period 1800–1920. The entire system was decidedly skewed to upper socioeconomic strata and served to produce and to maintain a *Bildungsbürgertum*—what Fritz Ringer denoted as the German Mandarins, borrowing a term from Max Weber—an intellectual elite in governing circles and universities serving as standard-bearers of a national cultural ideology. Although these Mandarins have peculiarly German traits, to North American eyes they clearly have relatives in Great Britain and France.

Before turning across the Atlantic to the American republic, a brief consideration of the gymnasium, the university institute, and the *Technische Hochschule*. The first was overwhelmingly classical. Despite glowing praise by Germans and others, its work in mathematics and natural sciences was modest until late in the last century. The closest recent analogue is the English grammar school on the arts side after the eleven-plus examination of the post–World War II years. That is, both are examples of premature specialization. As the *Abitur* was essential for university entrance and, consequently, for admission to many positions, accounts of German education are filled with the struggle to gain acceptance for other secondary schools that emphasized science and modern languages.[14]

The university institutes were directly funded by the *Kultusministerium,* not out of the regular university budgets. They were research entities, not part of the formal instructional process. Pending future research in all academic fields, not only the sciences, it is an open question to me how great a share the institutes actually contributed in preparation for the doctorate and the *venia legendi.* They were clearly not like the U.S. university departments, which, in general, controlled both instruction and research. In U.S. academic circles today, the unity of teaching and research is a widely accepted cliché and loosely credited to a German precedent, which may not exist in the American sense.[15]

The institutes acted as a device for separating new, specialized knowledge from the mainstream of formal education. The separate institutes helped keep the generality of knowledge from being overwhelmed by specialization, particularly in the sciences. Despite every noble phrase about philosophic unity, specialization was on the increase in the German university in the ninetenth century. Discovering particulars was acceptable, but many German academics apparently regarded specificity as akin to and leading to vocationalism; realia were déclassé.

Of all the vocations, engineering most troubled the Mandarins, some later writers say because of the French example and also in reaction to the cameralism of the eighteenth century. The *Technische Hochschulen* developed independently of the university, attaining the right to award degrees only at the end of the last century. Somehow, engineering as an applied profession violated some Germans' class sensibilities and cultural ideology.[16] Even academic chemists such as Liebig, strongly involved with applications, carefully and with apparent sincerity proclaimed their allegiance to purity, to the generality of knowledge. In the German university, as in the larger society, word and deed often did not match.

In nineteenth-century America, the problem was different: reality and aspiration were believed to be incompatible. During the early decades the four-year residential colleges provided education to a small portion of the population. A few of the colleges were associated with professional schools of law and medicine. Throughout the century and afterward conventional wisdom equated the first two college years with the work in the gymnasium, the grammar school, and the lycée, and the last two years, hopefully, with the European university level. Elevating the entire college course to university level often appeared to be beyond attainment.

Colleges are still important in American education. Even into this century, they retained their early purposes: moral inculcation; development of mental faculties (originally from the classics and mathematics); and provision of the knowledge regarded by the society as necessary for both cultivation and practical needs. The pattern was quite like what Nicholas Hans has disclosed for an eighteenth-century England in which preparation for adult life combined elements of both liberal education and transmission of specific vocational knowledge. In both eighteenth-century Britain and early nineteenth-century America, the assumption and reality were that hardly anybody would attend a real university. For that reason, the colleges and Hans's schools had to provide what each society required in the way of both cultivation and preparation for selected occupations.[17]

Movements for upgrading the colleges in the United States usually involved adding courses and professors to accommodate new definitions of cultivation and emerging vocations—from law, medicine, and theology, for example, to science and engineering. From Stanley Guralnick's work, we know of the significant improvements within the colleges in the sciences during the years 1830–60.[18] In many institutions there was a desire to expand the higher courses and the professional schools further in emulation of Europe, particularly Germany.

The phenomenon of lower-level educational entities raising themselves, some even to true university status, is a persistent aspect of U.S. life. During the pre-Civil War era a number of academies—the analogues of Hans's "private schools"—aspired to and attained collegiate status; a few even aimed higher. After World War II, some normal schools, junior colleges, and even community colleges "bettered" themselves, a few to university level. The lack of any national regulations or standard lured academic entrepreneurs into higher education. The coming of the graduate school, 1847–1920, was simply an instance of a general continuing trend.

The motivation of some of the U.S. professoriat for higher courses and for research opportunities is usually stressed in the historical literature. Perhaps as important, in most instances, was the relation to undergraduate education. The resultant historical development presents the following typology:[19]

a. Institutions in which the undergraduate college, later commonly designated as of "arts and science," was the funnel through which students passed before entrance into the graduate and professional schools. Only the college provided education to the bachelor's level. Harvard and Yale are the best examples, but their paths to that position were quite different. Princeton is an interesting variant. Yale's is by far the most important single U.S. influence on university development, particularly graduate education.[20] Three notable, influential new universities of the 1860–1900 era, Cornell, Johns Hopkins, and Chicago, were founded by Yale graduates—Andrew D. White, Daniel Coit Gilman, and William Rainey Harper, respectively. The Yale experiences entered into the actions of all three men significantly. Both White and Gilman studied in Germany; Harper, a Yale Ph.D., did not.

b. Cornell is the prototypical U.S. university, and consequently, Andrew D. White is perhaps the most significant of the university builders in the United States. Cornell was a private university, but one that received some of the Morrill land grant funds. While most of White's policies had earlier precedents, his example influenced the adoption of the pattern now typical of state universities and most others. Cornell was notable in its acceptance of women and the early place accorded alumni and faculty among the governing trustees. White favored the equality of fields in the university—whether pure or applied—and the pattern of "breadth and depth" now common in undergraduate education. Unlike Harvard and Yale, Cornell and the state universities did not exalt *the* undergraduate college of arts and science but allowed coequal schools to issue bachelor's degrees. Only in graduate education did Cornell—and the state universities—lag as an innovator and as a model. Although Cornell's first Ph.D. was awarded in 1872, to Henry Turner Eddy, who became a notable engineer and engineering educator, leadership in graduate studies was elsewhere.[21]

In 1910 a writer reported that 3,471 doctorates had been awarded during the preceding dozen years. A sample of fourteen leading universities produced four-fifths of that total, but nine of the private universities in the sample, including Cornell, accounted for 2,634, and the five state universities for only 89. Only after 1910 did the state universities expand graduate output, a process still going on.[22] The usual explanation for the lag of the state universities is their emphasis on vocationalism, reflecting the viewpoints of state legislators. This explanation rings false, not only because of the expansion of graduate education in many of the state universities after 1910, but also because of the pre-1910 pattern, in which efforts were made in many institutions to give an education beyond the narrow confines of vocational preparation.

c. A third category of institution are those founded with deliberate slants toward the university level, whether graduate or postgraduate, such as Johns Hopkins, Clark, and Chicago.[23] In all three unsuccessful efforts were made to bar or to diminish undergraduate education. Only a few, rare examples exist today in the United States—the Rockefeller University, for example, formed after World War II by adding a formal educational function, Ph.D. and postgraduate, to an existing research entity, the Rockefeller Institute.

To return to Yale, in 1847 it formed its Department of Philosophy and the Arts, which included the School of Applied Chemistry; in 1852 the School of Engineering was added. The department was the first in the United States to give earned master's degrees, and in 1861 it awarded the first U.S. doctorate to a physicist, A. W. Wright. In 1863 J. Williard Gibbs received his, not a bad start for an educational tradition.

The two schools were undergraduate entities, which merged to form the Yale Scientific School in 1854, renamed the Sheffield Scientific School in 1861 in honor of a wealthy donor. Sheffield, not Yale College, received Connecticut's land grant allotment under the Morrill Act. Not only was Sheffield in competition with Yale College on the undergraduate level, but it and its parent department embodied a quite different educational view, stressing the primacy not of the college but of the philosophical faculty. It provided an undergraduate education in which the sciences, modern languages, and what we now call the social sciences had pride of place. Sheffield even organized a "Select Course" designed as preparatory to nontechnical pursuits for future businessmen, offering a three-year B.S. Sheffield pioneered into newer fields. The study of English language and literature at Yale originated in Sheffield, not the college. To James Dwight Dana and others around 1870, the department and its Sheffield School provided an opportunity to form a real university. In retrospect Dana's view overlaps to some extent the idea of a polytechnic university, influencing W. B. Rogers in his contemporary founding of the Massachusetts Institute of Technology. The selection of Noah Porter as president of Yale killed Dana's initiative. A good part of Yale's subsequent history is the attempts of Yale College's adherents to develop graduate education while stifling Sheffield, which continued to exist in an attenuated state as late as the post–World War II era.[24]

An analogous policy existed at Harvard under its great president, Charles W. Eliot. Harvard's Lawrence Scientific School never attained the strength or other characteristics of Sheffield. Like the Yale administrators, Eliot favored the college and stifled any possibility of a rival undergraduate body by placing the science teaching labs under the control of the college in the interest of economy.[25] Yale followed suit. Eliot, a chemist, and the Yale leaders had no intention of dropping or diminishing the sciences. Perhaps on the same principle as the Oxford Examination Statute of 1850, which required the Literae Humaniores before entrance to the schools of natural science, law, and medicine, Eliot and his peers at Yale wanted the American equivalent as a requirement for entrance into graduate and professional work.[26] Eliot successfully attenuated undergraduate engineering at Harvard during the last century; at Yale that goal was reached only in 1962.

Both universities initially favored a slow growth of graduate education during the period after the Civil War but followed different strategies. In New Haven, fending off Sheffield to maintain the primacy of Yale College produced a policy of regarding the graduate school as something added on, not intrinsically related to the college. Arthur Twining Hadley, Porter's successor, thought that pure research belonged in research institutes while the college fostered good citizenship, not expertise. Training teachers, not advancing research, was the aim of the graduate work.[27]

Eliot is a fine example of an American patrician liberal of the nineteenth century. He too saw the college as training for leadership in the society. Clearly influenced by *Lernfreiheit*, Eliot developed a complete elective system for undergraduates, a form of laissez-faire applied to the formation of intellect and character. The graduate program was to develop naturally out of the elective system under the assumption of a continuity of identity of undergraduate and graduate education, perhaps by analogy with the German model. Not only did Eliot favor the college, but he initially doubted the viability of full graduate work in the United States.[28]

In 1876 a former professor at Sheffield shattered the assumptions that prevailed at Cambridge, Massachusetts, and at New Haven.[29] Daniel Coit Gilman and his trustees deliberately launched Johns Hopkins University as a graduate-level institution. A Berlin Ph.D., he called his creation the "philosophical faculty" and reluctantly had a nominal undergraduate enrollment, which later grew under community pressure. The small, high-quality faculty avoided formalities of structure and procedure. Throughout its golden age under Gilman, Johns Hopkins was a great success. In the years 1878–1889, while Yale produced 101 doctorates and Harvard 55, Hopkins granted 151 Ph.D.'s.[30] As so often remarked, a great part of the intellectual history of the United States during the subsequent quarter of a century is the result of the labors of these men. One even became president of Princeton and then of the United States.

On the English model, Hopkins established fellowships as a necessity for a successful graduate program. The German precedent produced a short-lived version of the *Privatdozent* system. But Hopkins had an unmistakable American flavor. As Hugh Hawkins puts it, the European aspects were subtly Americanized by their earlier stay at Sheffield. Like Hadley, Gilman did not see Hopkins as a research institution but as a place for teaching research. Most of its graduates—184 out of 212 by 1891—joined college and university faculties, spreading the word that real universities could and should exist in the United States.[31]

Gilman went on to establish the Medical School of Johns Hopkins, of great influence in subsequent developments in the United States. Even before Gilman's retirement, in 1901, Johns Hopkins was in trouble and went into something of a decline. The endowment proved inadequate to the needs. Expanding the undergraduate program did not lead to the acquiring of new funds, only to new problems and a loss of the singularity of the original conception.[32] More important, other universities with greater resources expanded their graduate programs.

By far the most spectacular new academic presence was the University of Chicago, which opened in 1891. It was well endowed by Rockefeller money, but its greatest asset was its president, William Rainey Harper, a specialist in Semitic languages. He was a wunderkind, a Yale Ph.D. at eighteen, given to grand conceptions, elaborate administrative schemes, and great ability in day-by-day operations. Like Eliot and like Nicholas Murray Butler of Columbia, Harper had the same skills visible in such men as Rockefeller. Conventional wisdom correctly describes Chicago as a great university from the moment its doors opened. The success challenged the conventional wisdom at Harvard and Yale about the necessity and desirability of universities developing from or around *the* college.

Chicago was very American in some ways; in others it clearly reflected Europe. Of the original faculty, most had earned U.S. degrees, with Yale in the lead—sixteen, of whom nine were Ph.D.'s. Even at the doctoral level there were fourten German degree holders to twenty-one or twenty-two from the United States. Only five of the faculty were Hopkins graduates. Chicago rejected the extreme elective system of

Eliot and chose the American norm of breadth and depth—that is, a pattern of distribution in all areas plus a concentration in one, the whole designed to provide both general cultural and specialized skills.

From its inception the graduate school was described as nonprofessional for pure research on the German model. There was a sharp distinction made between the first two undergraduate years, the junior or Academic College, and the last two, the senior or University College. Harper tried unsuccessfully to banish the former from his campus. Harper had two graduate schools—sciences; arts and literature—reflecting an effort to have broad subject-area faculties rather than departments as the unit of governance. As in the similar developments in Columbia University, the German influence is clear, but in the long run the departments rather than faculties became the sites of action and power. Another possible German influence is the complete absence of engineering, unique among major U.S. universities.[33]

Hopkins and Chicago, like Cornell, clearly influenced the then modest state universities. More important for the development of graduate education before 1920 was the effect on the older private establishments such as Harvard, Yale, and Princeton. Eliot has advised the Hopkins trustees not to go into graduate education. Spurred by Gilman's achievement, Eliot deployed Harvard's more extensive resources into improving the faculty so as to expand the graduate degree program. That success was linked in Eliot's mind to the elective system, the one justifying the other. Samuel Eliot Morison was probably correct in his typical Harvard mixture of complacency and self-mockery when describing his university as the "premier American Ph.D. mill" by 1900.[34]

In contrast, the Yale situation is best described as modesty and anguish. The faculty improved in quality, and the graduate program expanded accordingly. But the Yale College–Sheffield split hindered a full commitment to a university program. Among the Yale College faculty, the alumni, and the trustees (the Yale Corporation), strong allegiance to the ideal of the college and hostility to Sheffield with its Select Course and its science orientation persisted.

Although also dating from before independence, Princeton, earlier the College of New Jersey, was smaller than Harvard and Yale and notably different in lacking professional schools in law and medicine.[35] But for the successful ambitions of some faculty, it might have remained a four-year liberal arts college, as did Amherst College in Massachusetts. In 1868 James McCosh became president of Princeton. He was a Scot who was both a Presbyterian clergyman and a philosopher in the tradition of commonsense realism. McCosh instituted graduate study to the master's level in 1877–1881 and established fellowships, citing the precedent of Edinburgh University. The early program consisted largely—and curiously—of physics and philosophy. Despite lukewarm sentiment among trustees and alumni, the doctorate was authorized in 1887.

Earlier, in 1872, the School of Science was formed, requiring Latin for admission. Two years later the School of Engineering opened. (It is a peculiar anomaly that, to this day, only Princeton of the Ivy League trinity has a continuous history of undergraduate professional engineering.) By the end of the century Princeton began resembling the Yale College–Sheffield situation in at least one respect: the lesser classical requirements for admission to Sheffield and to the School of Science at Princeton—coupled with the attractiveness of the Select Course and the more "modern" subjects—produced a rising enrollment in both, to the dismay of the adherents of the traditional college. Having successfully attenuated the Lawrence Scientific School and having relaxed the rigidities of entrance and bachelor degree requirements, Eliot looked complacently on the problem of his colleagues to the south.

In actuality all three institutions were in trouble by 1900. The crux of their problems was the social and intellectual status of the undergraduate college. Those problems and their resolution had significant implications for graduate education. Earlier, McCosh was distressed by signs that Princeton was becoming a fashionable school for the wealthy with a rising stress on sports, eating clubs, and other undergraduate social activities. He wanted to attract talented, poor youths to the ministry, as Princeton had in earlier days. Graduate study, with its fellowships, was a way of restoring a more serious tone to Princeton.[36]

Yale graduates had a group memory of a golden age of the college as a republic of learning and morality with a simplicity of life-style for both faculty and students—the early Roman republic infused with Congregational piety. Whether true or not, by the latter years of the nineteenth century Yale had become fashionable, sports conscious, and filled with examples of undergraduate frivolity. Despite Eliot's genuine desire to have Harvard truly open to talent from whatever social source, Harvard came in time to present a picture at variance with his rather austere ideals. It had a "Gold Coast" and a student body a significant element of which was unconcerned with patrician ideals of leadership, let alone vocational and intellectual motives. At both Harvard and Yale around 1900, the reality of undergraduate education was a student body largely skilled in finding easy courses in order to graduate with a minimum of disruption to a preferred anti-intellectual style of student culture. A similar pattern was in evidence at Princeton. An ambiguous situation, if not a clash, existed between the emerging culture of the graduate school and the culture of the undergraduate college—specifically, the pattern of behavior of those undergraduates whose goals were not the same as those of the minority going on to the Ph.D. and the like-minded entering the traditional professional schools.

When Eliot's forty-year tenure ended in 1909, he was succeeded by A. Lawrence Lowell, who taught government at Harvard. Lowell was primarily concerned with the undergraduate college and did away with such Eliot reforms as the complete elective system. He worked successfully to raise the caliber of Harvard undergraduate education by instituting honors work, tutorials, and examinations—all clearly showing English influences. Like the Cambridge reformers discussed by Rothblatt in *The Revolution of the Dons,* Lowell sought a sense of community, leading to the establishment of a common freshman year, followed by later years in residential houses.[37]

Lowell's preference was decidedly for the humanities and the social sciences, but he was not necessarily hostile to the natural sciences. The sciences were included in his vision of community if performed in the proper spirit, an American version of old Göttingen and old Oxbridge. His role in the McKay bequest eventually led to an applied science program at Harvard. At Harvard and elsewhere there were complaints about the specialization and vocationalism of the Ph.D. No less a personage than William James wrote against the "Ph.D. Octopus."[38] (Although the degree was commonly associated with the natural sciences, from 1873 to 1928 only 25 percent of the Harvard doctorates were awarded in that sector.) Harvard cheerfully demolished the Union, used by graduate students, to make way for its new scheme of undergraduate housing.[39] At the same time, the departments pushed graduate education, which became a principal concern of James B. Conant, who succeeded Lowell in 1933. Where Lowell stood was manifest by the founding that year, at his instigation, of the Harvard Society of Fellows, a collegial body giving three-year appointments to those not more than twenty-five years of age who would take no courses nor study for any degree. It was old Oxbridge transmuted to Cambridge by the Charles River.

At Princeton the classicist dean of the Graduate School, Andrew F. West, had a parallel vision of a residential graduate college that would provide a community for graduates and postgraduates at Princeton.[40] At the same time Woodrow Wilson embarked on an upgrading of the undergraduate progam, stressing tutorials and a sense of intellectual community. When West won the struggle over the graduate college, Wilson entered politics. Veysey is probably correct in stressing the basic similarity of the two men, but Wilson's position was closer to Lowell's than was West's. Despite his speciality, it was West, the champion of graduate education, who received and refused an offer to head M.I.T., which was by then improbable for either a Wilson or a Lowell.

At Yale College the faculty introduced honors work and moved to eliminate the easy courses. Resistance to change in both Yale College and Sheffield prevented many major reforms before World War I. In 1907, the historian George Burton Adams proposed a radical solution. A proponent of Yale College, Adams feared that the popularity of the Select Course and the pressure to relax the traditional entrance requirements would lead to its demise. Not at all hostile to science or to other newer fields, Adams wanted Yale to have three undergraduate colleges: the time-honored Yale College, a scientific and technological school, and a school for the social sciences and modern languages. It was a solution in the spirit of Cornell and the state universities in that all fields were treated equally and undegraduates had more than one path to the bachelor's degree. Perhaps it failed for that reason.[41]

In 1919, taking advantage of faculty absence in war work, the Yale alumni, with aid from elements in the administration, pushed through a "great reorganization" despite a later near revolt from the faculty. Sheffield lost its separate undergraduate program. The college still had its prideful place. A common freshman year was followed by a community experience in residential colleges. Entrance and graduation requirements were reformed. Sheffield still existed in the form of the science and engineering departments at the undergraduate level and at the graduate level in certain arcane senses. Freed of the worst aspects of the old struggle, Yale's graduate program expanded.[42]

Sheffield's last dean, Charles H. Warren, according to Furniss's short history of the Yale Graduate School, was "wary of the purposes of high university authority, suspecting that policies injurious to the sciences might be incubating in secret"—this in the university of Silliman, Dana, and Gibbs! It was only after World War II that Sheffield became completely a legal fiction, its undergraduate program merging completely with the college while the graduate school became the holder of the historical tradition.[43]

Dean Furniss recognized the historical reality that much of Yale's graduate training and research traditions came from Sheffield, not the college, even a field such as English language and literature. And his brief account discloses a different spirit from that evident in Pierson's history of the college. Furniss notes that the great reorganization was supposed to overcome the reluctance of Yale's traditional families to send their sons into science. At first without comment, Furniss called attention to the emphasis on "inspirational teaching" and the claim "that advancement to the highest faculty rank would be awarded on teaching excellence alone." Although a Yale College graduate, Furniss represented a different worldview. He rejected the idea of "broad cultivation of the mind." He saw the doctorate as a professional degree, leading in most instances to employment in higher education. Believing in research as the function of the true university and of its graduate school, Furniss writes: "These young people [the graduate students], putting first things first, seek to qualify for remunerative employment.

Having obtained this primary objective they are prepared to espouse lofty non-utile aims. The Professor, fortunately, has a secure grip on his meal ticket and is in a position to rise above mundane things in his avowal of professional purposes." The graduate school was a vocational institution and "training in research was [its] inescapable duty." Long before Furniss wrote, in the era during which graduate instruction became established, his views represented the reality of higher education, not the increasing rhetoric of the exponents of the liberal culture ideal, as represented by Lowell and Wilson.[44]

That reality reinforced a policy based on a misapprehension about the German doctorate. Many United States academics assumed it was *the* prerequisite for teaching at the university level. That the German doctorate served as a prerequisite for gymnasium teaching also reinforced the high United States opinion of that institution which stressed its overlap with the college. The point about its relation to habilitation and the *venia legendi* usually vanished from academic arguments. Nor was it common for American university reformers to note the basic structural distinction between the single Ordinarius, the Professor, and the multiprofessorial departments becoming common in United States colleges and universities from 1890 to 1920. If a "doctorate" was to be required for teaching beyond the high school, the United States required a graduate system that would produce more neophyte professors and by a faster procedure than habilitation but clearly differentiated from both the liberal arts bachelor and the increasingly numerous products of the normal schools.

Originally, doctoral requirements in the United States extended the time beyond the B.A. for advanced study modestly, perhaps with an eye to the realities of resources and to the German tradition. In 1887 Princeton asked for two years of study beyond the bachelor's degree—a major and a cognate study—plus a thesis. If the two last years of undergraduate work are considered to be university level, the formal requirements were not much different from those in Germany. By 1899 a minimum of two years of graduate work was fairly standard, with the proviso—obviously German-inspired—of not more than one year in residence. The National Association of State Universities in 1908 specified three years of graduate study. By 1912, the prestigious Association of American Universities noted "never less than 2 years." By 1916 the latter association declared: "The amount and character of the work should be such that the degree rarely could be attained in less than 3 years following the attainment of a bachelor's degree or equivalent." That is, continuing the previous thrust, the doctorate now required a minimum of five years of university work, while only three were required in Germany.[45] Ultimately this structural change, aided by other trends, helped transform once colonial or provincial universities into the peers of the leading European schools.

What is being hypothesized here is not a claim that professors were superior, that students were more intelligent, that specific courses were more detailed or deeper, that theses were better or more original, or even that administrators were more enlightened and farsighted. It eludes me how anyone could properly make such comparisons between the United States and Germany, let alone all Western Europe. Any such claims are made in disregard of the obvious point that educational systems fit national traditions, social structures, and intellectual aspirations. The more sweeping the comparisons, the more futile the effort.

The claim here is more modest and arises out of a consideration of circumstances within the United States. The longer period specified for the doctorate made possible a formal requirement for a more comprehensive educational preparation, particularly in the leading schools. This produced a growing number of individuals trained, on the average, in a broader range of specialized topics, reflecting the increasing size and intellectual diversity of the academic departments. These individuals provided a human resource, a surplus of trained talent, for the further expansion of higher education, industrial research, the government science bureaus, and the independent research institutions during the years 1900 to 1940.

Particularly striking in retrospect is the way the revival of liberal culture figured in this outcome. The seekers after community and a common intellectual tradition often looked to Oxbridge for inspiration—or, rather, to an idealized vision of Oxbridge. The introduction of tutorials is an obvious example, as are the expressions of hostility to specialization and vocationalism, often accompanied by affirmations of intellectual purity. But the reformers of schools such as Harvard, Yale, and Princeton—and their imitators elsewhere—could only turn for yardsticks to the graduate school; attempts to foster other modes of education have had only limited success to this day. When another Yale wunderkind, Robert M. Hutchins, came to Chicago before World War II, his Great Books educational strategy had its effect, but a limited one, even in his own institution.

To illustrate the improvements in Yale College before World War I, Pierson proudly listed honors theses, as if to show that near-graduate-level work was achieved. Improvers of the undergraduate college often ended up supporting improvements in graduate programs in order to get better teachers. Adherents of the research-oriented graduate school and disciplinary department could readily assent to improvement of undergraduate education in order to get better-prepared doctoral candidates. No one wanted a college substantially involved in remedying the deficiencies of the high school, a nearly universal complaint throughout the last century.

Both groups could agree that abstract knowledge, now defined as "pure," was the necessary preliminary to both culture and utility; many Americans, in fact, had trouble differentiating the two. It was an old belief, present even in antebellum America, but not exactly British or German, by the turn of this century. Perhaps settlers in what was formerly British North America south of Canada carried European mirrors across the ocean in which to look at themselves. In the westward translation the mirrors may in time have acquired a distinct distortion. In the United States, liberal culture never had the same degree of antiscience bias present in Germany and Great Britain. Like the undergraduate college, the graduate school was also of "arts and science."

Notes

1.   Laurence R. Veysey, *Emergence of the American University* (Chicago, 1965); Richard J. Storr, *The Beginnings of Graduate Education in America* (Chicago, 1953); W. C. Ryan, *Studies in Early Graduate Education . . .* (New York, 1939); W. C. John, *Graduate Study in Universities and Colleges in the United States* (Washington, D.C., 1935).

2.   The best known statement on the importance of the multiprofessor department is Joseph Ben-David, "The Universities and the Growth of Research in Germany and the United States," *Minerva* 7 (1968/69):1–35.

3.   Perhaps the best place to start are the writings of Friedrich Paulsen, *The German Universities and University Study* (New York, 1906) and *German Education, Past and Present* (London, 1908). Of recent historical treatments, I am indebted to Charles McClelland, *State, Society and University in Germany, 1700–1914* (Cambridge and New York, 1980) and Fritz K. Ringer, *The Decline of the Mandarins: The German Academic Community, 1890–1933* (Cambridge, Mass., 1969).

4.   Paulsen, *Universities* p. 9.

5.   Paulsen, *Education* p. 184.

6.   See also W. H. Buford, *The German Tradition of Self-Cultivation: "Bildung" from Humboldt to Thomas Mann* (Cambridge and New York, 1975).

7.   Paulsen, *Education* p. 179.

8.   Ibid., pp. 294–295.

9.   Ibid., pp. 188–193. Interestingly, in the years before World War I, the "classic" German university was changing, and Paulsen saw this as part of the influences of the coming of the masses who had the option of free immigration to the New World. He also saw the Anglo-American college as a possible model for an institution between the gymnasium and the university. Idem, pp. 179, 204, 297; *Universities*, 278–279. McClelland, *State*, p. 137, notes the American example as influencing discussions on technology.

10. Paulsen, *Universities,* p. 3.

11. Storr, *Beginnings,* p. 79.

12. McClelland, *State,* pp. 104, 201.

13. Ibid., pp. 104, 166–167. This position was, in part, a reaction against the French example.

14. Ibid, p. 110.

15. Ibid., pp. 285–286; Paulsen, *Education* p. 137f.

16. McClelland, *State,* p. 104.

17. Nicholas Hans, *New Trends in Education in the Eighteenth Century* (London, 1951).

18. See his *Science and the Ante-Bellum American College* (Philadelphia, 1975) and "The American Scientist in Higher Education," in N. Reingold (ed.), *The Sciences in the American Context: New Perspectives* Washington, D.C., 1979), p. 99–14. For an account of German influence on the humanities, see Carl Diehl, *Americans and German Scholarship, 1770–1870* (New Haven, 1978).

19. This typology is indebted to but differs from those given by Storr, *Beginnings,* pp. 131–132, and by George W. Pierson, *Yale College: An Educational History, 1871–1921* (New Haven, 1952), pp. 44–45, basically because I am not that impressed with the essentiality of the factors differentiating Harvard and Yale.

20. In addition to Pierson, *Yale College,* see R. H. Chittenden's *History of the Sheffield Scientific School* (New Haven, 1928), 2 vols; E. S. Furniss, *The Graduate School of Yale* (New Haven, 1965); and D. C. Gilman, *The Relations of Yale to Letters and Science . . .* (Baltimore, 1901).

21. See Andrew D. White's *Autobiography* (New York, 1905) and Morris Bishop, *A History of Cornell* (Ithaca, 1962). I am not arguing for primacy, let alone downgrading other influential universities. What developed at Cornell—not wholly White's doing—is what most U.S. universities tended toward then and even now if one thinks of the belated admission of women to Princeton and Yale undergraduate life.

22. E. E. Slosson, *Great American Universities* (New York, 1910), p. 317.

23. Hugh Hawkins, *Pioneer: A History of the Johns Hopkins University, 1874–1889* (Ithaca, 1960); Richard J. Storr, *Harper's University: The Beginnings* (Chicago, 1966).

24. Pierson, *Yale College,* pp. 55–59, 377f.

25. Hugh Hawkins, *Between Harvard and America: The Educational Leadership of Charles W. Eliot* (New York 1972). Still useful is S. E. Morison, *The Development of Harvard University since the Inauguration of President Eliot, 1869–1929* (Cambridge, Mass., 1930).

26. Michael Sanderson. *The Universities in the Nineteenth Century* (London, 1975), p. 75. This and his *The Universities and British Industry* (London, 1973) are extremely valuable for comparative studies of higher education and necessary correctives for the limited vision of Sheldon Rothblatt's *The Revolution of the Dons: Cambridge and Society in Victorian England* (New York, 1968).

27. Pierson, *Yale College,* pp. 126–127.

28. Hawkins, *Pioneer,* p. 13; Robert A. McCaughey, "The Transformation of American Academic Life: Harvard University, 1821–1892," *Perspectives in American History* 8 (1974):239–332.

29. Veysey, *Emergence,* p. 160.

30. Hawkins, *Pioneer,* p. 122.

31. Ibid., pp. 37, 64, 291.

32. Ibid., pp. 239–241.

33. Storr, *Harper's University,* passim.

34. S. E. Morison, *Three Centuries of Harvard, 1636–1936* (Cambridge, Mass., 1936), p. 335.

35. T. J. Wertenbaker, *Princeton, 1746–1896* (Princeton, 1946) and W. Thorp et al., *The Princeton Graduate School: A History* (Princeton, 1978) are the best general sources.

36. A new study of McCosh came out too late for use in this paper: J. David Hoeveler, Jr., *James McCosh and the Scottish Intellectual Tradition* (Princeton, 1981).

37. For A. Lawrence Lowell, see the excellent *Dictionary of American Biography* entry by Hugh Hawkins.

38. W. James, "The Ph.D. Octopus," *Harvard Monthly* 36 (1906):1–9.

39. The percentage is from Morison, *Development,* p. 458, and his *Three Centuries* pp. 335, 370–371, 479. Even in Germany the percentages for science in the Ph.D. population was overestimated by those against vocationalism and specialization who blamed "science" as the cause of undesirable events and trends. Nor was Harvard unique, witness the statistics for the University of Pennsylvania in 1908/09 given in Slosson, *American Universities,* p. 366. Whether to stress reality or express his own inclination, Gilman at Hopkins took pains to repudiate charges of overstressing science (Hawkins, *Pioneer,* 50). A modern historian, disenchanted with the consequences of science, found only a small percentage of the early Ph.D.'s avoiding the stain (Veysey, *Emergence,* p. 173). Perhaps Gilman repudiated such charges to placate those in his day who had positions analogous to Veysey's.

40. For Andrew F. West, see the excellent *Dictionary of American Biography* entry by Laurence R. Veysey.

41. Pierson, *Yale College,* pp. 377f, 480.

42. Furniss, *Yale Graduate School,* and Pierson, *Yale College,* passim.

43. Furniss, *Yale Graduate School,* pp. 83–89.

44. Ibid., pp. 20–30, 90–91, and passim. Yale's history is anything but typical of the U.S. situation. Despite some points reflecting the peculiarities of that history, Furniss gives what was both the reality of graduate education and the accepted wisdom, whatever the rhetoric used on formal occasions. Of particular interest is his last two chapters.

45. Thorp, *Princeton Graduate School,* p. 19; Wertenbaker, *Princeton,* p. 379; John, *Graduate Study,* pp. 28, 35, 46f; Storr, *Beginnings,* pp. 158–159.

10 □ National Science Policy in a Private Foundation: The Carnegie Institution of Washington

In the early 1970s I was exploring the sciences in twentieth-century America, work leading to the volume I coedited, *Science in America: A Documentary History, 1900–1939* (1981). This article and the "Disappearing Laboratory" paper that follows were among the by-products of that research. My interest in the Carnegie Institution of Washington (CIW) went back to the period when my nineteenth-century documentary history was in gestation. Superficially, CIW seemed just what Joseph Henry would have wanted the Smithsonian Institution to be if only the Smithson will, the Congress, many of his colleagues, and the public had given him a free hand. Of course, I was partly wrong in that view. Henry was part of the problem and part of its eventual solution, if that is how we can characterize the present cultural conglomerate on the Mall.

On closer examination of its splendid archives, the CIW simply proved irresistible to a historian with my predilections. A key figure, Robert S. Woodward, turned out to belong to the geophysical tradition. Here was an organization where concepts and personalities resonated with contexts of the larger society. Some of my favorite contexts, in fact: public policy on research and development and the relation with the university world. The most obvious new factor was the scope of private philanthropy.

Here I noted a curious phenomenon present in Henry's day, in the early years of CIW, and in the period after World War II. Instead of stressing the formal aspects—a private gift to a quasi-public body, the Smithsonian; a multimillionaire's endowment of a private institution; and federal funds for research and development largely in private bodies—I considered the effect of scale. In each historical period, the coming of the new funds produced great expectations. Existing difficulties would disappear. Great consequences would follow. Both Henry and Woodward quickly realized the resources in their hands could not meet all demands and expectations. Even with increased funding after

1945, many federal science administrators had similar reactions. The Smithsonian developed an applied mission—popular diffusion. Being a private body, CIW under Woodward and his successors (including Vannevar Bush) resisted all pressures to change its nature. CIW even avoided the federal grants and contract route for many years after World War II. Inflation eventually eroded that resolve. After World War II, the way to attain the scale of support required by a growing scientific community also required linkage with applied missions. To use my previous terminology, the scientists often preferred research enclaves but historical circumstances in the United States fostered the spread of broad missions in both the private and public sectors.

This paper was originally published in *The Organization of Knowledge in Modern America, 1860–1920.*

Introduction

Within a small circle near the end of 1901 there was much excitement. A great gift to American intellectual life was coming from the Laird of Steel, Andrew Carnegie. In December 1901 the transfer of $10 million in United States Steel Corporation bonds to the newly formed Carnegie Institution of Washington (CIW) was publicly announced and tension spread widely within the communities of American scientists and scholars who might potentially benefit. A distinguished Board of Trustees was designated by Carnegie, and Daniel Coit Gilman, lately retired from the presidency of Johns Hopkins University, was named the Institution's first president.[1]

To aid them in wisely disbursing Carnegie's gift, the trustees appointed committees of experts to deliberate and to advise on which intellectual fields should be supported and how each of these areas could best be advanced. These reports were supplemented by solicited and unsolicited advice of individuals given in print, correspondence, and conversation. The committee reports are curious reading today, being a presumably authoritative survey of the state of many disciplines in America and their future needs. They range from the detailed to the laconic, from modest reasonableness to the rodomantade of special pleading. For example, the anthropologists' committee, which included Franz Boas, gave as one of the aims of the field "to discover the principles and laws of human development with the view of utilizing them to regulate the present and to mould the future of the race." With such advice, a wide variety of options were open to the trustees.[2]

Today, some seventy-five years later, the Carnegie Institution of Washington is a private center for research in the physical and biological sciences; its divisions include a department of terrestrial magnetism and a geophysical laboratory in Washington, D.C.; a department of plant biology at Stanford, California; a department of embryology at Baltimore, Maryland; a genetics research unit at Cold Spring Harbor, New York; and the Hale Observatories (owned and operated jointly with the California Institute of Technology).

Unlike such European countries as France, Germany, and the Soviet Union, America has relatively few independent institutions devoted solely to basic research. In the United States, most research establishments are associated with, or exist in juxtaposition to, larger entities in government, medicine, industry, agriculture, and higher education. In whole or in part, their research is justified by such missions as national defense, public health, agricultural and industrial production, or even popular enlightenment (in the case of museums). Thus, the process by which the Institution established its purpose and determined its future is of particular interest to historians.

Retrospection coats events in the Institution's development with a patina of inevitability. In fact, great uncertainty about its nature and function existed from the formal launching until well into the administration of its second president, Robert S. Woodward (1904–1920).[3] Conceptual issues, personalities, ideologies, institutional rivalries, and many other factors openly entered into the discussions of the trustees and their interchanges with interested parties outside the Institution.

When the Carnegie Institution was established at the turn of the century, there were no governmental bodies with wide concern for intellectual life; thus CIW necessarily dealt with what are now considered to be national policy questions, principally the establishment of priorities for the support of research areas. The Rockefeller philanthropies were then largely involved with public health and medical research. Only toward the start of World War I did they edge into the broad concerns manifested so significantly between 1920 and 1940. Nor did such programs as the Guggenheim fellowships exist in 1902. In the absence of alternative bodies, the support not only of science but of research in the broadest terms became the concern of the officers and trustees of the Institution. Perhaps this is what Andrew Carnegie really wanted, but the record of his motivations is quite obscure.

Emphatically, Carnegie asserted that the first aim of the new body was the promotion of original research, in which the country was so deficient. Clearly, he was reacting to the turn-of-the-century literature about American indifference to basic research.[4] But Carnegie's assertion left much unanswered; quite deliberately, the formal founding papers were both sweeping in their statements of scope and obscure in their particulars. For the purposes of this chapter, the origin of the gift and the nature of Carnegie's views are of interest only insofar as they entered into the policy discussions. Fiduciary responsibility led the trustees to invoke their interpretation of Carnegie's intentions from time to time; the possibility of increased endowment was ever present in the early days, possibly influencing solicitude for the founder's motives. Once the Institution was established, Carnegie was rather conscientious about letting the trustees manage it. When Gilman in 1903 asked Carnegie to confirm that the funds could support non-Americans, Carnegie agreed, but added it was not proper for him to interfere.[5] Only two instances of Carnegie's pressure on the Institution are known to me. In 1910 Carnegie tried unsuccessfully to have Woodward use simplified spelling in Institution publications.[6] More significant is his role in the case of Luther Burbank, an important incident dealt with later.

The Gilman administration was marked by a highly interesting clash between the president and two influential trustees, C. D. Walcott, then the head of the U.S. Geological Survey and the secretary of the Carnegie Institution, and John Shaw Billings, who early became chairman of the Board of Trustees. A physician, Billings is best known as the man who developed two great libraries, the Surgeon General's Library, now the National Library of Medicine, and the New York Public Library. Ostensibly, the dispute was on questions of governance, the president vs. the trustees, or more specifically, the Executive Committee dominated by John Shaw Billings.[7] However, more than administrative prerogatives were involved; as will be evident, Gilman, Walcott, and Billings had strong differences on matters of policy.

As far as governance is concerned, Gilman's successor, Robert S. Woodward, achieved the strong executive leadership Gilman sought but could not attain. Woodward came to the Institution from Columbia University, where he was Dean of Pure Science. Originally trained as an engineer, Woodward was an old Washington hand with service in the Lake Survey of the Corps of Engineers, the Coast and Geodetic Survey, the Geological Survey, and the Naval Observatory. From the pre-graduate-school era, Woodward was a highly competent applied mathematician or classical physicist who saw the Earth as *the* great object of study. His principal scientific work occurred during his years with the Geological Survey, notably a finding contrary to Kelvin's estimate of the age of the Earth. After three years with the Coast and Geodetic Survey, Woodward became Professor of Mechanics and Mathematical Physics at Columbia University in 1892; in 1895 he was named Dean of its College of Pure Science.

Billings and Walcott knew Woodward from his days with the surveys. Before his elevation to the presidency of Carnegie, Woodward headed two of the advisory committees. When Billings saw that Walcott and other favored candidates were not acceptable for the position of president, he pushed Woodward forward, confident in having a willing collaborator.[8] As president, Woodward quickly and smoothly edged Walcott away from control of the paperwork that constituted much of the vital life of the organization.[9] Smoothly but not so quickly, Woodward reduced Billings and the Executive Committee to a subordinate role. In the last years of Billings's life, a bitter enmity existed between the two men,[10] and an observer in the Rockefeller Foundation could write of the autocratic control in the Institution.[11] Whether this characterization is true or not, Woodward evolved a policy uniquely his own—with points of agreement and disagreement with Gilman, Billings, and Walcott.

Certainly the most fundamental issue in the early history of the Carnegie Institution was its relationship to the university world. At the first meeting of the trustees, both Andrew Carnegie and Daniel Coit Gilman presented statements identifying the Carnegie Institution of Washington as a research organization, distinguishing between the Institution and a university, and very carefully disentangling the new body from its origins in the national university idea and in the movement to aid graduate students coming to Washington.[12] The concept of a national university was not new, of course; its lineage traced back to George Washington. In the decades before the turn of the century, the national university was pushed with great vigor by John W. Hoyt. Although he succeeded in gaining considerable support, Hoyt's moves were frustrated first by the obvious disinclination of some to increase federal activities and second by the opposition of those in higher education who claimed their institutions now filled many of the roles destined for the national university and looked with trepidation on a possible federally funded rival. Still others, eager to expand research and advanced training, had qualms about the ambitious but vaguely defined institution being promoted by Hoyt. Such individuals, including key scientists and educators, wanted suitable action on a national level.[13] Among these were Andrew D. White, the president of Cornell, and Daniel Coit Gilman.

According to standard accounts, the origin of the Carnegie Institution was a conversation in Scotland in the spring of 1901 during which White and Carnegie discussed the possibility of a national university.[14] Hoyt had previously raised the issue with Carnegie with no success. On Carnegie's return to America, Gilman took up the task of promoting the idea.

Simultaneously, a sequence of related events occurred. Apparently independent of Hoyt, a group of women formed the George Washington Memorial Association before the turn of the century with the avowed purpose of establishing in the District of Columbia "a Washington Memorial University." So far as one can judge from a distance, the women were sincerely motivated by a combination of patriotism and a desire to advance education. In the years 1900 and 1901, the Association worked out a promising alliance with a group of influential scientists and educators.

In 1898 the Washington Academy of Sciences came into existence. Despite its name, the Academy aspired to a national role and was so viewed at least by some outside its ranks. (The National Academy was in a quiescent state.) Leading the Washington Academy in these early years was C. D. Walcott, John Wesley Powell's successor as director of the U.S Geological Survey. Connecting the Washington learned community with institutions of higher learning in the country was a major point in Walcott's program for the Academy. That point fitted in well with the inclinations of a number of educators like Nicholas Murray Butler; they wanted to avoid a national university while obtaining access for their students to the facilities and resources of the Washington area.[15]

Shortly afterward (May 20, 1901), the Washington Academy and the George Washington Memorial Association joined forces to incorporate the Washington Memorial Institution; its purpose was to bring students to the nation's capital to utilize its scientific and cultural resources. To house the Institution, the George Washington Memorial Association hoped to construct an edifice which would be suitable as the administrative headquarters of the national university that would hopefully emerge from this nucleus. Walcott and his female allies turned to Andrew Carnegie to give substance to their blueprint.[16]

By the fall of 1901, however, Carnegie was definitely not interested in any university but in aiding research and in developing exceptional men. Despite the opinions of other historians, I doubt Carnegie's interest in the national university was more than a momentary one or ever at a high voltage. At the November 16 morning meeting of Carnegie, Gilman, and Billings that launched CIW, Gilman still urged a national university in Washington. In contrast, Billings "advised instead of a university [that] the institution be founded for the promotion of original research, advanced teaching, etc., to be located in Washington, to be so original as not to interfere with the work of any existing university, but on the contrary to assist them, and to grant scholarships, etc., for work in laboratories and institutions outside of Washington as well as in it." Billings prevailed, and the resulting plan of November 20 won Carnegie's assent two days later.[17]

Gilman had no illusions about the fate of the national university and the Washington Memorial Institution. Writing C. W. Eliot on December 3, 1901, he thought the Carnegie gift killed both.[18] Nor did the women of the George Washington Memorial Association have any doubts about the effect. CIW was "exactly along the lines with identical aims, and with a charter embodying the very same features" as their proposals: CIW "rendered forever impossible of fulfillment our purpose that there should be here a worthy memorial of Washington's deep interest in science and learning."[19]

No doubt Gilman sincerely wanted to help the universities, but, in general, there was an aloofness from, or a coolness to, the university world on the part of Carnegie and the trustees—an aloofness rooted in the determination of CIW's leaders to separate teaching from research. Today, universities are generally regarded, in the words of the title of a recent book, as "the home of science."[20] Yet Walcott, Billings, and Woodward did not accept that view. Their governing model was not the German or some other form of university but the Royal Institution of Davy and Faraday. In a February 1902 memorandum, Billings called for the establishment of laboratories like those of the Royal Institution, adding "they do no teaching." The only other precedent cited by Walcott and Billings in those early days was that of the Smithsonian Institution.

Given the spectacular personalities and the fascinating byplay between them, there is a strong temptation to structure historical explanations in terms of human dramas: in the beginning, Carnegie coyly playing off Andrew D. White and Gilman against Billings and Walcott, then Walcott teaming with Woodward against Billings, with Woodward first using Billings and Gilman to supplant Walcott and later isolating Billings with the help of William H. Welch, Dean of the Johns Hopkins Medical School. And there are others in the cast—members of the Carnegie Board of Trustees including the banker Henry L. Higginson, the lawyer Elihu Root, and, toward the end, the senator Henry Cabot Lodge—not to mention such extraordinary concerned outsiders as the astrophysicist George Ellery Hale and James McKean Cattell, the psychologist who edited *Science*. But the participants' continuing explicit awareness of issues renders quite suspect explanations largely in terms of dramaturgy.

The tension between teaching and research, between the concept of the university and that of the research institute was a pervasive factor underlying the three areas of contention that dominated the formative years of the Carnegie Institution: the support of individuals, the allocation of funds among intellectual fields, and the relation of the Institution to higher education. Quite often, disputes in one area were linked to differing viewpoints in another, but all of the disputes occurred more or less simultaneously.

"The Exceptional Man"

Following the primary objective—support of original research—the deed transferring $10 million of United States Steel Corporation bonds to the newly formed Carnegie Institution of Washington gives as the second aim of the trust: "To discover the exceptional man in every department of study whenever and wherever found, inside or outside of schools, and enable him to make the work for which he seems specially designed his life work."[21] Billings, Gilman, and Walcott undoubtedly agreed with that aim, even though they knew exceptional men were not always readily found. Being scientists with university connections, they assumed exceptional men were largely at or from universities, so the "inside or outside" phrases did not, at first, appear vexing. But they and their fellow trustees quickly took a step that was to have serious consequences for universities. By the time of the first annual meeting of the trustees on November 27, 1902, there was agreement with Billings's general principle of aiding individuals, not institutions, in the conduct of research.[22] With this seemingly modest act, universities were cut off from direct aid.

Even before the formal launching of CIW, the leading figures were seeking and receiving advice on this question. Billings, who was intent on aiding individuals, received two letters in January 1902 from the Johns Hopkins University that are a good introduction to the problems discussed by the trustees over the next decade or so. The author of the first letter, William H. Welch, would become a CIW trustee in 1906; in 1902 he wrote from the perspective of a trustee of the newly founded Rockefeller Institute as well as Dean of the Johns Hopkins Medical School. Although the Institute was initially giving research grants to investigators, Welch looked toward a general policy of concentrating on in-house research—exactly what occurred in the Carnegie Institution of Washington. As to fellowship recipients, Welch saw two categories: those with demonstrated ability, and those recommended by laboratory directors to work under their supervision. Better still, freeing established men from the burden of teaching would increase the output of American laboratories. But the number of those with "any real fitness for original research is small, and these are likely to come to the front under any circumstances." On the other hand, Welch saw a danger—not from CIW—"of encouraging those who have no genuine capacity for such work and who had better be at something else."[23]

Gilman's successor at Hopkins, the chemist Ira Remsen, agreed that only a few individuals were capable of significant research and not many of these were languishing for want of funds. The few really good men were snatched up by the universities. Referring to large-scale investigations requiring substantial funding, he stated, "I do not think there are many such." (That presumably meant the kind of installations CIW did fund subsequently.) "What we lack is men," wrote Remsen, who doubted CIW could enlarge the supply of exceptional men. Instead he urged CIW to increase the efficiency of the demonstrably productive men in universities by providing funds for research assistants. As to fellowships for the untried, "that is now overdone . . . the danger in it is that it tends to develop a lot of men up to a certain point and then drop them." Only a few fellowships were needed for those with clear signs of an ability to carry out independent research.[24]

Billings, Gilman, and the other trustees agreed to aid individuals immediately and not wait for the careful formulation of a program based on the reports of the expert advisory committees. At the outset, at least, Billings and Gilman were in agreement on the two principal forms of aid: what we would now call fellowships to as yet unrecognized individuals, and what we now designate as project grants to the more established investigators. By the end of 1905, Woodward could inform Nicholas Murray Butler, the president of Columbia University, that 260 individuals in 89 institutions had benefited from CIW funds.[25]

Right from the outset, two semantic problems were created for the Institution, most likely by Walcott—not the most auspicious move for a future secretary of the Smithsonian Institution and a future president of the National Academy of Sciences:

First, much to Billing's distress, the young investigators were tagged with the title of "research assistants." Billings had originally suggested "junior associate." Walcott's term immediately confused those young neophytes being tested "in special lines of work, for which they profess to have a special aptitude," with assistants provided to a professor as "a careful, accurate, serviceable human tool."[26] That confusion only confirmed the fears of some that professors would use the "research assistants" to do their drudge labor, not to launch careers. To some of the young investigators, the term carried unflattering connotations.

The second semantic problem arose from Walcott's attempt to distinguish between the two principal objects of research that fell within the purview of CIW. In his 1902 instructions to the advisory committees, he differentiated between objects of broad scope for the discovery and utilization of new forces to benefit man—exemplified by the CIW laboratories—and objects of lesser scope such as filling gaps in known areas, acquiring knowledge of specific phenomena, or undertaking research in restricted fields.[27] Given Walcott's personal support of laboratories over short-term research grants, perhaps there was a pejorative connotation from the start. More likely, the choice of the term "minor grants" to identify the latter category was unfortunate; it clearly lacked an inspirational ring. Yet a harassed Robert S. Woodward later would point out that T. W. Richards of Harvard received America's first Nobel Prize in chemistry for the kind of work supported by his minor grants.

The Institution advertised the research assistantships. Applications were sent to "expert advisors" for report and awarded on the judgment of the Executive Committee. Billings pointedly called attention to the geographic diversity of the research assistants and their origin in small colleges as well as large universities; "their connection with this place or that" was not to be the determining factor. Awarding minor grants was more vexing, given the difficulty of evaluating the competing projects. A Charles Sanders Peirce did not impress the Executive Committee. Despite grants to the "wrong men" or grants "used not to a very good purpose," the Executive Committee was satisfied in 1903 to have found "about a dozen really exceptional men." Observing the proceedings, Carnegie expressed great gratification: "As the Chairman [Billings] has said, if you have three or four exceptional men, that is a great deal. We do not judge by the number that do not produce, we judge by the number that are successes."[28]

Enter the Dean of Pure Science, Robert S. Woodward, filled with ideas for cooperation in science and somewhat leery of the founder's "happy phrase" about the exceptional man (the characterization is Woodward's). To his dismay, some scientists accepted Carnegie's formulation. Richards of Harvard, for example, doubted the value of cooperation in physics and chemistry, whatever the merits in astronomy:

On the other hand it seems to me that the making of a great original discovery is not unlike the writing of a great poem or the painting of a great picture. The thought and its execution must be hammered out by genius alone. . . . I agree with Mr. Carnegie entirely in his belief that from the individual exceptional man alone is any great addition to be expected to the sum of our original conceptions. Who can imagine Faraday as cooperating? In short it seems to me that cooperation may be highly productive of routine work and of a general rounding off of already acquired knowledge; but might be equally destructive of great advance in entirely new directions.[29]

Within less than a year after Richards wrote, Woodward and the trustees began wrestling with an exceptional man, quite unlike Michael Faraday, who found cooperation very difficult—the horticulturalist Luther Burbank. In 1904 the Executive Committee tried to have the Department of Agriculture examine and report on the methods and results of Burbank. Popularly acclaimed as a "wizard," Burbank was the object of considerable contention in the scientific community and among his fellow horticulturalists. CIW had given funds to Burbank and wanted to place before the public and its own biological stations the knowledge acquired by this extraordinary man who kept few, if any, records. In Woodward's words: "He is like a mathematician who never has to recur to his formulas; all information he possesses he can summon in an instant for his use." Complicating the task of extricating Burbank's "formulas" from his mind was the problem of commercial rivalry with competing horticulturalists. Nevertheless, Woodward was impressed with the man, his work, and the potential benefit to humanity:

> A word or two as to his [i.e., Burbank's] personality. He is not a trained man of science; he lacks knowledge of the terminology of modern science. He often expresses himself in a way quite offensive to many scientific men, if due allowance is not made; but he is a man who unconsciously works by the scientific method to most extraordinary advantage. I think anybody who goes to his orchards and sees what he has produced and who studies Mr. Burbank as I have done will admit at once that he is a most unusual man. But along with his unusual abilities he has unusual peculiarities. Let any ordinary man of science go to his orchards and enter into a discussion with him and they will be at loggerheads in fifteen minutes. It is our duty to make allowances for these things. If we can make such allowances then what we can get out of Mr. Burbank will be extremely valuable to humanity; if we cannot make that allowance we shall fail.[30]

With those words in the trustees' meeting of December 12, 1905, Woodward precipitated unusually animated exchanges. His idea was to place one or two bright young scientists with Burbank to extract the underlying findings and general method. The subsequent discussion took an unexpected antiscientific turn that Woodward would implicitly counter in later years. First, Carnegie worried about men of science interfering with the workings of a genius: "a body of professors acting in a professorial role will look askance at the idea of such a man . . . revolutionizing us . . . with his experiments." Like Carnegie, a number of the trustees had some reservations about science, its methods, and its practitioners. Elihu Root contrasted Burbank's "extraordinary and exceptional faculty" with the "careful, scientific way." Billings observed, "The scientific language is rubbish; plain simple langue is what is needed." As to whether Burbank is in accord with Darwin, Mendel, or de Vries, Billings thought "it is of no importance."[31]

Only S. Weir Mitchell expressed doubt about the nature of Burbank's work. The argument in support of Burbank "has taken a very practical turn." He added, "We did not quite begin that way." Although Mitchell supported Woodward's proposal, he differed from him in regarding Burbank as no more than "an exceedingly careful observer." Woodward was right, in Mitchell's opinion, in seeking the laws underlying Burbank's work.[32] On July 26, 1906, CIW's young scientist, George Harrison Shull, completed a report on Burbank and his work.

Fascinated by Burbank, even fond of the man, Shull's report clearly indicated the limits of Burbank's empiricism.[33] But CIW continued Shull's work with Burbank; as late as 1911, it still hoped to publish his full study.[34] By one of the ironies that makes history interesting, Shull was one of the generation of pioneer geneticists who played an important role in the development of hybrid corn.

In 1908 the trustees once more clashed over Luther Burbank. Woodward now spoke quite differently: "It should be understood also that from a scientific point of view there is nothing mysterious or occult in the work of Mr. Burbank. His methods are as a rule neither unique nor unknown, and his work turns out to be of less value to biological science than is generally supposed." Having translated Burbank's work into scientific language, its stature was diminished. Woodward— despite bad publicity and despite Burbank's commercializing of his findings—recommended continuing support of Shull's work and of Burbank's. At the same time, however, he indicated that "much to our disappointment, our association with him [Burbank] does not appear to have been at all effective in raising him to the level he might deservedly occupy"; that is, Burbank did not wholly appreciate the possibility of becoming a scientist.[35]

Shortly afterward, Andrew Carnegie made one of his few incursions into the proceedings:

Will you let me interrupt you, gentlemen, for just a moment? Would you like the view of a rank outsider, and a most ignorant man compared with you learned people? My friend here [Woodward] said the only thing we expect is a scientific report, and my other friend here [John L. Cadwalader, a trustee] said that the only thing we would get would be an economic result from Mr. Burbank. I would like to know what our scientific reports avail us if the end be not economic gain, that we shall get plants which will yield revenue, which are now useless. If we can sustain Mr. Burbank in his work so that in the end we will have economic gain, I would be in favor of increasing the amount given him.

Although Woodward and Carnegie prevailed, this was one of the rare occasions in which some trustees voted in the negative.[36] By 1909, the Institution withdrew further support "to avoid entangling alliances with Mr. Burbank and his numerous exploiters."[37]

Shortly after Woodward began his disillusioning relationship with Burbank, he declared open war against the two principal CIW mechanisms for aiding individuals, the research assistantships and the minor grants. One aspect of his hostility remains shrouded in obscurity. In print in the Institution's *Yearbooks* in 1905 and in 1911, Woodward leveled serious charges against the programs. In the first year he condemned them as a scientific "spoils system"; in 1911 he described at least some participants as motivated by "dreams of avarice."[38] Writing in 1920 to his friend, the geologist T. C. Chamberlin, Woodward referred cryptically to the "sinister aspects" of the Institution's history,[39] presumably a reference to the "spoils system" and to the "dreams of avarice." To the best of my knowledge, these words were never challenged in print nor were they elaborated on by Woodward. If true, they not only involved appicants for funds but also distinguished referees. More important, implied is the involvement, knowingly or unwittingly, of the Executive Committee, which up to 1905 meant Billings and Walcott.

Opposition to the minor grants and the research assistantships did not merely spring from maladministration, real or alleged. Woodward favored support of research, certainly over any educational function. And the best way to advance research was by supporting proven people, either in CIW laboratories or in their home institutions. Such a policy avoided sheer waste (which had not bothered Carnegie) and the danger of involvement with cranks.

At the trustees' meeting of December 11, 1906, Woodward outlined his reasons for opposing the existing program: "At least three quarters of my time is absorbed by the business appertaining to these small grants and to the business of research-assistantships. . . . At present we are conducting a species of Havana Lottery, with monthly drawings, in which the inexperienced and the inexpert man is almost as likely to receive a prize as the expert and the experienced man."[40] Further in the meeting, Woodward asserted that only one in seven of the research assistants produced results worthy of publication. In the next year, similar statements about the unproductiveness of the short-term minor grants would embroil Woodward in a dispute with his former colleague James McKean Cattell.

What Woodward proposed instead was a system of research associateships: commitments of long-term support to proven investigators who could remain at their universities. The research associates would also serve as expert advisors to the president in their areas of competence. S. Weir Mitchell, speaking for himself and for Billings, defended the original program as in accordance with Carnegie's intention:

I wish to say that in Mr. Carnegie's original gift to us he especially dwelt on the desirability of experimenting on the finding of competent men. He did not expect to do what the President desires to have done, to limit our gifts only to those who had through years proven their capacity. . . . I will say also that his request for research associates and none others, to have these grants, would be cutting off a certain amount of valuable future human material. I feel strongly about this subject and Dr. Billings and I have gone over the statistics of success and failure in these minor grants and we believe that we have had a sufficient amount of success in these gifts to entitle us to continue for some years longer this search for competent men. I feel very strongly in regard to it, because there have been times in my young life when certainly, under the President's rule, I would not have been a proper person to give a grant to . . . and when I should have longed to have one, and applied to institution after institution without getting it.[41]

Although Woodward did not prevail in 1906, in time his views became accepted. The pressure of inflation was on his side, not only the economic kind, but the steadily rising demands of the successful departments and laboratories. And such great scientists as Thomas Hunt Morgan became CIW research associates. Persistence also paid off. Recognizing the key role of the relationship with universities in the minor grants and research assistantships, Woodward told the trustees in 1906 that he had in the past two years had more than 2,000 interviews on the subject. Methodically, he would delimit the contacts with higher education. Just as methodically, he would review the support of fields of learning and of the "exceptional man." Between 1906 and his retirement he carefully restated again and again what amounts to a theory of research, not very original but striking in its certitude.

As to the "exceptional man," by 1916 he had clearly become the proven specialist within an established discipline. As Woodward noted, "It has been very commonly supposed . . . that the chief business of your humble servant should be hunting by means of a lantern in the bushes and the tall grass for exceptional men. That happy phrase of our Founder has worked out very unhappily for the Institution. . . . [W]e have had our experience with wizards—and . . . that Burbank was by no means the only one, nor the worst of them."[42] Yet many people continued to think otherwise. Woodward characterized Edison's Naval Consulting Board in the World War I period as the mobilization of genius "and that meant, if it means anything, the stimulation of cranks." In Woodward's view, science was not a matter for abnormal minds, and Woodward's friend Richards agreed, forgetting his earlier comment on the artistic genius theory of scientific advances.[43] In 1918 Woodward reiterated his belief in supporting scientists of proven ability by contrasting the work of the Naval Consulting Board, now out of Edison's control, with that of George Ellery Hale's National Research Council:

On the one hand, the National Research Council has proceeded on the supposition that discoveries and advances may be most reasonably expected to arise with those who have already shown capacity to make them. The theory of the Council has been that the best advice to the Government in cases of emergency is most likely to come from experts of repute in their various fields of research. The initial theory of the Naval Consulting Board, on the other hand, was that discoveries and advances are about as likely to come from untrained as from trained minds and that, since the number of amateurs is very large, the best way to secure advances is to set experts at work examining the suggestions and inventions of inexperts. In addition, initially, the Naval Consulting Board was also encouraged to believe that discoveries and advances are developed chiefly by abnormal minds and that it is therefore worthwhile to set men of proved efficiency and capacity at work scanning the horizon for the scintillations which might otherwise emanate unperceived from exceptional men, who are supposed to be in hiding, or at best more or less concealed behind books and bottles in dingy laboratories. . . .

Happily for the reputations of the members of the Naval Consulting Board, this initial and popular theory was subjected to the tests of plain experience, which proved what is well known in the history of science and what has been demonstrated on a grand scale in the experience of the Institution, namely: first, that revolutionary discoveries, advances, and inventions do not arise suddenly or in necromantic fashion; and, secondly, that the poetic process of winnowing vast quantities of intellectual chaff with the hope of securing good grains of truth is the sheerest of futilities.[44]

Intellectual Fields

While the search for the exceptional man was difficult, choosing fields for support was a deeper, more fundamental problem. The sums in the control of the trustees were finite; the demands seemingly infinite. Each time a choice was made—laboratories rather than minor grants, this field rather than another—the Institution made enemies among the disappointed. Supporting every field, no matter how worthy, was impossible. Somehow, priorities had to emerge.

After Billings met with Carnegie and Gilman on November 18, 1901, a memorandum was prepared on the nature of the proposed institution. Its subject scope obviously included science "in the ordinary sense of the word." To this was added history, political science, economics, philology (especially "Chinese and Oriental"), and library economy and management. (The last clearly reflected Billings' devotion to developing great libraries.) After listing a number of topics in the physical sciences, the text surprisingly called attention to the need for instruction in statistics and actuarial sciences and in Chinese and other Oriental languages; it also noted the importance of establishing an advanced school of bibliography and related topics. Nowhere else in the CIW context did Billings suggest support of teaching programs; in fact, teaching programs vanish from the surviving record of later discussions. A further Billings memorandum of February 1, 1902, noted that the objective of CIW was the same as that of the Smithsonian: "the increase and diffusion of knowledge, not limited to any field of inquiry."[45]

No doubt Gilman agreed with these sentiments. But by the first annual meeting of the trustees on November 27, 1902, Billings was qualifying his Smithsonian-like scope. Medical research was excluded, being a territory occupied by the Rockefeller interests. Despite that limitation, Billings's influence resulted in CIW support of *Index Medicus* from 1903 through 1926. Billings purged literature, music, and the fine arts from the Intstitution's purview. At least publicly, Gilman agreed. Emphatic as usual in his views, Billings was quite unimpressed by the nonscientific applicants: "We have had a number of requests for funds, to enable a man to abandon all his other business and write a treatise on logic or a treatise on philosophy, or a treatise on the history of religion, or a treatise on some particular points in theosophy or metaphysics. I have not considered that the urgency of those demands or that the character of the men who proposed to do it was so exceptional."[46] But where the exceptional man existed, like Ewald Flügel, the Middle English scholar at Stanford, the trustees provided support. Perhaps more important, the "Bureau," later the Department of Historical Research, was organized in 1903. The Department of Economics and Sociology became active in 1904. Individual archaeologists and anthropologists, like Raphael Pumpelly, received CIW funds.

Despite these actions, the bulk of CIW support went to science "in the ordinary sense of the word." Twelve of the eighteen advisory committees were concerned with scientific fields; of the remainder, engineering was not viewed as an applied field and the rest—economics, history, psychology, anthropology, and bibliography—were as far as the trustees cared to venture outside the established physical and biological fields.

In the sciences proper, the Institution encountered considerable skepticism and even hostility. Doubts were rampant about the ability of the trustees to act wisely. For example, in correspondence with George Ellery Hale, the astronomer Lewis Boss expressed reservations about the trustees' interest in big reflectors: "Dr. Walcott is an old friend of mine. I have known him ever since you were a small boy. . . . But I cannot bring myself to attach much weight to his judgment upon a technical matter in astronomy."[47]

Distrust of the trustees, largely nonscientists, produced the one major setback of the Institution's early years. After a majority of the trustees of the Woods Hole Marine Biological Laboratory voted to join CIW in August 1902, C. O. Whitman, its director, mustered enough support to thwart the merger. Whitman believed scientists should run their own institutions.[48] That experience, and some relations with universities, undoubtedly hardened CIW's belief in the need to have its own laboratories.

Woodward and Walcott had key roles in this development, perhaps best illustrated by the fate of the advisory committee reports on geophysics, physics, and chemistry. Woodward chaired the first two committees. Serving also with him on both were the physicists A. A. Michelson and Carl Barus. Joining these three on the geophysics committee was the entire geology advisory committee: T. C. Chamberlin, C. R. Van Hise, and C. D. Walcott. Chamberlin and his colleagues maintained that since geology was in good shape, the future was in geophysics. Woodward produced two symmetrical reports, calling for a special research laboratory in geophysics and another in physics.

What followed was probably Walcott's greatest moment as a policymaker in CIW. Word of the reports of the various advisory committees spread quickly in 1902. By late summer and early fall, James McKean Cattell, the Columbia psychologist and editor of *Science,* was on the attack, particularly against the geophysical laboratory. Earlier, he had given the trustees unsolicited advice; at Woods Hole, Cattell had sided with Whitman against absorption by CIW. Cattell favored laboratories for physics, chemistry, and psychology in addition to a very liberal program of fellowships and grants. He was decidedly against tying up an appreciable portion of Carnegie Institution funds for geophysics. No amount of assurances by Chamberlin, Walcott, and presumably Woodward could convince him the subject merited so much cash.[49] In addition, Cattell was against the growing inclination of the trustees to favor the expensive plans of the astronomers, notably those of George Ellery Hale to construct an observatory on Mount Wilson.

In 1902, Walcott purposefully moved to commit CIW to a program of laboratory construction. Locating tracts of land in Washington, he presented plans to the trustees on November 25. One hundred and ten acres seemed excessive before the Institution had decided on its policy: as Billings declared, "that is practically an endorsement of the centralized large laboratory scheme for Washington." Following that assertion, the trustees had a heated, if rambling, debate. Putting resources into land and then into bricks outraged those interested in finding and developing the exceptional man. To others, a good buy in real estate was simple prudence looking to the future. Alexander Agassiz wavered, tempted by the possibility of a biological laboratory in northwest Washington. Still others argued for giving the exceptional men physical homes for their talents. Walcott's particular site proposal failed, but the net effect was to harden opinion among the trustees for acquisition of land in Washington, particularly for a geophysical laboratory.[50]

Woodward's other proposed laboratory scheme for physics quietly expired. No exceptional man appeared, at least not one interested in leaving academe. Presumably, if someone like Michelson had evinced interest, then the Institution might have erected another laboratory in Washington to go along with those for geophysics and terrestrial magnetism. Significantly, the trustees, including Woodward himself in later years, made no effort to find such an exceptional investigator. By 1906 at least some leading physicists were convinced CIW support for their field was a lost cause.[51]

The advisory committee on chemistry submitted majority and minority reports. Remsen of Johns Hopkins chaired a committee of three that included T. W. Richards and E. F. Smith. The majority favored giving funds to established chemists at universities so they could hire suitable research assistants. Yet in their initial concept of research assistantships, the trustees in effect disregarded the substance of the majority report.

The attitude of T. W. Richards regarding CIW support of chemistry is of particular interest. While signing the majority report, Richards issued a minority report calling for an alternative rejected by his two colleagues—the establishment of Carnegie Research Professorships "to relieve university professors from part of the routine work which they are now doing." Richards and Woodward tentatively considered forming a laboratory of physical chemistry, but their idea foundered on the ambiguous attitude of the trustees toward universities. In general, however, Richards's views on the pursuit of science differed significantly from those of Woodward and other trustees. For example, when Woodward's appointment was under discussion. Richards initially opposed it because, in his view, chemistry had greater possibilities of promoting the "welfare of mankind" than physics or astronomy. On the basis of no evidence known to me, Richards stated that Carnegie favored experimental work over mathematical or abstract investigations. Science, Richards asserted, is more likely to advance along inductive rather then deductive lines. Yet Woodward himself had no doubt that he was an inductive scientist; his own work included instances of careful observation and experimentation.[52]

With the exception of the abortive merger of CIW and Woods Hole, the biological sciences did not represent an area of contention with the Carnegie Institution. The Desert Laboratory opened in 1903; the Station for Experimental Evolution and the Department of Marine Biology were established in 1904; the Nutrition Laboratory followed three years later; and the Department of Embryology in 1914.

By contrast, the Carnegie policy toward the "newer" sciences led to several disputes, with Cattell once again at the center of the opposition. Beyond Cattell's great skepticism of the need for a research facility for geophysics was a strong desire to obtain funds for experimental psychology. To him the issue was favoritism for older established fields over the newer ones. As Hale's plans progressed, Cattell shifted his criticism from geophysics to astronomy. His way of looking at the matter was quite different from that of Hale, Woodward, and the trustees. In 1905 he wrote Hale: "Any distribution of the funds of the Carnegie

Institution is likely to cause criticism. . . . As a psychologist I naturally feel that we are as much in need of money as the astronomers, and I do not see just how the account can be settled, except on the assumption that as there are about an equal number of workers in astronomy and psychology, they should receive equal consideration."[53] Assuring Cattell of his desire to advance psychology, Hale argued that his proposals were unique, valuable, and timely.

Cattell's criticism roused Woodward, particularly when points made in correspondence subsequently appeared in print. Being a methodological and conscientious person, Woodward thought out and presented an answer. After citing "the practical and commercial value" of astronomy, he cited "a still higher value":

Astronomy, for example, has done more than all the other sciences put together up to date to straighten out the kinks in the minds of human beings, to remove superstition and to enable us to think straight on questions in general. . . . It is much easier in general to get money to carry on astronomical researches than to carry on other researches. . . . There is good reason for it. Many people have asked me how it is that, if we speak of shares amongst the different sciences, astronomy is getting the lion's share. I have tried to find out why it is, and it is this: It is the oldest and most highly developed science. . . . Look over the applications we have for projects [and] you will find that the certainty, the definiteness of these applications is almost directly proportionate to the age of the sciences represented. Roughly speaking, the order of precedence, which is fixed not by us but by nature, is this: Astronomy, chemistry and physics second, and zoology third. . . . One of the aims [of CIW] should be to bring up the other sciences to the level of the older sciences.[54]

Woodward's certainty about the order fixed by nature arose from the clear, self-evident truths of the history of science—at least as seen by him.

Given the limited funds, the trustees understandably favored the more established sciences. Being sympathetic with his former colleagues Cattell and Franz Boas, Woodward was not wholly comfortable with his position and the prejudices of his trustees:

It is not alone sufficient to show the possibility of taking up work in new directions. This is in fact the least of the difficulties. The greatest difficulty is to overcome the prejudice and inertia of the society in which we live. Some lines of work which appear to me to be of the greatest importance to the future of humanity are looked upon with unmitigated contempt by many of our most eminent contemporaries. The Institution is severely criticised for pursuing the obviously advantageous sciences. What could be the stream of criticism and abuse received if we were to take up work in lines less obviously advantageous?[55]

An outsider like Cattell did not realize that Woodward and the trustees were often genuinely interested in moving into new fields. As proposals for new thrusts encountered the limitations of funds, they reacted sometimes out of prejudice, sometimes with genuine concern to think out issues and do right. The openness to expansion of the Institution's scope is well illustrated by the new departments that were under consideration in 1907: municipal government, classical philology and archaeology, legal research, hydromechanics, and anthropology. While Woodward's sole reservation on the five was fiscal, only anthropology became a research area of the Institution.[56]

Despite Woodward's interest, he was very cautious about entering into anthropology. His 1901 presidential address to the American Association for the Advancement of Science spoke approvingly of science entering and exploring "the domain of manners and morals."[57] Woodward had great respect for Franz Boas, his former colleague at Columbia, and consulted with him on this issue. Boas's initial training was related to Woodward's, and one can speculate that he might have served as the "exceptional man" in anthropology. But Boas preferred to remain at Columbia. In correspondence with Cattell in 1906, Woodward summed up his problem with the field: "The fact that there are numerous tribes of the human race now rapidly disappearing has been stated to me at least once a day on the average during the past year, and yet I fear we could hardly get two men to agree on a practicable method of dealing with the problems presented by these tribes."[58] Woodward was noting the absence of a consensus, one of the characteristics of a science, a point later rediscovered by Thomas Kuhn.

These misgivings aside, by 1911 Woodward seemed ready to find an exceptional man for a department of anthropology. One of the trustees, William Barclay Parsons,[59] avidly pushed Central American archaeology, and other trustees enthusiastically supported him. But at the next annual meeting, Woodward balked. In his view, there was still no single person exceptional enough to take a broad view of the entire field. Instead, he recommended going forward in human embryology, "a field which lies at the basis of anthropology." For Woodward, Central American archaeology was too narrow, yet Parsons persisted. The result was that, in 1914, CIW began support of Sylvanus G. Morley's archaelogical work in Central America. Eventually, a separate program for the study of Central American archaeology was developed.

The Carnegie Institution's experience with two other social sciences, economics and sociology, was completely different. The Department of Economics and Sociology was organized in 1904—one of the earliest to be established in the Institution. It differed from the other CIW departments in two respects. First, its director from 1904 to his death in 1909 was Carroll D. Wright, a statistician and one of the trustees.[60] Perhaps for that reason, little attention was paid to the department at trustees' meetings during Wright's lifetime.

After Wright's death, Woodward and the Executive Committee looked carefully into the work of the department and were appalled. Wright had ambitious plans for a series of studies constituting a comprehensive social and economic history of the United States.[61] Yet despite sizable allocations by the standards of the day, few results were produced nor was there optimism about future works reaching CIW for printing. At fault, in Woodward's opinion, was a totally pernicious manner of organization unlike that of any of the other operating departments of the Institution. Woodward pointedly reminded Billings that the defective procedures were specifically approved by the Executive Committee in January 1904.[62] But he was careful to point out, at a trustees' meeting in December 1909, that the faults were not inherent in economics[63] and in sociology:

It appeared, moreover, that most of the work which has been done under the auspices of this department had been carried on or executed by young men or women who were, at the time, candidates for the higher degrees in colleges and universities. It appeared, on further examination, that many of these candidates for degrees had been supported in going through colleges and universities, and had actually had their dissertations and other papers printed at the expense of the Institutions. . . . It appeared from the confessions of these collaborators of Colonel Wright, all very able and eminent men, that nearly all of the work which had been collected by their assistants was regarded as untrustworthy, defective, much of it defective to such an extent that it could not be used at all.[64]

To Woodward, it was vindication for his belief in the full-time pursuit of knowledge. By 1916, he was more than a bit smug in reminding the trustees of his opposition to assistantships and to minor grants and how opinion inside and outside the Institution opposed him by a margin of six to four, the majority preferring to aid university men.

With the discovery of conditions following Wright's death, Woodward tried to rescue something from the debacle. To the trustees he displayed, on December 11, 1916, "the first fruits from that department." It was too late; the trustees voted for its abolition. Woodward hoped to revive the group in five years when funds might become available, but this proved impossible.[65] In retrospect, the department did achieve a measure of success; its products eventually included important contributions, some not wholly supplanted by later work.[66]

At the same trustees' meeting that revealed the bad news about Carroll Wright's creation, a pointed question was asked about the quality of the work carried out by the Department of Historical Research. Staunchly defending the department, Woodward pointed to its productivity—a result of full-time devotion to research so unlike the university-type arrangement in Wright's group. While the Department of Historical Research survived from 1903 until 1930, its career within CIW was not always smooth. Walcott and Billings strongly opposed it. In 1903 Walcott decided to remain on the Executive Committee "to prevent any development of the Institution on the lines of an educational organization. I fear we are drifting into it in the historical work."[67] A number of years later, Billings unsuccessfully attempted to abolish the Department of Historical Research.[68] Woodward stopped that move and in 1916 defended the department, then under John Franklin Jameson, against the charge of not producing history but simply "materials for historic research" in the form of guides, catalogs, and indexes.[69] The products of Jameson's group were valuable and influential in their day; in addition to guides to sources, a number of important documentary collections eventually appeared.[70]

Full-time professionalism rated high in Woodward's scale of values. That concept had much to do with the demise of CIW support of a program in classical archaeology and philology. At the very same trustees' meeting of December 13, 1904, in which Woodward was elected president, Gilman introduced an appeal for funds for the schools of classical studies in Athens and Rome. The sum was modest, $3,300, and the sponsors impeccable—T. D. Seymour of Yale, James R. Wheeler of Columbia University, and Andrew F. West of Princeton. Billings voiced his objections; the work was not under the direction of CIW but of outside organizations. Gilman prevailed largely because many of the trustees regarded the schools as admirable examples of self-help.[71]

Woodward was also sympathetic. Linking classical studies to his support of anthropology, he observed in 1908:

It seems to me that it would be quite practicable now to raise classical philology and archaeology from the plane of amateurism and dillettantism on which they have rested, perhaps for centuries, to the plane of anthropology. The ablest philologists and archaeologists, I think, in this particular field are engaged in work which is as truly scientific as that carried on in the physical sciences. It should be remembered that a very large majority of our fellow citizens of the world, especially among educated men, are men whose education and whose interests are almost wholly confined to those lines of work, and it appears to me that one of the best ways to get them to appreciate other lines of work, which now certainly receive the preference in educational institutions and in the world generally, would be to give our classical friends an opportunity to come in on the same ground with the physical sciences.[72]

But in 1912, Woodward withdrew CIW support for the American School of Classical Studies in Rome: "I visited it last summer, spent nearly a month in Rome, and looked into it very carefully. . . . [It] is an admirable school for elementary education, or perhaps for the dilettante or amateur who aspires to a modicum of culture; it is a school . . . somewhat of the character of an international afternoon tea . . . but it does not lead to research." Loath to desert the field, Woodward made two scholars from the Rome School research associates: Elias A. Lowe (later at the Institute for Advanced Study) and Esther Boice Van Deman. Unfortunately for the humanities, Woodward's observations of "an international afternoon tea" occurred in the period when he and CIW were under pressure to expand into that intellectual area.[73]

At the same meeting of December 14, 1909, in which Woodward disclosed the shortcomings of the Department of Economics and Sociology, the trustees received a petition from ten national societies calling for a redress in the balance of CIW funding for science and the humanities.[74] The petition was endorsed by the presidents and former presidents of nine universities. Although it was sent to the Executive Committee for consideration, no significant change occurred in CIW's allocation among fields. Despite Woodward's sympathetic words a year earlier, neither he nor the trustees were inclined to adopt a change in policy, especially in view of a fairly tight budget.

What the petition did cause was a reexamination of the status of the humanities within CIW, a topic that concerned Woodward almost to the end of his tenure. Earlier in his presidency, Woodward sided with Gilman against Billings and Walcott on support of the humanities but with only modest effects.[75] After the receipt of the 1909 petition, Woodward began a systematic study of how CIW "may best promote research and progress in the humanities." Naturally, the results confirmed his long-held beliefs. "About thirty distinguished authors" replied to a circular letter. Very little agreement existed as to the definition of the humanities; asked to indicate which CIW publications were in the humanities, the respondents reached no consensus.[76]

Woodward had no qualms about where he stood. While sympathetic to the humanities, he regarded science as both intellectually and morally superior. In public addresses while Dean of Pure Science at Colombia, he made his position quite explicit. In his presidential address of August 27, 1901, before the American Association for the Advancement of Science, he sketched a scheme for the history of science which was, to him, *the* history of mankind. To extend the sway of man and to diminish the scope of the supernatural, three distinct methods were developed in the past and applied down to the present: the a priori, the historico-critical, and the scientific.

The second . . . depends, in its purity, on tradition, history, direct human testimony and verbal congruity. It does not require an appeal to Nature except as manifested in man. . . . And in the serenity of his repose behind the fortress of "liberal culture," the reactionary humanist will prepare apologies for errors and patch up compromises between traditional beliefs and sound learning with such consummate literary skill that even "the good demon of doubt" is almost persuaded that if knowledge did not come to an end long ago it will soon reach its limit. In short, we have learned, or ought to have learned, from ample experience, that in the search for definite verifiable knowledge we should beware of the investigator whose equipment consists of a bundle of traditions and dogmas along with formal logic and a facile pen; for we may be sure that he will be more deeply concerned with the question of the safety than with the question of the soundness of scientific doctrines. . . . I would not disparage the elevated aspirations and noble efforts of the evangelists and the humanists who seek to raise the lower to the plane of the higher elements of our race; but it is now plain as a matter of fact, no matter how repulsive it may seem to some of our inherited opinions, that the railway, the steamship, the telegraph, and the daily press will do more to illuminate the dark places of the earth than all the apostles of creeds and all the messengers of the gospel of "sweetness and light."[77]

Two years later Woodward tilted at the humanities in a commencement address:

Thus, even at the present day, many of the older schools of education hold, tacitly, if not openly, that studies may be divided into sharply defined categories designated as "liberal," "humanistic," "scientific," "professional," "technical," etc.; and men and women are said to have had "a liberal training," "a professional training" or "a technical training," as the case may be. They say, by implication at least, that mathematics, when pursued a little way, just far enough to make a student entertain the egotistic but erroneous notion that he knows something of the subject, is an element of liberal training. On the other hand, if the student goes further, and acquires a working knowledge of mathematics, his training is called professional or technical. Similarly, studies which include the memorabilia of Xenophon and Caesar, the poetry of Homer and Virgil and Dante and Shakespeare, or, in short, the so-called polite literature of ancient and modern times, are said to lead to breadth and culture; while studies which include the works of Archimedes, Hipparchus, Galileo, Huygens, Newton, Laplace and Darwin are said to lead to narrowness and specialism; as if the first class of authors were somehow possessed of humanistic traits, and the other class of demoniacal tendencies. So far, indeed, are these distinctions carried that higher moral qualities are not uncommonly attributed to the young man who studies Latin and Greek in order that he may earn a living by teaching them than are attributed to the young man who studies engineering in order that he may earn a living by building bridges which will not fall down and kill folks.[78]

Above all, the attribution of "higher moral qualities" to humanists outraged Woodward. For example, in 1906 he had a dispute about John Tyndall with one of T. W. Richards's Harvard colleagues, C. L. Jackson. Woodward's defense of Tyndall unexpectedly brought in the humanities:

We Americans, I think, are especially indebted to Tyndall for helping us to get a start in our colleges for the pursuit of postgraduate studies. We should remember also that while Tyndall left the proceeds of his "Lectures in America" to found fellowships at Harvard, at Columbia, and at the University of Pennsylvania, the prince of humanists, Matthew Arnold, refused to go on the lecture platform in some instances before receiving a check in payment for his address.[79]

In addition to the claims of the academic humanists, Woodward had to consider the views of a number of his lay trustees, some of whom favored support for poetry. As Senator Lodge stated: "Literature and the humanities are pretty nearly dead, science has pretty nearly killed them, but I would like them to live and I would like to see something done to keep them alive." One gathers that there was little sympathy among the trustees for Woodward's attempt to convert the humanities into research fields.[80] His last public words on the subject were: "It appears to be the duty of the Institution to proceed . . . in a spirit of sympathy and equity based on merit toward all domains of knowledge, with a full appreciation of the necessary limitations of any single organization and with a respectful but untrammeled regard for the views, the sentiments, and the suffrages of our contemporaries."[81]

Here George Sarton's dedication to volume 3, part 1 of his *Introduction to the History of Science* becomes a clue to Woodward's behavior. Extravagantly, it calls Woodward the "second author" of the work, pointing up a bookish, learned element in the man who wanted Archimedes, Hipparchus, Galileo, Huygens, Newton, Laplace, and Darwin to be part of liberal training. Such a man gloried in the publication of a translation of a revision of Ptolemy and could proudly call attention to the publication of the Vulgate version of the Arthurian romances, Charles W. Hodel's translation of the "Old Yellow Book," and concordances to Spenser, Horace, and Keats. And he could even approve the publication in 1913 of Morgan Calloway's statistical study, *The Infinitive in Anglo Saxon.*

If one looks at the pattern of CIW support of fields of learning as a whole—not only at the areas of contention—two factors emerge as decisive for a major commitment of its funds. In each and every case there was an exceptional man (in the eyes of the trustees) and a field already familiar to the American scene. Where a field, such as physics or chemistry, was marginally established in the hands of a small number of university practitioners, even eminent ones like A. A. Michelson, CIW support of a laboratory was most unlikely. Even the two nonscientific departments—History, and Economics and Sociology—represented activities and intellectual styles already familiar to Americans with any reasonable cultural awareness. Walcott and Chamberlin properly stressed the originality of the proposed geophysical research. Concern with the physical and chemical properties of the earth's constituents was, nevertheless, an intrinsic part of a widespread, successful research enterprise in geology in nineteenth-century America. A special laboratory was a logical extension of this tradition. Terrestrial magnetism had attracted American physical scientists in the nineteenth century. The early interests of Ferdinand Rudolph Hassler and Alexander Dallas Bache embedded the subject in the Coast and Geodetic Survey. By 1900 Americans had invested considerable intellectual and economic capital in astronomy; even the newer specialty, astrophysics, was fully assimilated.

Much the same pattern exists within the biological installations; the Station for Experimental Evolution (later the Department of Genetics), the Desert Laboratory (later the Department of Plant Biology), and the Department of Marine Biology all had American antecedents. At first glance the Nutrition Laboratory (1907) and the Department of Embryology (1914) do not quite fit the pattern. Both had exceptional men as directors, Francis G. Benedict and Franklin P. Mall, respectively. Yet the fields are not part of the conventional historical landscape of the sciences in America. If, however, one remembers R. H. Chittenden's history of physiological chemistry in America and the earlier work of Wilbur Atwater, the Nutrition Laboratory does fall within the pattern.[82] On the other hand, Franklin P. Mall's focus on human embryology required a fetus collection analogous to the anatomical and pathological collections of the Army Medical Museum. I am inclined to view CIW's interest in embryology as an outgrowth of concern with medical pathology and histology, now extended to include physiology.

Citing the impact of precedent and of exceptional men is not to deny nor to diminish the value of the research issuing from CIW installations. Quite often the laboratories and the Mt. Wilson Observatory did fresh and influential work. What is at issue here is the range within which the trustees were willing to make long-term obligations in contrast to short-term fellowships and specific grants. Understandably, even enlightened laymen would hesitate about venturing too far from the familiar. Moreover, the scientific trustees were not the sort likely to lead the CIW into revolutionary ventures. Billings and Mitchell were elderly physicians not inclined to stray from the safety of the reports of expert advisers. Although he was great paleontologist, Walcott's intellectual scope was limited; after the Geophysics Laboratory was assured, his influence diminished. As a man from a rising, bustling university, Woodward was more knowing and sympathetic to new trends; countering that characteristic were his administrative caution and his ideological set. Yet, despite this cautionary approach, CIW in its early years was a notable increment to the intellectual life of the United States.

Higher Education and the Carnegie Institution

After support of original research and the discovery of exceptional men, the trust deed establishing the Carnegie Institution of Washington set forth three aims related to institutions of higher learning:

3. To increase facilities for higher education.
4. To increase the efficiency of the Universities and other institutions of learning throughout the country, by utilizing and adding to their existing facilities and aiding teachers in the various institutions for experimental work, in these institutions as far as advisable.
5. To enable such students as may find Washington the best point for their special studies, to enjoy the advantages of the Museums, Libraries, Laboratories, Observatories, Meteorological, Piscicultural, and Forestry Schools, and kindred institutions of the several departments of the Government.[83]

These three aims plus the two prior ones could easily lead to the conclusion that the bulk of the Carnegie gift was destined for universites, their faculty members, and their students. Indeed, many drew that conclusion—to their bitter disappointment as the Institution's program unfolded. Despite the promising language of the trust deed, the origins of aims 3, 4, and 5 practically guaranteed only limited aid to higher education either directly by grants to institutions or indirectly to faculty and students. To this birth defect (from the standpoint of the universities) must be added the resolve of Carnegie and the trustees (with the notable exception of Gilman and White) to avoid the enmeshment of CIW with the world of higher education.

As has been shown in the preceding pages, evidence of this determination to keep a careful distance from the university world became increasingly apparent as the Carnegie program developed. But even in the very beginning there were indirect indications of this intent. For example, a letter from Henry Smith Pritchett, then president of the Massachusetts Institute of Technology, reveals that in the early planning stages of CIW five of the trustees were to have come from institutions of higher learning.[84] Yet when the Institution actually came into existence, an unwritten rule barred active university administrators and faculty from the Board of Trustees. President Charles W. Eliot of Harvard was asked to join only after his retirement; Welch of Johns Hopkins was the first to breach the rule. University men by definition were interested parties. The trustees feared their funds might blight the initiative of the universities; they also worried about the diversion of funds to support routine educational activities, not research or outstanding individuals. When a revision of the CIW charter occurred in 1904, the trustees, without any qualms, dropped those clauses on education objected to by the Congress.[85] Finally, innumerable encounters with university men seeking funds only strengthened an existing bias for in-house research. Writing to Billings only a few months after Woodward assumed office, Walcott was "somewhat amused to hear of the experiences Dr. Woodward is having with university and other men. I told him it was a duplication of what you and I have been through and that he would probably come to the same conclusions before long."[86]

At the very end of his tenure as president of the Institution, Woodward was still complaining that university men misunderstood the nature of CIW. They disbelieved or resented the effort to make the Carnegie Institution more than a mere "disbursing agency," in fact, a real participant in research. Nor was Woodward wholly successful in conveying his rationale, even though he wrote several statements on his "theory of research" in the Institution's *Yearbooks,* emphasizing his belief that research should be conducted on a full-time basis at an institution designed specifically for that purpose.[87]

A good example of this lack of successful communication can be found in Woodward' long and friendly relationship with T. W. Richards. Despite an extensive exchange of correspondence over many years, neither really understood the other nor succeeded in convincing the other. As late as 1914, after Richards had spent several years as a CIW research associate, the gap was still wide. As a notable researcher and a university man, Richards continued to favor the endowment of research professorships at universities to relieve outstanding faculty from the routine of teaching. Although the possibility of a CIW laboratory for physical chemistry came up several times, Richards always insisted that the laboratory be at Harvard.

Richards's reasons for endowing professorships and building laboratories at universities were in part highly practical. Using existing staff, buildings, and funds at universities would enable CIW funds to go further. In addition there was the "great mass of advanced students . . . held partly by their desire for the degree . . . and by sentimental considerations" available as cheap labor. This argument based on economy was flatly opposed by Woodward; CIW experience was to the contrary. In addition he pointed to the irritations and frictions almost invariably arising between grantor and grantee. Nor was Woodward impressed by Richards's call for CIW to expand the small portion of funds available for research at Harvard.

The raising of the flag for the universities by Richards brought forth rebuttal from Woodward:

Curiously enough, it seems to be tacitly assumed by many eminent men that there is something about a university that will prevent investigators from stagnating more completely than similar influences in institutions organized for the express purpose of research. This argument applied to the Royal

Institution of London, for example, would lead us to suppose that Davy, Faraday, Tyndall, and Dewar ought to have undergone deterioration immediately on being given life positions for the purpose of devoting their entire energies to investigations.

An eminent man of science wrote me a few days ago that he viewed "with alarm" the tendency of our institution to build departments of work, for . . . the men at work in these departments will tend to stagnate. . . . Everyone knows . . . that most men tend to stagnate rather early in life; but I fail to see why picked men should tend to stagnate in an institution whose atmosphere is that of research, any more than men in academic institutions, who are, as a rule, to a much less extent subject to stimulus to original productivity.[88]

That elicited an interesting rejoinder from Richards: "What I meant to say was this:—no man can predict what a man is going to do in the future; some increase in productiveness; others diminish. If a research institution engages a man in a life position, and his originality diminishes, both the institution and the man are in an unfortunate position. If the university does the same thing, the man may throw his energies into teaching."[89]

In the 1912 *Yearbook* of the Institution, Woodward gave a number of "inductions" from CIW experience in supporting research. Full-time professionalism loomed large. One induction was aimed at the universities. "Many investigations of real scope require men untrammeled by other occupations. . . . The common notion that research demands only a portion of one's leisure from more absorbing duties tends to turn the course of evolution backwards and to land us in the amateurism and the dilettantism wherein science finds its beginnings."[90]

Walcott and Billings supported Woodward in his determination to create an organization devoted purely to research. Although so different in personality and intellectual interests, the three men shared at least two experiences. First, since they entered science before the widespread establishment of graduate education, they had followed other career routes. Walcott never went to college, being largely trained by apprenticeship and self-study. He is probably the last American scientist of consequence to lack formal academic training. Billings (if viewed as a scientist) entered the world of research from medical school, a path fairly common in America during the first half of the century before the growth of science in the liberal arts colleges, in the scientific schools, and in the graduate schools. And Woodward was a mathematically talented engineer who turned to pure science. Not having passed through a graduate school, they did not consider the university as necessarily the site of research. What they saw was some research conducted at a few of the outstanding schools and very little investigation, indeed, performed at the mass of colleges and universities. A second point in common was a career in the federal service.

Each had engaged in full-time professional work in the government. Their support of an independent installation specializing in research was a reasonable extension of their personal experiences. In no way could they agree with the tacit assumption of their critics that persists to this day: the support of research must be necessarily linked to the support of higher education. By implication, Walcott, Billings, and Woodward were advocating policies at the Carnegie Institution of Washington that called for universities to specialize in teaching. Lurking among the minutes, memoranda, and correspondence of the CIW may be a moral.

Yet even at the founding, such a moral was futile because it came too late. The Carnegie Institution was not alone in attempting to separate teaching from research. In the same period, similarly motivated organizations appeared in the United States and in Europe—the Rockefeller Institute for Medical Research in New York, the National Physical Laboratory in England, the Kaiser-Wilhelm-Gesellschaft in Germany.[91] However, by the turn of the century the leading American universities, as well as those institutions just below the first rank, had come to include research in their definition of the essence of higher education. Perhaps as important, universities and colleges with pretensions to excellence scrambled down the same path. As circumstances permitted, university after university with little hesitation and considerable support from their faculties incorporated more and more aspects of the research institute. A T. W. Richards of Harvard, after all, vied with Woodward, Billings, and Walcott in his admiration for the Royal Institution of Michael Faraday. And there was no shortage of university faculty who regarded students as obstacles to the pursuit of knowledge.

Notes

1. By far the best account of the founding of the Carnegie Institution is in chap. 9 of Howard S. Miller's *Dollars for Research: Science and Its Patrons in Nineteenth-Century America* (Seattle: University of Washington Press, 1970). Joseph Wall's *Andrew Carnegie* (New York: Oxford University Press, 1970), the latest biography, adds only a few details; see pp. 858–863. A valuable article bearing on the founding is David Madsen, "Daniel Coit Gilman at the Carnegie Institution of Washington," *History of Education Quarterly* 9 (1969):154–186.

2. Carnegie Institution of Washington, *Yearbook,* 1901, Appendix A, pp. 1–238.

3. See the entry "Robert S. Woodward" in the *Dictionary of Scientific Biography,* in press at this writing.
4. Nathan Reingold, "American Indifference to Basic Research: A Reappraisal," in George H. Daniels' (ed.), *Nineteenth-Century American Science: A Reappraisal* (Evanston, Ill.: Northwestern University Press, 1972), pp. 38–62 [Paper No. 3 in the present volume].
5. Gilman to Carnegie, Apr. 13, 1903, with reply of Carnegie. Carnegie Institution of Washington Archives (hereafter CIW).
6. Woodard to Billings, Nov. 30, 1910, Billings Papers, New York Public Library.
7. As this matter is adequately handled in Madsen, "Daniel Coit Gilman," it will not be discussed in detail here.
8. Billings to H. L. Higginson, Nov. 23, 1904, Billings Papers, and Woodward to T. C. Chamberlin, July 11, 1904, Chamberlin Papers, University of Chicago.
9. Billings to Walcott, Oct. 25, 1905, Billings Papers, in which Billings agrees to Woodward acting as secretary of the Executive Committee.
10. W. H. Welch to Woodward, Nov. 30, 1911, and Woodward to Welch, Dec. 21, 1911, CIW.
11. File DR 75, May 24, 1916, p. 3, Record Group 3, Ser. 915, Box 1, Folder 6, Rockefeller Archive Center.
12. While Carnegie's remarks are in the trustee's minutes, Gilman's are not but survive separately at CIW.
13. The standard account is David Madsen, *The National University: Enduring Dream of the U.S.A.* (Detroit: Wayne State University Press, 1967); chap. 5 deals with Hoyt.
14. For White's optimistic reaction, see his letter to Gilman of June 21, 1901, CIW.
15. An 1892 statute had opened government facilities to students and investigators for institutions incorporated by Congress or under the laws of the District of Columbia. In March 1901, this privilege was extended to all.
16. Miller, *Dollars for Research,* has a good account of the events. In the Walcott Papers, Smithsonian Archives, are Walcott's files on the Association and the Institution. An interesting account in the context of the later events is Philip C. Ritterbush, "Research Training in Governmental Laboratories in the United States," *Minerva* 4 (1966):186–201. Walcott hoped to have Gilman head the Washington Memorial Institution. Walcott to Gilman, Nov. 18, 1901, CIW.
17. Billing's memo of the meeting, dated Nov. 22, 1901, is in Box A of his papers in the folder "Memos–Bibliographies."
18. Gilman to Eliot, Dec. 3, 1901, CIW.
19. In the Walcott file cited in n. 16. See also Trustees' Minutes, Mar. 1, 1902, 12f, and the undated letter (1902) of Walcott to N. M. Butler about CIW developing as a research institute rather than promoting "education in the usual sense" (Walcott Papers). Butler headed the Washington Memorial Institution in these early days. By 1903 it was fading; Walcott thought that Columbian University in Washington would take up the function by becoming purely a graduate organization. But although Columbian changed its name to George Washington University, nothing came of this. Walcott to Billings, July 13, 1903, Billings Papers. As to the Memorial Association, almost down to World

War I it persisted in trying to get a memorial hall constructed suitable for a national university.

20. Dael Wolfle, *The Home of Science* (New York: McGraw-Hill, 1972).

21. *Yearbook,* 1902, p. xiii.

22. Billings' copy is in Box A of the Billings Papers.

23. Welch to Billings, Jan. 16, 1902, Billings Papers.

24. Remsen to Billings, Jan. 24, 1902, Billings Papers.

25. Woodward to Butler, Nov. 23, 1905, Butler Correspondence, Columbia University Archives.

26. Billings to Walcott, Dec. 10, 1902, CIW.

27. Walcott's memo, dated May 28, 1902, is in the Billings Papers.

28. Trustees' Minutes, Dec. 8, 1903, pp. 167–168, 179–180, 216. The Carnegie quote is from the last.

29. T. W. Richards to Woodward, Mar. 28, 1905, Richards Papers, Harvard Archives.

30. Trustees' Minutes, Dec. 12, 1905, pp. 470–471.

31. Ibid., Dec. 20, 1905, pp. 445, 452, 469–471, 474, 486.

32. Ibid. Mitchell was referring to the stress on the practical application of Burbank's work. But Woodward, who later usually noted that practicality should never be raised in considering the best research, also noted at that meeting the possible application of the work in geophysics.

33. Shull's report, dated July 26, 1906, is in the Shull Papers, Library of the American Philosophical Society.

34. Trustees' Minutes, Dec. 15, 1911, p. 210.

35. Ibid., Dec. 8, 1908, pp. 736, 754–756, 761–763.

36. Ibid. Carnegie quotation appears in p. 761.

37. Ibid., Dec. 14, 1909, pp. 25–29.

38. The first is on p. 29; the second is on p. 6 of the respective *Yearbooks.*

39. Woodward to Chamberlin, July 10, 1920, Chamberlin Papers.

40. Trustees' Minutes, Dec. 11, 1906, pp. 581–582. A statement of Woodward's views is in his letter to George Ellery Hale, Dec. 12, 1906, Hale Papers, California Institute of Technology Archives; see also Hale's reply of Dec. 20, 1906. For an exchange on minor grants, see Woodward to Cattell, Mar. 6, 1907, and Cattell to Woodward, Mar. 9, 1907, Cattell Papers, Library of Congress.

41. Trustees' Minutes, Dec. 11, 1906, pp. 583–584.

42. Ibid., Dec. 15, 1906, pp. 611–612.

43. Woodward to Richards, Mar. 22, 1918, and Richards to Woodward, Apr. 6, 1918, Richards Papers.

44. This discussion, including a restatement of Woodward's "theory of research," is in the Trustees' Minutes of Dec. 13, 1918, pp. 759, 760–761.

45. Both memos are in the Billings Papers.

46. Trustees' Minutes, Dec. 8, 1902, pp. 192–195.

47. Lewis Boss to Hale, Aug. 27, 1902, Hale Papers.

48. In addition to the references in the Trustees' Minutes to Woods Hole, there is a very interesting run of letters of E. B. Wilson to Billings in the Billings Papers, as well as a scattering of related letters in the same collection.

49. Walcott to Cattell, Sept. 17, 1902, and Chamberlin to Cattell, Sept. 10, 1902, Cattell Papers, Library of Congress.

50. Trustees' Minutes, Nov. 25, 1902, pp. 63–68, 86–106, 113–119, 127–

128. Before the organization of the Geophysical Laboratory in 1906, CIW supported the researches of the Geological Survey's George F. Becker (an old school chum of Henry Adams and C. S. Peirce) and Arthur L. Day. Under the latter's directorship, the Geophysical Laboratory had a notable record in research.

51. E. P. Rosa to A. G. Webster, Mar. 12, 1906, Webster Papers, Archives, University of Illinois (Urbana).

52. Richards to Higginson, Dec. 1 and 3, 1904, Richards Papers.

53. Cattell to Hale, Apr. 15, 1905; Hale to Cattell, May 2, 1905, Hale Papers.

54. Trustees' Minutes, Dec. 12, 1905, pp. 471–473.

55. Woodward to Cattell, Feb. 2, 1906, Cattell Papers.

56. Woodward to Billings, Oct. 18, 1907, Billings Papers.

57. "The Progress of Science," *Science* [n.s.] 14 (1901):305–315. The reference to anthropology is on p. 312.

58. Woodward to Cattell, Jan. 24, 1906, Cattell Papers.

59. Parsons (1859–1932) was an eminent civil engineer. Woodward apparently regarded the proposed archaeological program as too much like a conventional museum program. In modern terms, he leaned to a more "behavioral" approach. Trustees' Minutes, Dec. 15, 1911, pp. 206–207, 250–255; Dec. 13, 1911, pp. 301–305, 333–339.

60. From 1885 to 1905 Wright (1840–1909) headed the Bureau of Labor, the predecessor of the U.S. Department of Labor.

61. Wright to Billings, Feb. 26, 1904, Billings Papers. Billings was concerned that the products have more than data and be analytical.

62. Woodward to Billings, Apr. 15, 1909, and Feb. 14, 1910, Billings Papers.

63. For example, see Woodward to E. R. A. Seligman, April 28, 1908, Seligman Papers, Columbia University Library, with its discussion of aid to economics in the context of Woodward's opposition to minor grants and aid to universities.

64. Trustees' Minutes, Dec. 14, 1909, pp. 16–18; Dec. 15, 1911, pp. 202–203.

65. Ibid., Dec. 11, 1916, pp. 540–541, 666.

66. For example, Emory Johnson et al., *History of Domestic and Foreign Commerce of the United States,* 2 vols., Publ. No. 215A (Washington, D.C.: Carnegie Institution, 1916); Victor S. Clark, *History of Manufactures in the United States,* 3 vols., Publ. No. 215B (Washington, D.C.: Carnegie Institution, 1929); Balthasor Meyer et al., *History of Transportation in the United States,* Publ. No. 215C (Washington, D.C.: Carnegie Institution, 1917); P. W. Bidwell and J. I. Falconer, *History of Agriculture in the Northern United States to 1860,* Publ. No. 358, ser. 2 (Washington, D.C.: Carnegie Institution, 1933).

67. Walcott to Billings, July 13, 1903, Billings Papers; Trustees' Minutes, Dec. 14, 1909, pp. 25–26.

68. Woodward to Welch, Dec. 2, 1911, CIW.

69. Trustees' Minutes, Dec. 11, 1916, p. 542.

70. For example, only recently have historians picked up the work done in the 1920s on cases involving slavery and on the African slave trade.

71. Trustees' Minutes, Dec. 13, 1904, pp. 452–459.

72. Ibid., Dec. 8, 1908, pp. 736–737.

73. Ibid., Dec. 13, 1912, pp. 336–337.

74. Ibid., Dec. 14, 1909, pp. 29–32.

75. Gilman to Woodward, Nov. 16, 1906, CIW.

76. Referred to in the *Yearbook*, 1917, pp. 19–20. The correspondence apparently no longer exists.

77. "The Progress of Science," *Science* [n.s.] 14 (1901):305–315.

78. "Education and the World's Work of Today," *Science* [n.s.] 18 (1903): 161–169. The quotation is from p. 163.

79. Woodward to C. L. Jackson, June 29, 1906, Richards Papers.

80. Trustees' Minutes, Dec. 15, 1916, pp. 544, 608–609, 666–675.

81. This is from a discussion of the "claims of humanists" in the 1917 *Yearbook*, pp. 16–21.

82. Russell Henry Chittenden, *The Development of Physiological Chemistry in the United States* (New York: Chemical Catalog, 1930).

83. The text of the trust deed and the original articles of incorporation appear both in the *Yearbook* for 1902 and the minutes of the first meeting of the Trustees on Jan. 29 and 30, 1902.

84. Pritchett to Carnegie, December 13, 1901, Carnegie Papers, Library of Congress.

85. Madsen's article ("Daniel Coit Gilman") is quite good on this point.

86. Walcott to Billings, May 26, 1905, Billings Papers.

87. A convenient summary of this view is in Woodward's address, "The Needs of Research," *Science* [n.s.] 40 (1914):217–229. See the *Yearbooks* 1912, pp. 11–12; 1914, pp. 16–17; 1915, pp. 11–17; 1917, pp. 21–26.

88. Woodward to Richards, Mar. 30, 1906, Richards Papers.

89. Richards to Woodward, Apr. 2, 1906, Richards Papers.

90 *Yearbook*, 1912, pp. 11–12.

91. This is discussed in Loren R. Graham, "The Formation of Soviet Research Institutes: A Combination of Revolutionary Innovation and International Borrowing," *Social Studies of Science* 5 (1975):303–329, esp. pp. 303–306.

11. The Case of the Disappearing Laboratory

I had occasion, early in the 1970s, to visit the warehouse in New York City then housing the archives of the Rockefeller Foundation. Proud of his holdings, the archivist in charge challenged me to ask for specific documentation. Remembering my reading of the *Autobiography of Robert A. Millikan* shortly after I came to Washington, I asked for the 1918 correspondence leading to the establishment of the National Research Council Fellowships. On opening the folder he brought, I realized I had stumbled onto something significantly different from Millikan's account. I did not then realize what complex, extensive research awaited me in order to understand just how different the past had actually been.

The Disappearing Laboratory paper is an account of successive choices between concepts of science and the scientific life, between ideological preferences for particular kinds of institutions over other kinds—the university vs. the government bureau and/or the independent research institution, and the private university vs. the state university. The story also hinged on a conscious preference of some in the scientific community for replicating themselves even over the increase in research output. An unexpected ideological strand was the preference for downgrading the role of medicine in retrospective accounts in favor of a none too subtle exaltation of pure physics and chemistry.

A favorite historical subject of mine, the National Academy of Science, provided the stage setting within which all these possible choices were acted out. The nature of the Academy itself entered into the competition between the alternatives. Looked at in perspective, the events of 1918/19 had resonances with both the world of Joseph Henry and Alexander Dallas Bache and the world of Vannevar Bush and his successors as the Pooh-Bahs of federal science.

This paper was originally published in the Spring 1977 issue of *American Quarterly,* copyright 1977, the American Studies Association.

Robert A. Millikan remembered it as a great event.[1] As recounted in his *Autobiography* (1950), a letter arrived in February 1918 from George E. Vincent, the president of the Rockefeller Foundation, and from this came the National Research Council Fellowships, so influential in developing physics and chemistry in America. From an important participant, Millikan's recollections became the standard account—the official mythology—a half-truth around which all could rally.[2]

A "suggestion" was made to the Foundation, Vincent wrote, "that a research institution to deal with physics and chemistry ought to be independently established and endowed." Next follows a syllogism in support of the suggestion:

1. Industrial competition after the war will require greater scientific efficiency.
2. Undoubtedly great industrial establishments will increase their research activities. It is true, however, that practical research seeking immediate returns tends in time to defeat itself.
3. An institution, therefore, devoted to pure research, unhampered by obligations to teach and uninfluenced by commercial consideration, is needed for leadership in American progress in the physical sciences.

Then Vincent has a paragraph of questions. Is such a new research center justified? Is there another device for organizing science "for national service"? Will the National Research Council survive the war? What about the possibility of the federal government creating such a research body by analogy with activities in the Geological Survey and in the Department of Agriculture? Could the Bureau of Standards become "a national research institution"?

Millikan recounts how he showed the letter to the astrophysicist George Ellery Hale[3] before taking soundings. Later Simon Flexner[4] enters the office, announces that he—the head of the independently endowed Rockefeller Institute for Medical Research—had promoted the idea among his colleagues in the Rockefeller philanthropies for the progress both of medicine and the two disciplines. Millikan next relates how he convened a caucus of sixteen scientists who voted nine to seven against an independent institute for physics and chemistry. The only point unanimously agreed upon was the desirability of the fellowships which came into existence in 1919 and continued to 1954. Replying to Vincent on February 18, 1918, Millikan argued against centralization and for a number of regionally distributed research centers at universities. Fellowships did not loom large in the reply.

Millikan's account is incomplete and misleading in three important respects. First, the Council presented two applications to the Foundation, both containing provisions for research institutes for physics and chemistry not connected with universities.[5] The earlier proposal was rejected. Second, while Millikan may be truthful in saying the sixteen scientists unanimously favored fellowships, it is not true that he initially intended them to be modeled on the British 1851 Exhibition Scholarships.[6] Until 1919, a different aim persisted. Millikan asserts that the fellowships were to aid domestic institutions, but they were established primarily for the benefit of promising young scientists. Nineteen percent actually studied outside the United States, a practice starting with the first awards. Yet, Millikan's version reflects important American values. In retrospect, Millikan is fairly characterized, I think, as a Babbit who became a Nobel Laureate in Physics.[7]

This hotly contested decision about a proposed research institute can explain much about the thinking of a small, influential group. To do so requires consideration of a number of topics Millikan passed over lightly, if he mentioned them at all. Why was the laboratory intended for physics and chemistry only, and why was it ultimately rejected? And why, when the fellowships came, were they initially restricted to these two fields?[8] What influenced research-minded individuals to support fellowships rather than a research institute? Flexner's institute was a success; even Millikan replied in his letter to Vincent that a body like the Kaiser Wilhelm Institute was "the most direct and obvious solution" to the need for better research. How did government science figure in the process? The role of the universities is at issue, as are the conceptions of science and research in the minds of influential individuals like Hale. How did the Rockefeller foundation come to this tempting offer of a great research institute, and why did it ultimately accept another mechanism for aiding physics and chemistry? Millikan knew the decision was close in Washington. He did not know that the letter he received was just one of five identical letters sent by Vincent. Only John Zeleny of Yale joined Millikan in opposition. Even he, however, took the institute as a foregone conclusion and suggested desirable modifications. A. A. Michelson and the chemist Julius Stieglitz, both from Millikan's own university, Chicago, supported the idea, as did the chemist Alexander Smith of Columbia.[9] Nor is this result surprising. There were two prior proposals to advance the physical sciences in the United States by establishing a research institute.[10]

George Ellery Hale, not Millikan, was the principal actor in this boardroom drama. He was fertile with ideas and ambitions for science in America. If Millikan was a "Nobel Babitt," Hale was the J. P. Morgan of the scientific community. The setting for the demise of the idea of an institute of physics and chemistry was his creation, the National Research Council of the National Academy of Sciences. How and why

Hale formed or did not form intellectual pools, trusts, and conglomerates is very much to the point. Hale's ambitions for the Academy, his wartime experiences, and his postwar plans provide the context for considering how he and others perceived the scientific community, the government, and the universities. And these perceptions determined the outcome of the negotiations with Vincent.

Almost from the year Hale joined the National Academy of Sciences (1902), its revival had engaged his ingenuity. A nationally chartered honorific body, the Academy was in decay.[11] Yet no other organization in the United States was comparable to the great academies of Europe, the Royal Society and its analogues in Paris, St. Petersburg, Berlin, and elsewhere. Europe and the history of its science loomed large in Hale's mind.

Hale culminated his public campaign for reform of the Academy in November 1913 with a paper on its future.[12] The text was sent to each member asking for comments. Before its appearance in *Science,* Hale reviewed the replies and made revisions. The original survives, as do most of the comments. Taken together the letters of the academicians say much about the viewpoints of a key group of men.[13] Analysis of Hale's ideas and the responses will provide a backdrop to later events. Of 130 members, 75 responded and are tallied.[14] An older group, the 75 do not match either the population of the scientific community or the most active research areas. Only two scientists, the astronomer Edward C. Pickering and the paleontologist Henry Fairfield Osborn, are rated as in complete disagreement,[15] with 15 in total agreement. The remaining 58 have some qualms on specific points in Hale's program. Average affirmation totaled 27 percent; average negative stand, 9 percent; and 64 percent of the academicians were rated as neutral. Unless neutrality meant assent, this was no landslide for the Hale program.[16]

Five of Hale's points merit discussion: specialization, popularization, laboratories, government relations, and the comparative status of fields. The dominant contemporary role of specialized professional societies, universities, and even local scientific societies bothered Hale. Soothing words about cooperation by no means masked his desire to supplant these organizations. Pickering thought most of Hale's suggestions fitted local societies better than the National Academy. S. J. Meltzer of the Rockefeller Institute dismissed the idea of the Academy gaining any influence on research. Recent progress was due to universities, research institutes, and "the awakened spirit in the intellectual set of the younger men." Academies were, and should remain, honorific bodies. Nothing said by Meltzer and other critics was new or even alien to Hale's own thinking. But Hale favored the revival of the National Academy because he opposed specialization's harmful consequences. Specialization deprived scientists of a view of the "unity of knowledge" leading to perceptions of "large relationships." "Intermediate unexplored territory" was overlooked. The Academy could correct this and also encourage interdisciplinary research. Hale's vision of a science directed by great academies stressed cooperation within and between fields. Astrophysics provided an excellent example.

Research efficiency alone would not have sparked an attack on specialization. "Unity of knowledge" to Hale was not merely instrumental. Earlier in 1913, writing to Elihu Root, he cited Alexander von Humboldt's *Kosmos* as a model instructive to both scientists and humanists. The latter rarely saw science as cultural, only as a confusing complexity of details. But Hale asserted that science's "cultural value is no wise inferior to that of the humanities," by which he meant it was integrative and not merely practical. By courses showing the evolution of each field and the "meaning of evolution in every department of human activity" such instruction would help the nation: "The United States must play a large part in the politics of the world, and take a deeper interest in foreign affairs. I believe such instruction would be a useful means to this end." This message, conveyed earlier to Elihu Root, was repeated in November 1913.[17]

The evolutionary college course—by Humboldt out of Darwin and Spencer—was for Hale an "entering wedge" to reach the nonscientist. Particularly, Hale wanted to convince manufacturers of the value of science for industry. They presumably were in the humanities, not the sciences. The published speech cited the German example: writing to Andrew Carnegie in May 1914, Hale pointed to depletion of resources as the reason for the need to emulate the Reich.[18] Hale knew the matter was sensitive. In the printed version a reference to Michael Faraday, the great British scientist, appeared immediately thereafter; Faraday was misused as a paragon of purity. Hale also wrote a new passage about keeping industrial needs and basic science in proper perspective. Faraday's attitude toward applied science was different from the position of Hale and others. Faraday contributed to the development of physical theory and had no qualms about involvement in applied work. Nor did his younger contemporary, Lord Kelvin. But the belief in the primacy of pure research with an implicit downgrading of applied work became a matter of ideology among some scientists in the latter half of the nineteenth century.

Mindful of the needs of nonscientists, Hale carefully proposed methods of disseminating the latest in science to them. A new Academy building would house scientific exhibits, both current and historical. From its lecture hall, leading scientists would address the public. Hale proposed, in fact, to endow a series of lectures in honor of his father with evolution as its unifying theme. Abstracts of new research results given in Academy sessions and in the *Proceedings* would go to magazines and newspapers. But Hale's purposes were far from popularization. He was aiming for a very elite audience—indeed, one capable of rewarding the Academy "ultimately in larger endowments for research." To be fair, Hale thought he was dealing with "the average American citizen" who was "well-acquainted with the name of the Paris Academy though press reports" but knew not the National Academy. News of discoveries of the National Academy would similarly reach this average citizen through abstracts given to the Associated Press and "also through certain conservative newspapers and magazines."

Michael Faraday counted greatly in Hale's thinking. In Hale's correspondence in February and March 1913 with Root and Henry Smith Pritchett, Faraday's precedents were carefully cited. A building, more members, and a splendid journal were not enough to revivify the National Academy. Pritchett, of the Carnegie Foundation for the Advancement of Teaching, pointed to Faraday as a great investigator and a leader of the scientists. Someone had to live in Washington and gain public recognition of the Academy's leadership. Picking up the theme, Hale wrote to Root of Faraday's lectures and of the example of the Royal Institution. A Home Secretary is needed, "a man who would be as nearly as possible a counterpart of Faraday," a good lecturer and a good investigator. "To attract and hold an investigator, the principal essential is a well-equipped laboratory." Hale proposed two in the projected building, one for a biologist and one for a physicist.

He amplified this idea in the talk sent the membership and later in the printed article. Here Hale gave a building plan, stressing the laboratories and calling attention to research conducted by other national academies.[19] Only six replies backed Hale on the laboratories; sixteen opposed him.[20] Three arguments were offered in rebuttal. One group of academicians objected to the Academy competing with existing organizations: "I do not approve of the Academy undertaking the functions of a great university." Many objected to the Academy pursuing a course favoring one or two fields over others. Some noted this disapprovingly as a form of controlling or directing the conduct of science. Still others favored raising endowments to support a grant program. Reviewing the responses at the Academy's business meeting on April 21, 1914, Hale cited reaction to the laboratory proposal as the "most important divergence." Reiterating the Faraday theme, he regarded the laboratories as "Perhaps the chief point of view in the scheme."

In preparation for his 1913/14 reform move, Hale had asked a number of academicians to compare current American and European research in their fields. All but one replied that American work equaled or surpassed the best in Europe. But the respondents were from the biological and earth sciences. The one naysayer, the Johns Hopkins physicist J. S. Ames, described Americans as good in exact measurements but contributing little to theory or new discovery in physics.[21]

Hale's 1913 talk noted parity in certain unnamed fields. But the printed text only notes American contributions to astronomy. To justify the increase in members of the Academy, Hale stated in print that American "investigators of ability" were more numerous than in Britain. To Carnegie, in private, Hale noted the real strength of many fields; his public stance had to be otherwise. Some literature upheld a belief in American indifference to basic research, a special case of the conviction about American anti-intellectualism.[22] If a significant number of scientific fields were on a par with the best in Europe, crying poor would not work as it had in the past. More important, such equality implies that lagging fields could and might improve by the very processes responsible for the productivity of the flourishing specialties. Because he had a fresh program involving the Academy with a specific vision of how science functioned as well as a nationalistic role, Hale could not publicly avow what his correspondents asserted.

In view of the statutory role of the Academy as a possible adviser of the federal government, Hale's original text of 1913 was notably silent on relations with the Congress and the Executive Branch. Very few of the academicians questioned such relations. Pickering wanted to use the endowment for lobbying. The Yale chemists L. B. Mendel and T. B. Osborn thought the government would help when the Academy won a place in the scientific world. Hale's published version reversed their argument. When the Academy has developed its "standing and prestige, then it will influence and help the government."[23] Hale saw science almost as an autonomous or equal body. Good science was not to be subordinated to or dependent on the politics of the American republic. Hale (and Millikan) were conservative, narrowly circumscribing government's role. Real science, the best and basic science, was largely for the private sector.

For the governance of the scientific community, Hale was promoting an academy model. A self-perpetuating elite of elites, the National Academy's stature and effectiveness were to arise from the achievements of the members. Collectively, the Academy was to exemplify research and to symbolize the greatness of the scientific tradition of western civilization. A "Faraday," after all, was to serve as Home Secretary. By setting standards, the Academy could encourage and reward the promising young. Research eminence of the national scientific community would enable the Academy to deal with similar national bodies on equal terms. Domestically, a revitalized Academy could represent science as an autonomous force in negotiations with government, industry, and various professional groups. And it could negotiate as a peer with these other segments of society.

After World War I started, Hale had no doubts about his position. He was for the Allies and eager to have the Academy demonstrate what science could accomplish for national preparedness. As early as 1915, he had unsuccessfully urged the Academy to reform. Even more serious from Hale's viewpoint were the moves of rival groups in 1915 for national power and position. The Naval Consulting Board,[24] headed by Thomas Alva Edison, was seen initially as eliminating part of Hale's war program. Although engineers and scientists served with Edison, the initial purpose was to stimulate inventors, not to mount a research effort. If inventors could meet the needs of the armed services, the Academy would have no significant role in the war. Hale contemplated starting his program, therefore, in the areas of medicine and surgery where any aid was valuable and "no such thing as failure" existed.[25]

Potentially a greater threat were the activities of James McKean Cattell, Pickering, and their Committee of One Hundred. Sharing some of Hale's points, the Committee was both within the scientific community and more widely based than the Academy. In April 1913, when Hale was formulating the program presented that November, Cattell rejected the notion of the National Academy of Sciences as a fourth branch of the government: "A self-perpetuating corporation electing its limited membership for life on the grounds of scientific eminence cannot be such a body." The Academy was old, conservative, timid, and inefficient. Instead, Cattell proposed a representative academy with delegates from both the specialized scientific societies and the local societies.[26] That was unacceptable to Hale. By 1915, Cattell had settled on the American Association for the Advancement of Science as his chosen instrument. The elaborate subcommittee structure of the Committee of One Hundred represented not only the older, established disciplines but also the social sciences, an array of biomedical specialties, engineering, and even some nonscientific fields. The Committee had one serious weakness: it was a pressure group, not an administrative body. Later, Hale co-opted its stronger subcommittees into his war program. While the Committee of One Hundred remained alive on paper, an attempt to revive it after World War I failed. Hale's creation, the National Research Council (NRC), dominated the area. But Pickering and Cattell had stirred the interest of the Rockefeller philanthropies.

Before the Vincent letter, none of the Rockefeller philanthropies had undertaken to support general scientific research.[27] Areas of concern were conventional—aid to health, education, and social welfare. But with so large a scale, even conventional charitable activities tended to new shapes and directions, and the trustees were often responsive to innovation.[28] The earliest discussions on broad support of scientific research involved Jerome D. Greene, Secretary of the Foundation, and Pickering. Despite the differing title, Greene was George Vincent's predecessor as executive head of the foundation since its inception in 1913. Before coming to the Rockefeller Institute in 1910 as its general manager, Greene had been secretary to President Eliot and, later, the Harvard Corporation. In 1912 Green became an adviser to the elder

Rockefeller.[29] In 1913 Pickering was nearing the end of a notable career as director of the Harvard College Observatory when he broached the matter of general support for scientific research. For decades he had worked to raise money both for the Observatory and for research in all fields. Pickering's field and his promotional efforts were parallel to Hale's, allowing for generational differences. Success came in aiding the Observatory; despite ingenuity and effort, Pickering's other fundraising efforts yielded only modest results. Eventually, Hale's exertions surpassed Pickering's.

When Pickering in 1913 tentatively raised the possibility of aiding "men of genius . . . isolated" with "no other means of carrying on his [sic] work," Green declined, stressing the concentration on public health and social welfare.[30] In 1915 Pickering, now allied with the Committee of One Hundred on Scientific Research,[31] appealed to the Rockefeller Foundation as perhaps the last hope for support. Earlier, the Carnegie Institution was the expected source of grants.[32] By 1915, the Institution was concentrating on its own research installations. But isolated in the universities were men of talent requiring but small amounts, less than $500 in fact, to produce new knowledge. A single appropriation ranging from $20,000 to $50,000 would suffice. And the Committee, with its experience, could wisely disburse the funds with due regard for fiscal prudence.

Greene favored support for increased endowments. Both the Carnegie Institution and the Rockefeller Institute initially had a program of research grants. Both felt the results were meager, especially in light of administrative problems. Both then turned to support of in-house research.[33] More important, the Foundation was not yet committed to the support of research.[34] Pickering's application in 1916 stirred Greene to state his position and to present this subject to the Foundation's trustees.[35]

In nearly every branch of philanthropic effort . . . the doubt is bound to rise now and then as to whether such artificial nourishment as may be given by a rich Foundation is likely to be of permanent and unqualified good. It will certainly not be so if it fails to draw forth from the community its own effort and its own financial resources to an extent immeasurably far in excess of the money received from the bounty of a single institution or individual. . . . The danger of diverting a great Foundation from its true function of experiment, discovery, initiative and demonstration, to being a mere bag of money, has been guarded against with unremitting vigilance, a vigilance, the necessity for which has been proved by daily experience.

There is one kind of philanthropy which is not open to this particular role, although it has difficulties and dangers of its own, namely, the extension of the bounds of human knowledge. The discovery of new facts and new laws covering the world in which we live, biological, physical, economic, is in itself a service which may be called thoroughly good without drawback or qualification of any kind; and experience has shown that all branches of scientific knowledge are worthy of equal respect, so numerous and unpredictable are the interrelations of knowledge in the various branches. A good example of this truth is found in the enormous usefulness of the science of entomology to medicine and agriculture, a usefulness which could never have been attained had not the characteristics and habits of various bugs interested for generations men whose sole incentive to study was their love of nature and truth without any practical utilization.

At Greene's initiative, the Committee's request was set aside pending a report from a committee of the Board of Trustees. But after Greene's resignation on September 1, 1916, the appointment of the committee was indefinitely postponed.[36]

In 1916 Hale wrote: "I am planning a National Service Research Foundation to tie together research in universities, Government Bureaus, manufacturing establishments, medical schools, etc."[37] Formally called into existence by a presidential order, the National Research Council was the operating arm of the National Academy. On the Council sat delegates from universities, government bureaus, industry, and professional societies. Hardly the representative academy proposed by Cattell, the Council presented a new model for the governance of the scientific community—a corporate model, more particularly a vertical trust. In the cause of efficiency, diverse interests, not merely basic science, met at the Council to avoid wasteful competition and to promote production and cooperation. By setting and enforcing standards, marginal research efforts were shunted aside. It was a traditional American voluntarism, powered by moral suasion and braced by the muscle of the state.

In the summer before his resignation, Greene had asked Willis R. Whitney of the General Electric Research Laboratory if the National Academy of Sciences was capable of coordinating scientific research. Aware of Hale's efforts, Whitney endorsed the Academy.[38] At the same period, Hale's stirrings were reaching the attention of the General Education Board's energetic Abraham Flexner, Simon Flexner's brother, who responded to Hale's call for more research.[39]

While the Rockefeller group pondered its role, Hale and his associates moved to be useful to the armed services. Only one element was lacking, a full-time head of operations for the Council. As the academy model was not dead, Hale looked for and found his "Faraday."[40] And, on February 16, 1917, Robert A. Millikan arrived in Washington, eventually joining the Signal Corps. Millikan was a great success as the administrator of the World War I scientific mobilization. Apparently, only with the arrival of George Vincent's letter did Hale realize Millikan wanted neither the academy model nor the corporate model but something different—a university model.

During the months Hale and Millikan were developing the National Research Council the Rockefeller philanthropies remained interested. On December 5, 1917, this concern became explicit in President Vincent's "Forecast of Policy." At the end of the document he wrote:[41]

National efficiency will be tested by intense competition after the war. The Foundation's Health Work will be an important contribution. It has been suggested that an *Institute of Physics and Chemistry* might be made a source of enormous National value. Modern industry depends upon the development of the physical and chemical sciences. Merely "practical" research tends to defeat itself. The cultivation of pure science is essential to the best practical results. The officers are planning to investigate this proposal with the utmost care.

Writing in December to Simon Flexner, the source of the suggestion, Vincent proposed a discussion of the new institute in the new year.[42]

Oblivious to the preference of his chosen instrument and unaware of the intentions of the Rockefeller Foundation, Hale presented a postwar plan to the meeting of the National Academy's Council on December 19, 1917.[43] Like the 1913/14 proposal, Hale wanted Carnegie funds for a building and endowment.[44] The arguments offered in 1917 were similar to the earlier presentation. A "Michael Faraday" was needed, a permanent representative of science in the District of Columbia. Such a man—and Hale stopped just short of naming Millikan—could inspire and coordinate investigations within and without the government. Hale was against taking a "Faraday" away from research. Therefore, the building should not display instruments like a museum but should show the current conduct of basic science. Laboratories were essential.

The presentation to the Academy's Council had two features different from the 1913/14 text. Biology no longer rated a laboratory in the projected building. Like Flexner and Vincent, Hale wanted facilities only for

The organization of researches in physics and chemistry, the fundamental sciences underlying all others, for the advancement of knowledge and the discovery of new principles applicable to the arts. . . . Their cultivation has been left almost entirely to the universities where funds for equipment and for the payment of assistants are limited or altogether lacking, and where the entire burden of research falls upon men whose energies are sufficiently taxed by their work of instruction.

Given this position, Hale could agree to Vincent's suggestion of an institute and accede to Millikan's thrust for decentralization. Understandably, he initially held out against placing the laboratories at universities. By the 1917 presentation Hale could safely drop biology and push solely for physics and chemistry, since biology was, after all, in good shape. War experiences had brushed aside all of the previous concerns. While the mobilization program used talents from many fields, physicists and chemists were strategically located. Almost by definition certain problems, such as submarine detection and gas warfare, were physical and chemical. The wartime environment seemingly confirmed the positivistic belief in the basic nature of the two fields.

The second novelty was the recognition of a possible federal role. Even in wartime Hale maintained certain fictions about the autonomy of NRC. Whatever power it exercised, however, stemmed from the government. The NRC was the department of research of the Council of National Defense. Personnel, funds, and programs shuttled between NRC and the War and Navy Departments. The scientific community had more power and prestige than ever before. Hale wanted this state preserved without subordinating the Academy or any principal part of the research world to governmental control. The postwar plan adroitly sidestepped the most visible threat—the ambitions of S. W. Stratton, director of the Bureau of Standards. Neglecting the possibilities of growth, Hale smoothly described the Bureau as limited in scope; its director was already involved in the NRC, which should supplement the Bureau.[45]

Vincent's letter had asked about Standards and the role of federal agencies. As Jerome Greene put it, being a "mere money bag" was not enough; the community's "own efforts and its own financial resources" had to come forth. That meant proper government action, and no foundation or academy could assume anything about the continuing propriety with which the state exercised its coercive power. In 1917 the Foundation, nevertheless, retained its faith in government by experts. Vincent's "Forecast of Policy" calmly assumed the coming of a federal Department of Health, exactly two weeks before Hale ruled out Stratton's Bureau.

But not one of the replies to Vincent's letter of February 5, 1918, favored any governmental role. Federal science meant practical, limited-scope research. Millikan placed his opposition on a higher plane: "An institution which is free from all political influences is better adapted to the stimulation of research in pure science." Stieglitz elaborated Millikan's point: "In the Government departments mentioned problems are to a certain extent limited by the relations of the government to the people and to specific work for the people." Strong doubts existed of the ability of physicists and chemists to persuade the people to support good research.[46]

At the time of the arrival of Vincent's letter, somehow Hale had to ensure the dominance of the Academy, to avoid entanglement with government, to promote the development of physics and chemistry, and to downplay the universities. While the elevation of Throop Institute into the California Institute of Technology was very much in his mind, Hale still thought of an academy model in these months of early 1918.

Undoubtedly the biggest concern at that time was governmental action which even threatened to impinge on higher education.

For example, visiting Millikan in February 1918, perhaps right after the reply to Vincent, Whitney argued with him abut how to advance research in physics and chemistry. But Millikan undoubtedly agreed when Whitney consequently wrote of the two fields as the "great bottleneck." For the sake of physics and chemistry, Whitney wanted to improve the conditions at universities. Citing the achievements of German universities, Whitney called attention to Germany's "broad system of federal aid to scientific education."[47] Earlier, Hale and Millikan had supported the Newlands Bill to form engineering research stations in the states. It stalled in Congress; a competing bill, the Smith–Howard, downgraded NRC's role.[48] But by February 1918 Millikan and Hale had soured on government aid to state universities for applied science. Apparently, that caused the dispute between Millikan and Whitney. The later proposals to the Rockefeller Foundation for laboratories at universities excluded state universities.[49]

Vincent's offer of an institute for physics and chemistry had stimulated Hale. As late as February 9, 1918, four days after Vincent's letter, he wrote that the NRC Executive Committee favored "providing a small number of laboratories" for the research of men coming to Washington to work for the council. By the end of March, Hale believed Rockefeller funds would provide three branch laboratories for the National Research Council, not for the universities.[50] Before receiving Vincent's letter, Hale had approached the Rockefeller Foundation for support of the activities of NRC's Division of Medicine and Related Sciences. He needed $50,000 to coordinate American research in pure and applied sciences in both war and peace. (While the context was medical, the literal sense was quite unrestricted.) Supported by Flexner, Hale applied on February 19, 1918; the Foundation's trustees awarded the funds nine days later.[51] Writing to Vincent earlier, Hale had noted the availability of other funds, private and federal, for the nonmedical work. But there was uncertainty about postwar support. The Carnegie Corporation funds for the building and endowment were not in hand; nor was there assurance of continued governmental support when the war ended. On March 10, 1918, when Hale, Vincent, and Millikan discussed matters, Hale probably saw the institute of physics and chemistry in a new light. With Rockefeller support, NRC could also coordinate nonmedical research in America. The plan now was for the National Research Council to have three branch laboratories supported by Rockefeller funds. First, the laboratories would work on war problems; in peace they would do basic physics and chemistry,[52] presumably for the benefit of industry. As Hale noted, there was a "difference of opinion." Undoubtedly both Millikan and Vincent had reservations. Vincent wrote to Flexner of a "complicated but pressing problem."[53] Hale, as chairman of the National Research Council, and Millikan, as its executive officer, would compose the grant application.

Vincent knew of the division in the NRC ranks from Simon Flexner, one of the sixteen. Their original intention was strengthened by the majority support from their five correspondents. Flexner found the opposing arguments familiar but felt they could be taken into consideration. Hale was under pressure from the Rockefeller interests to consider an independent, central research institute of physics and chemistry, a possibility kept alive until the very launching of the fellowships. At the same time Millikan and his allies in NRC pushed for the founding of research laboratories at private universities.

Formidable opposition required compromise; Hale backed down adroitly but probably only accepted the university model later. To minimize the NRC aspect, the first (and unsuccessful) proposal had the Rockefeller Foundation awarding the fellowships and assigning the recipients to particular laboratories. Hale may have also backed down because of the belief that physics and chemistry differed from other fields. Hale and Millikan disagreed on the desirability of cooperative efforts in those two disciplines. To counter the one institute idea, the proposal distinguished between "large" experimental and observational institutions like those of the Carnegie Institution of Washington and laboratories in physics and chemistry generating new ideas. Hale was "not in the least concerned about some of the details . . . provided only that some such plan be carried out."[54]

The salient feature of the resulting June 13, 1918, proposal was not the fellowships but a call for regional research institutes for physics and chemistry connected with universities. A central independent institute unconnected with education was offered as an alternative. Such a body, it was argued, would not train new researchers for still other laboratories. Regional dispersal provided wider recognition and stimulation to research; smaller laboratories were likely to yield more research and, especially, new ideas, than one large institute. The regional laboratories were better investments than small research grants or additional fellowships. In addition, two categories of fellowships were proposed. By far the larger category ($75,000) was for assignment to the as yet unestablished regional research laboratories. Only $10,000 was asked for "traveling" fellowships, the final form of the program.[55]

Almost immediately, Flexner wrote he was "uncertain about the precise plan submitted." Hale reported that Flexner still wanted a central laboratory analogous to the Rockefeller Institute but felt Flexner could be converted to the first proposal or something similar. The two men went overseas in September to promote international cooperation in research. But Flexner remained unconverted. Despite initial optimism, the Council by early October tried unsuccessfully to withdraw the plan. The Foundation's deliberations went inexorably to a rejection on October 22, 1918.[56]

While Flexner stuck to his position, British advice changed Hale's attitude. J. J. Thomson, Ernest Rutherford, Joseph Larmor, and H. F. Newall all supported placing the institute or institutes at universities. While Larmor and Rutherford conceived of a national center in conjunction with local groups, all favored pluralism and a university environment.[57] It was impressive support, and necessary in light of doubts about the universities in many quarters of the scientific community.

In the National Research Council were documents indicating the poor equipment, heavy teaching loads, and general lack of support for research in universities.[58] Hale had expressed the viewpoint at the end of 1917. Carnegie Institution's reluctance to fund university science was well known. Replying to Vincent's letter, Smith, Michelson, and Stieglitz endorsed separating the Institute of Physics and Chemistry from pedagogic responsibilities. Stieglitz and later the chemist G. N. Lewis argued that a separate institute would stimulate the universities. In arguing for an independent research body, these individuals were not merely giving vent to antiuniversity feelings. They were reporting the lack of support for research in most institutions of higher learning; they recognized the teaching responsibility of colleges and universities.

But now, in early 1919, Hale and the Council's leaders started all over again. Hale's views had evolved under the pressure from Millikan and his cohorts, from the force of British advice, and perhaps with the future of Caltech in mind. The Foundation and the Council were exchanging comments and drafts. As Millikan wanted, the research laboratories were still at universities with advanced students assigned to the research professors at each, although Flexner objected it would "sterilize and render impotent the staff doing the regular teaching, besides putting too great a burden on the research professorships."[59]

On February 6, 1919, Hale sent Vincent a draft of the NRC plan. It provided for matching funds at a limited number of private universities to establish research professorships and laboratory facilities. Fellows were to be assigned to these laboratories. As the central laboratory idea was alive in the Foundation, Hale had to reassure Vincent on February 24, 1919, "I am not in the least opposed."

Two days later the Rockefeller Trustees learned of the National Research Council's ideas on two ways of supporting physics and chemistry. The first called for selecting five universities for support—to release men from other duties for research and for financing of buildings and equipment—and to create 50 research fellowships annually. NRC would administer both the fellowships and the aid to universities. The fellows were to go to the selected universities; a separate provision appeared for traveling fellowships only if the endowments fell through.

The second possibility was to create an endowed Institute of Research in Physics and Chemistry. Advantages and disadvantages of this possibility were noted—the former stressing increased research output, the latter underlining the possible injuries to the universities.[60] Authorized to seek a resolution of the differences, Vincent exchanged more views with the Council. On March 20 he approved a draft proposal, suggesting one new point. The amended text of March 22 was considered and approved by the Foundation's Executive Committee on April 9. Flexner was absent from the session. Now, only fellowship support was asked for, and the fellows were not restricted to working at a few laboratories.[61]

Underlying the compromise of the competing claims of the university and of the central research laboratory was a wide area of agreement between Hale and Vincent. While they recognized the anti-university points, counterarguments existed. Zeleny of Yale had argued that removing even four good organic chemists would ruin the universities. He urged the maintenance of the research ideal at the universities, both to provide fundamental knowledge and to yield a supply of new men. Whitney and those interested in applied research thought of the state universities. Unlike Stieglitz, opponents of central or independent laboratories saw them as stifling the universities. The Trustees noted on February 26 that the central laboratory would withdraw stimulating men from the university while "[d]isassociating research and teaching wholly, to the possible detriment of instruction in fundamental sciences."

That argument hit home. Many in the scientific community were interested not only in increasing research output but also in perpetuating the scientific community. Although like Millikan they favored a Kaiser Wilhelm Institute for the first goal, they willingly subordinated that to the second goal when faced with the choice. The university was both a home and an incubator of science. Replicating one's kind in a graduate school simultaneously advanced knowledge and assured perpetuation of the great tradition. Hale and his respondents in 1913/14 had agreed on seeking out the productive man; on giving awards, grants, and prizes to younger men not yet tenured; and on using such recognition to influence administrators and trustees.[62] Given his own inclinations, Hale could agree to the support for fellowships, especially as the Rockefeller Foundation was not quite ready to launch its aid to research installations.

But why would Vincent agree to maintaining the universities for replicating the scientific community? As a university man, he was undoubtedly sympathetic: earning his Ph.D. in sociology from Chicago in 1896, he served there as a dean from 1900 to 1911, and from 1911 to 1917 he was president of the University of Minnesota. More important, perhaps, is the medical strand throughout the events associated with the origins of the fellowship program. Flexner wanted basic physical science for medical progress. Greene supported pure research's worthiness because, among other instances, Charles Wardell Stiles showed how to combat hookworm. To launch NRC, Hale had used a medical gambit; to perpetuate NRC, he tried still another medical ploy. Men like Meltzer of the Rockefeller Institute show up as apostles of basic research. A clue is the one addition Vincent made in the final NRC proposal. He stipulated that all research results of the fellows become freely available in the open literature.[63] This was not a major point but probably underscored Vincent's uneasiness with Hale and Millikan's continuing emphasis on aiding industry. Trade secrets and proprietary information were in ill repute among the leaders of scientific medicine.

As president of the University of Minnesota, Vincent had pointed out the role of the universities in medicine, establishing postgraduate medical work there. Not only did the universities provide the fundamental scientific knowledge, but

the general atmosphere of investigation and professional teaching which characterizes the university environment was singularly favorable to the proper development of research and instruction in the medical sciences.

In the university community it was also easier to foster the altruistic ideals which ought to guide and inspire the medical profession. Important, too, was the fact that the university stands as a disinterested social agency, aiming primarily, not at offering special privileges to individuals, but at rendering service to the public. By creating standards and applying tests it tries to protect the community against incompetent and unscrupulous exploiters of public credulity and confidence.[64]

Vincent's words had an explicit medical context. They also echoed reformist views. Given this attitude towards universities and graduate education of physicians, Vincent could assent to an NRC position compatible with his organization's emphasis on public health.

In the *Autobiography,* Millikan credited Simon Flexner for much of the success of the fellowships—Flexner was chairman of the Fellowship Board for many years. Had he read Flexner's letter to Hale of September 18, 1919, perhaps Millikan might have written otherwise. Noting how well the Board was working, Flexner reminded Hale that both the Carnegie Institution of Washington and the Rockefeller Institute started out supporting individuals by grants. The central institute would come. Others continued to press for the idea. The notable Berkeley chemist G. N. Lewis urged Hale to move.[65] But Millikan was ever alert against basic research being placed outside the universities. When an independent institute for colloid chemistry was proposed in 1923, he wrote to the NRC recalling the vote on the Vincent letter.[66] The Foundation in the 1920s implicitly followed his idea of strengthening a few institutions.

The decision for a fellowship program, then, reinforced preexisting trends favoring a university model for science. In this model fundamental research occurred in university settings, performed by individuals or small groups. Independent nonprofit research bodies existed (e.g., the Rockefeller Institute and the Carnegie Institution). A few industrial research laboratories did basic research (e.g., General Electric, Bell Telephone). Here and there in government, enclaves of basic research persisted. But the university model returned science to the world of the Committee of One Hundred and Pickering's solitary investigator. It was a world minimizing entanglements with public policy.

Surviving the war with the dubious blessing of a presidential Executive Order, the NRC became a voluntary association of modest achievements. Hale's corporate model persisted in a curious and fateful manner; key individuals liked the idea of a body governing research and development, yet outside the workings of American politics. One example remained from the World War I period, the National Advisory Committee for Aeronautics, the precursor of the present space agency. Originating in small scale and outside of Hale's orbit, NACA was a mild irritant to Millikan and the NRC operations during World War I. It was an advising and coordinating committee which administered a modest research program. Like NRC, it brought together government agencies, industry, and the universities. Unlike NRC, a legislative enactment guaranteed NACA's survival. Shortly after World War II, NACA described itself as a research corporation outside of politics, an impartial source of data for defense and for industry. The Committee, as Board of Directors, represented all pertinent parties, the creators and users of the research.[67]

A young engineer in NRC's antisubmarine program, Vannevar Bush later came to Washington before World War II as president of the Carnegie Institution of Washington; he served briefly (1939–1941) as head of NACA. That position influenced his ideas about the organization of research.[68] Collegial bodies of experts from the producers and the users of research became a key feature in defense research. Above all, defense justified the great freedom of action of the research body, the Office of Scientific Research and Development. Peacetime plans, as far as possible, called for an equivalent autonomy for science. Even the National Academy was brought into an early bill for national defense research. And the proposed National Science Foundation was practically self-governing. President Truman's veto of the NSF legislation ended any illusion about the exemption of science from government.[69]

Hale was dead, but Millikan still lived during those post–World War II days. A querulous tone pervades his correspondence with an old pupil, Frank B. Jewett, the former head of Bell Telephone Laboratories, now president of the National Academy. The old order, Millikan's old order, was crumbling. Under the "influence of FDR's socialistic charade" prominent scientists like "Richards, Bronck, Bush, Conant, and Compton" were getting their snout into the federal trough for the benefit of basic research. Millikan preferred industrial support of science in the universities, despite industry's modest record of giving.[70] The changes bothering Millikan were far from random events but occurred in a pattern replete with symbolism and irony. In 1951 the Academy asked the Rockefeller Foundation for a renewal of the fellowship program. In the spirit of Jerome Greene, Warren Weaver declined, citing the coming of large-scale funds from the National Science Foundation.[71] While post–World War II science policy came into being, Robert A. Millikan remembered and wrote an autobiographic elegy to his days of power.

Notes

1. *Autobiography of Robert A. Millikan* (New York: Prentice-Hall, 1950), pp. 180–184. All subsequent references are to these pages.

2. See Myron Rand's early account, "The National Research Fellowships," *Scientific Monthly* 73 (1951):71–80. Later historical accounts are A. Hunter Dupree, *Science in the Federal Government* (Cambridge, Mass.: Harvard University Press, 1957), pp. 326–337; Daniel Kevles, "George Ellery Hale, the First World War, and the Advancement of Science in America," *Isis* 59 (1968): 427–437; Ronald C. Tobey, *The American Ideology of National Science, 1919–1930* (Pittsburgh: University of Pittsburgh Press, 1971), pp. 53–58.

3. For Hale, see Helen Wright, *Explorer of the Universe* (New York: Dutton, 1966), and H. Wright, J. N. Warnow, and C. Weiner, *The Legacy of George Ellery Hale* (Cambridge, Mass.: M.I.T. Press, 1972), as well as Kevles, "George Ellery Hale."

4. The best introduction to Simon Flexner is George W. Corner, *A History of the Rockefeller Institute . . .* (New York: Rockefeller University Press, 1964).

5. This is a simplification explained subsequently. The final version of the second proposal was solely on fellowships. Rockefeller Trustees had been given the option of an Institute right up to the formulation of the successful fellowship application.

6. These were established as a consequence of the Crystal Palace Exhibition. See R. M. MacLeod and E. K. Andrews, "Scientific Careers of 1851 Exhibition Scholars," *Nature* 218 (1968):1011–1016.

7. The best accounts of Millikan are Daniel Kevles's entry in the *Dictionary of Scientific Biography* and his essay, "Millikan: Spokesman for Science in the Twenties," *E & S: Engineering and Science* 32 (1969):17–22.

8. Medical fellowships began in 1922. Biological fellowships began the following year. Immediately afterward mathematics and astronomy were added to the physical science program.

9. Michelson already had a Nobel Prize; Millikan received his after the war. The three chemists were quite reputable, if not as eminent as Michelson and Millikan. The letters of reply to Vincent (Smith, Feb. 6, 1918; Michelson, Feb. 8, 1918; Steiglitz, Feb. 15, 1918; Millikan, Feb. 18, 1918; Zeleny, Feb. 20, 1918) are in Record Group 1.1, Ser. 200, Box 37, Folder 415, Rockefeller Foundation Archives.

10. Astronomer E. C. Pickering in 1877 called for a central research institute for physics, chemistry, and astronomy. American Association for the Advancement of Science, *Proceedings* 26 (1877):67–68. Prior to becoming its president, Robert S. Woodward advised the Carnegie Institution of Washington to found a research laboratory for physics. Carnegie Institution of Washington, *Yearbook,* 1902, pp. 12–24.

11. Dupree, *Science in the Federal Government.* The official account used by Hale is Frederick W. True, *A History of the First Half-Century of the National Academy of Sciences, 1863–1913* (Washington, D.C.: National Academy of Sciences, 1913).

12. "The Future of the National Academy of Sciences," *Science* 40 (1914): 907–919 and 41 (1915):12–22. Hale's articles in *Science* were republished later as *National Academies and the Progress of Research* (Lancaster, Pa.: New Era Printing Co., 1915). Further citations to the article are to pp. 94–167 of this volume.

13. In the file "Future of the National Academy of Sciences, 1913–1915" in Box 53, Hale Papers. All subsequent references to these documents are in this location unless otherwise cited.

14. Apparently by Arthur L. Day, Home Secretary of the Academy, who sent out the copies of Hale's address. Hale was the Academy's Foreign Secretary. (The tally was made under eleven different rubrics.)

15. Pickering to Day, Dec. 26, 1913, and Osborn to Day, Dec. 18, 1913.

16. Perhaps the most interesting comments were from S. J. Meltzer of the Rockefeller Institute. Meltzer to Day, Apr. 13, 1914.

17. Pritchett to Hale, Feb. 3, 1913; Hale to Root, Mar. 3 and 10, 1913, Hale Papers. Comments in the printed version are on pp. 99–102, 120–126. The quotes are from the Mar. 10 letter.

18. Hale to Carnegie, May 3, 1914, Hale Papers.

19. Hale, *National Academies,* pp. 145–148.

20. The tally says fourteen, but I have counted sixteen adverse replies in the Hale Papers. The quotation is from the Dec. 17, 1913, reply of Joseph P. Iddings, the geologist, who admiringly compared Hale's administrative fecundity to that of W. R. Harper.

21. The Ames letter of December 19, 1913, was in response to Hale's circular of Nov. 22. Statements about comparative performances of Americans, such as Conklin's letter of Oct. 29, antedate the circular. See Hale, *National Academies,* p. 151.

22. Hale to Carnegie, May 3, 1914, Hale Papers. See Nathan Reingold, "American Indifference to Basic Research: A Reappraisal," in George H. Daniels (ed.), *Nineteenth-Century American Science: A Reappraisal* (Evanston, Ill.: Northwestern University Press, 1972), pp. 38–62 [Paper No. 3 in the present volume].

23. Hale, *National Academies,* p. 164–165.

24. I have benefited from reading an unpublished paper on the Board by Thomas P. Hughes.

25. Hale to E. G. Conklin, July 21, 1915, Conklin Papers, Princeton University Library. Also see Hale to Conklin, Oct. 25, 1915, Hale Papers.

26. Cattell to W. H. Welch, Apr. 26, 1913, and Cattell to Hale, July 3, 1915, Hale Papers.

27. Two studies of the foundations in the interwar period are still well worth studying: Edward C. Lindeman, *Wealth and Culture* (New York: Harcourt, Brace, 1936) and Ernest Victor Hollis, *Philanthropic Foundations and Higher Education* (New York: Columbia University Press, 1938). For an account of pre–World War I foundation activity, see Jesse B. Sears, *Philanthropy in the History of Higher Education* (Washington, D.C.: Government Printing Office, 1922), Bulletin of the Bureau of Higher Education No. 26.

28. A. Flexner, *An Autobiography* (New York: Simon & Schuster, 1960). See also Raymond B. Fosdick, *Adventure in Giving: The Story of the General Education Board* (New York: Harper & Row, 1962).

29. For Greene, see Fosdick, *Adventure in Giving,* pp. 140–143. Prior to Vincent's coming in 1917, John D. Rockefeller, Jr., was president of the Foundation.

30. Pickering to Greene, Oct. 24, 1913; Greene to Pickering, Oct. 27, 1913. Record Group 3, Ser. 915, Box 1, Folder 1, Rockefeller Foundation Archives (hereafter cited as Record Group 3, RFA).

31. For a discussion of the Committee within the context of elitism, see Reingold, "National Aspirations and Local Purposes," *Transactions, Kansas Academy of Sciences* 71 (1968):235–246.

32. Discussed in an unpublished paper by Reingold on the Carnegie Institution of Washington presented at a conference on "Knowledge in American Society, 1860–1920," June 1975.

33. Corner, *History of the Rockefeller Institute,* pp. 43–46, is far more appreciative of the program than Flexner and Greene. In this period foundations strongly favored giving endowments and avoiding limited grants. Abraham Flexner later described grant programs, after the Rockefeller Foundation changed its policy: "Their main effect is to keep the recipients on their knees, holding out their hats from year to year." *Autobiography,* pp. 275–276. See also Wilder Penfield, *The Difficult Art of Giving . . .* (Boston: Little, Brown, 1967), pp. 294–296, for further comments on the grants vs. endowments issue.

34. Pickering to Greene, May 21, 1915; Greene to Pickering, June 14, 1915, Record Group 3, RFA.

35. From a draft dated Apr. 18, presumably prepared by Greene for one of his scientific correspondents, Record Group 3, RFA.

36. Minutes, Trustees, Oct. 25, 1916, Rockefeller Foundation Archives. The intention to form a committee occurs in the Minutes for June 1, 1916.

37. Hale to Walter Adams, June 1, 1916, Hale Papers.

38. Greene was responding to Whitney's address, "Research as a National Duty," *Science* 43 (May 5, 1916):629–637. Whitney, a chemist, was the first to head the General Electric Research Laboratory. Greene to Whitney, July 13, 1916; Whitney to Greene, July 15, 1916, Record Group 3, RFA.

39. A. Flexner to T. H. Morgan, Sept. 25, 1916; Morgan to Flexner, Sept. 27, 1916; Morgan to Flexner, Nov. 15, 1916; Flexner to Morgan, Nov. 16, 1916,

General Education Board Collection, Ser. 1, Subser. 5, Box 702, Rockefeller Foundation Archives.

40.   Millikan's *Autobiography,* p. 135, recalls a Hale talk identifying Millikan as the "Faraday."

41.   Minutes, Trustees, Dec. 5, 1917, Rockefeller Foundation Archives.

42.   Vincent to S. Flexner, Dec. 27, 1917, Record Group 3, Ser. 903, Box 2, Folder 17, Rockefeller Foundation Archives.

43.   Text in Hale Papers, Box 53. All future references are to this source.

44.   In 1919 the Carnegie Corporation granted $5 million to the National Academy.

45.   See Daniel Kevles, "Federal Legislation for Engineering Experiment Stations: The Episode of World War I," *Technology and Culture* 12 (1971):182–189.

46.   Flexner agreed, citing the experiences of Theobald Smith in the U.S. Department of Agriculture. Flexner to Vincent, Feb. 19, 1918, Record Group 1.1, Ser. 200, Box 37, Folder 417, Rockefeller Foundation Archives.

47.   Whitney to Millikan, Feb. 25, 1918, Chronological File, National Research Council, National Academy of Sciences Archives.

48.   See Kevles, "Federal Legislation."

49.   Hale queried this exclusion late in the process leading to the fellowships. See Hale to Vincent, Feb. 24, 1919; E. W. Washburn to A. A. Noyes, Feb. 4, 1919, Chronological File, National Research Council, National Academy of Sciences Archives. For the long-term effect, see the table in Hollis, *Philanthropic Foundations,* p. 274.

50.   Hale to G. W. Stewart, Feb. 9, 1918, Chronological File, National Research Council, National Academy of Sciences Archives. Hale to F. H. Seares, Mar. 25, 1918, Hale Papers.

51.   See the following, all in the Rockefeller Foundation Archives: Vincent to W. H. Welch, Jan. 26, 1918, and Flexner to Vincent, Jan. 26, [1918], both in Record Group 1.1, Ser. 100N, Box 65, Folder 647; and Minutes, Trustees, Feb. 27, 1918.

52.   Hale to F. R. Seares, Mar. 25, 1918, Hale Papers.

53.   Flexner to Vincent, Mar. 13, 1918, Record Group 1.1, Ser. 200, Box 37, Folder 417, Rockefeller Archives. Comments are in Hale to James A. B. Scherer, Mar. 14, 1918, Chronological File, National Research Council, National Academy of Sciences Archives; and in Vincent to Flexner, Mar. 14, 1918, Flexner Papers, American Philosophical Society Library.

54.   Millikan to Hale, June 15, 1918, and Hale to Millikan, June 21, 1918, Hale Papers; Hale to A. A. Noyes, July 5, 1918, and Hale to Millikan, July 10, 1918, Chronological File, National Research Council, National Academy of Sciences Archives.

55.   A. A. Noyes to Vincent, June 13, 1918, with enclosed proposal. The letter is in Record Group 1, Ser. 200, Box 36, Folder 415; the proposal is in Record Group 1.1, Ser. 200, Box 37, folder 417—both in Rockefeller Foundation Archives.

56.   Hale to Noyes, July 8, 1918, and John C. Merriman to Vincent, Oct. 14, 1918, Chronological File, National Research Council, National Academy of Sciences Archives. Minutes, Trustees, June 25 and Oct. 22, 1918, Rockefeller Foundation Archives.

57.   Thomson to Hale, Nov. 7, 1918; Rutherford, Nov. 13, 1918; H. F. Newall, Nov. 12, 1918; J. Larmor, Nov. 14, 1918—all in Hale Papers. In a letter of Feb. 6, 1919 to President Hadley of Yale, Hale also lists Lord Rayleigh; Chronological File, National Research Council, National Academy of Sciences Archives.

58.   For example, an undated NRC memorandum, 7 pp., on roll 50 of the Hale Papers microfilm.

59.   Flexner to Hale, Feb. 6, 1918, Chronological File, National Research Council, National Academy of Sciences Archives.

60.   Minutes, Trustees, Feb. 26, 1919, Rockefeller Foundation Archives. Hale's letter and the draft are in Record Group 1.1, Ser. 200, Box 169, Folder 2065, Rockefeller Foundation Archives.

61.   Walcott and Hale to Vincent, Mar. 18, 1919; Vincent reply Mar. 21, 1919; and final text to Vincent from Walcott and Hale, Mar. 22, 1919. Chronological File, National Research Council, National Academy of Sciences Archives. Executive Committee Minutes, Apr. 9, 1919, Rockefeller Foundation Archives.

62.   Hale, *National Academies,* p. 151, and Hale's Apr. 21, 1914 presentation.

63.   Vincent to Hale, Mar. 20, 1919, Chronological File, National Research Council, National Academy of Sciences Archives.

64.   George E. Vincent, "The Universities and Higher Degrees in Medicine," *Journal of the American Medical Association* 64 (1915):790–794.

65.   The Flexner letter is in the Hale Papers. Lewis urged Hale to support a separate institute, at least until 1920. See, for example, his letters to Hale of Sept. 12 and Oct. 15, 1919, Hale Papers. The idea was very much alive in 1921. See the Hale–Millikan correspondence in May, particularly Millikan's letter of May 21, Hale Papers.

66.   Tobey, *American Ideology,* 57.

67.   See NACA, "National Aeronautical Research Policy, March 21, 1946." Also based on my conversations with NACA officials while preparing an essay on the agency in 1953.

68.   For Bush's recollections, see his *Pieces of the Action* (New York: Morrow, 1970).

69.   The simplest and perhaps best introduction to this story is still J. L. Penick et al. (eds.), *The Politics of American Science* (Chicago: Rand McNally, 1965).

70.   These are in the Millikan Papers, at the California Institute of Technology, in particular the letters of Oct. 3, 15, and 21, 1949. The "charade" reference and the list of FDR dupes is from the last. A. N. Richards, Detlev Bronk, Karl T. Compton, Vannevar Bush, and James B. Conant hardly constituted a socialistic element in the scientific community. They were of a later generation which no longer regarded government aid as unthinkable.

71.   Weaver to Douglas M. Whitaker, June 19 and Sept. 15, 1951. "Board in Natural Sciences: Postdoctoral" file, National Academy–National Research Council General File, National Academy of Sciences Archives.

THE PERILS OF MATURITY

12. Refugee Mathematicians in the United States of America, 1933–1941: Reception and Reaction

For reasons I cannot wholly explain, the research and writing of "Refugee Mathematicians" gave me more pleasure than anything else I have done as a historian. When I started out on the topic, what I knew was from the papers of Oswald Veblen in the Library of Congress, a collection I had brought in years before. I have a very warm memory of talking to his widow, Elizabeth Veblen, the sister of one Nobel Laureate in physics (Owen Richardson) and the sister-in-law of another (C. J. Davisson). From the collection I had formed a very high opinion of her late husband.

At all costs I was determined to avoid the conventional celebratory piece on how all those wonderful people crossed the ocean and produced all that wonderful research. From casual reading I had a hazy knowledge of Greek scholars fleeing Byzantium to the West before the conquering Turks. In the literature I had encountered, again slightly, the anthropologists' fretting over diffusion. What I knew more about was the extensive literature on immigration to the United States. What I wanted was insight into the process by which individuals and groups from overseas settled into the environment of the American republic. But these mathematicians were hardly typical of the millions crossing the ocean to populate the lands between Canada and Mexico. Quite early, I concluded that the key to understanding was an intense scrutiny of the receiving professional population.

What I found was extraordinarily engrossing and not at all contained within the confines of this one paper. Running down detail after detail, thinking through issue after issue was simply marvelous. I cannot find a better word to describe how it was then and how it appears now in retrospect. Reluctantly, I had to push away from files dating back to the early years of the American Mathematical Society. Reluctantly, I had to remind myself to stop.

One point I noted with glee related to the indifference theme. The American mathematicians, as as group, were more adamantly committed to pure mathematics than the Germans—just the reverse of the conventional wisdom on the indifference theme I so strongly doubted. I again noted with interest how the rhetoric of science about value neutrality and meritocracy did not apply exactly to living scientists but to the bodies of knowledge. Real, live scientists could and did behave quite differently. Once again I noticed how ideologies influenced perceptions and actions. Somehow or other, despite the Depression, despite the strong qualms of many, the mathematicians as a group behaved generously and decently. But that did not occur simply and easily. How that result came about is the subject of this paper. While that may come through to readers, the prose quite properly does not convey the sheer fun I had doing this article.

This paper was originally printed in the May 1981 issue of *Annals of Science*.

Introduction

Immediately after the Nazis moved in April 1933 to expel non-Aryans and the politically tainted from German universities, concerned American institutions and individuals reacted by organizing efforts to aid these individuals. The Academic Assistance Council in Britain came into being earlier, influencing U.S. efforts; in relative terms, the United Kingdom ultimately absorbed more of these displaced scholars.[1]

American efforts were facilitated by the specific exemption of university teachers from immigration quotas in Section (4)d of the Immigration Act of 1924. The clause represented an unintended exception to the nativist, if not racist, character of that legislation. To take advantage of the exemption, a refugee needed the assurance of a post. The newly formed Institute for Advanced Study at Princeton and a few universities directly hired refugee mathematicians; many others (as in other fields) came for temporary employment under the aegis of the Emergency Committee for Displaced German Scholars (later Displaced Foreign Scholars), often with the aid of the Rockefeller Foundation.[2] Although the number of mathematicians involved is not very large, the migration is significant both for its consequences and for what it discloses about historical processes of human and intellectual transfer. By the end of 1935, forty-four mathematicians were dismissed by the Nazis from their posts, to be joined by others subsequently.[3] Through

1939 the number reaching America from the German language world reached fifty-one, plus others from elsewhere as Hitler's sway expanded. By the end of the war, the total migration was somewhere between 120 and 150. Many of these individuals remained permanently or for long periods in the United States, including a number we can characterize as being of, or near, world class in eminence. These numbers do not include younger mathematicians not yet in the professional community when forced to flee.[4] What follows is an account of how the American mathematical community received and absorbed their overseas colleagues up to U.S. entrance into the war. That event materially changed the situation.

The actions of the American mathematicians is a story of the influence of the ideology of the universality of science; of the hazards of Depression conditions; of the reactions to the policies of Nazi Germany; of the influence of nationalistic and anti-Semitic feelings in the United States; and of the persistence of the image of the United States as a haven for the oppressed. It is a story of a real world far removed from the certainty and elegance of mathematics as a monument to human rationality.

The programs of the Emergency Committee and the Rockefeller Foundation are important because they were designed as a mechanism leading to permanent posts. In the absence of a governmental position, these organizations formulated a *de facto* official policy. Mathematics loomed very large in both. Of the 277 individuals aided by the Committee, twenty-six were mathematicians, more than any other scientific field. Most were aided early in the period, before efforts switched to other disciplines. The Foundation supported twenty (some being aided by both).[5] What is most striking is that mathematicians were singled out for rescue early for three reasons: (1) scientists and others recognized the intellectual importance of the field for modern culture, as evidenced by the composition of the newly founded Institute for Advanced Study; (2) mathematicians were influential in both organizations—for example, the president of the Foundation, Max Mason, and the head of its natural science program, Warren Weaver, were mathematicians; and (3) leading U.S. mathematicians and their organizations became active participants in the reactions to Nazism.

Because the policies of the Committee and the Foundation reflected the difficulties in absorbing the displaced scholars and greatly influenced the rescue effort, I will first briefly sketch these policies in the next section. Then I will discuss pertinent aspects of the American mathematical community in the following two sections before giving a series of illustrative events from 1933 to 1939 in two more sections, concluding with an attempt to place these events in a larger framework in the final two sections.

The Committee and the Foundation

In April 1933 officials in the Rockefeller Foundation became concerned over the fate of displaced German scholars. Many were known to them because of prior contacts and support, such as the mathematician Richard Courant (1888–1972) and his colleagues at Göttingen. The Rockefeller Foundation officials were appalled to see a great European nation rejecting the ideal of the universality of learning and lapsing into barbarism. They encouraged the formation of the Emergency Committee in May with funds from other private sources.[6]

Both groups were most conscious of the effects of the Depression. In October 1933, Edward R. Murrow, the second-in-command of the Committee, penned a memo on "Displaced American scholars," noting that more than 2,000 persons had been dropped from the faculties of 240 institutions out of a total of 27,000 teachers.[7] The two organizations decided that universities could not use their refugee funds to displace existing faculty; that they had to avoid a nationalistic reaction to the coming of foreigners; and that, at all costs, the program had to avoid the danger of arousing anti-Semitism.

Specifically, the two bodies decided that they would aid scholarship, not provide relief to suffering. The selection of individuals was based on merit as measured on a worldwide scale. That largely meant mature, or at least recognized, scholars. Some younger mathematicians received fellowships from the Institute for Advanced Study. A later writer characterized the programs of the Committee and the Foundation as dealing with the "few, often well-off and well connected."[8] In practice, the distinctions were sometimes overlooked even at the start. The Committee negotiated with individual universities requesting a particular scholar. Like the Rockefeller Foundation, the Committee wanted to place scholars in research settings but ones hopefully leading to permanent placement—a principal difference from the British program.[9] Grants were often made for two-year periods, with local matching funds or a Rockefeller donation. The positions were not regular teaching posts but for research, perhaps involving an occasional gradute course. This was to reduce the perils of nationalism and anti-Semitism by avoiding regular posts and limiting calls on university funds.

The effect was often to the contrary. Some faculty members greatly resented giving special privileges to foreigners at a time when money was hard to get for research and when others were forced to carry heavy teaching loads. These Americans viewed the program as a way in which opportunities would be denied to young, promising native-born scholars because the top Europeans would be brought in under this program and then given a permanent place. It would cut off opportunities for young Americans "right at the top," words uttered by more than one person. But others in the Foundation and the mathematical community had different perceptions.

The American Mathematical Community

The two principal agents of the mathematicians in aiding emigrés were Oswald Veblen (1880–1960) and R. G. D. Richardson (1878–1949).[10] The former, a nephew of Thorstein Veblen, the great social theorist, was a distinguished topologist at the Institute for Advanced Study. For many years before, he had helped develop the mathematics department at Princeton, eventually attaining the Fine Professorship. Even before World War I, he had thoughts about expanding mathematical research in the United States. During the previous decade he had served a term as president of the American Mathematical Society (1923/24), turning some of his ideas into reality. As Richardson admiringly wrote on several occasions, "He is our master strategist." Veblen wrote about his efforts: "One of the greatest dangers . . . is the timid attitude which is taken by most of the scientific people who deal with these questions."[11] Veblen was not timid.

A few more events in his life may give the flavor of the man. In 1943 he wrote to the Secretary of War, Henry L. Stimson, protesting against the form he had to fill out at Army Ordnance's Aberdeen Proving Grounds. He had been a major in Ordnance in ballistics research in World War I and was now a consultant; he protested against filling out a form that had an entry for race. Veblen said it was like the Nazis, and he did not want to do it. In 1946 at Aberdeen he refused to sign a form that waived the right to strike. He said he would not do it. A few years later, during the McCarthy period, there was an attempt to deny him his passport on the grounds that he was a Communist, which, of course, he was not. He described himself then as an old-fashioned liberal.[12]

R. G. D. Richardson was chairman of the mathematics department at Brown University. Since 1926 he had been dean of its graduate school. He was also—most significant in this context—from 1921 to 1940 the secretary of the American Mathematical Society (AMS). In other words, he ran the Society; he was the establishment. He was born in Nova Scotia and had come to Yale to get his Ph.D. Never to my knowledge did he publicly mention in this period that he was an immigrant himself. In 1908 to 1909 he was in Göttingen to study. (Veblen had not studied overseas; he was a graduate of Chicago.) During World War II Richardson launched a program that would eventually produce a notable applied mathematics institute at Brown. In contrast to the pure mathematician Veblen, Richardson advocated and promoted the application of mathematics.

During the 1920s the mathematical research community in the United States was a small but active and expanding body, in large measure because of Veblen's fund-raising. For one thing, he got mathematicians added to the National Research Council fellowship program, a very important move much appreciated by his colleagues. He also launched an endowment drive for the Society. The mathematicians up to that time had had very little success in fund-raising, and they were impressed by his skills in talking to foundations and the wealthy. He obtained money to subsidize the publications of the American Mathematical Society so that more research could be published in the United States.

By the time the Depression began in 1929, in the United States a mathematical community that had been expanding modestly from the start of the century was undergoing a great period of growth partly due to the infusion of money. Richardson, writing to the Rockefeller Foundation in 1929, ascribed all of this to the AMS: "The atmosphere of scholarly devotion which has raised the sciences and arts of the European countries to a lofty plain is being cultivated by the SOCIETY."[13]

After the Depression came, this promising expansion was imperiled. In 1932 Richardson estimated that, at a minimum, 200 of the members of the Society were out of work. The Society passed a resolution readmitting members who had had to drop out because of economic stringencies without asking them to pay a new initiation fee. Richardson and his colleagues were trying very hard to find ways to get these people back in the Society.[14]

Although Richardson and Veblen operated through the American Mathematical Society, it was not the entire formal institutional structure of mathematics in the United States. There was another group, the Mathematical Association of American (MAA). The MAA was quite different in character and purpose. It was founded in 1915 when the AMS had refused specifically to take on a concern for teaching at the undergraduate level and below, and was largely concerned with teaching. There was a great overlap in membership, and the two organizations quite frequently met together.[15]

The split was very important because most of the jobs that might be available for mathematicians were for teaching undergraduates, not for research and graduate education. Veblen, of course, decried the overemphasis on teaching;[16] to counter that, he strived to develop the research-oriented Institute for Advanced Study. Richardson, as an academic administrator, had a greater sympathy for the problem of undergraduate education. Academic administrators hesitated about hiring foreigners for undergraduate teaching. Even more than any language difficulties, many emigrés were startled and troubled by the different methods and attitudes in teaching in American colleges. Very few realized, as one emigré later wrote: "It takes a long time for anyone not born or brought up in this country to realize . . . that . . . the primary aim of a college . . . is to educate members of a democratic society, that it includes among its functions the training of mind and character, of social attitude and political behavior."[17]

A 1935 survey of the job market for mathematicians concluded that, given the normal demographic turnover, there were more potential teaching positions than the annual estimated production of Ph.D.'s. This assumed only a slight relaxation of the economic conditions plus an upgrading of some posts not then occupied by holders of the doctorate.[18] But as late as 1940, the job market had not appreciably improved, in the opinion of many mathematicians. In a time of economic distress the always present conflict between teaching and research could and did become acute.

To cite one example, in 1934 the University of Michigan's College of Arts and Science had a greatly increased enrollment in mathematics. Significantly, the faculty member reporting the rise ascribed it to the perception of the importance of mathematics, "both culturally and practically." A lot more people were taking mathematics in 1934, at least in Ann Arbor. In the following year a young mathematician at Michigan wrote to Richardson that there had been an enormous struggle between the "research" and "non-research" groups over his tenure. He had won; he had a permanent job. He was not going to have to teach summer school any more in order to do his research and to earn a living; yet it was a real struggle, and it affected the kind of job opportunities for all mathematicians.[19] The refugees, being marginal men because of Nazism, were particularly vulnerable in clashes between culture (that is, research) and practicality (that is, elementary teaching). As long as they were viewed as "merely" researchers, not involved in the routine teaching, they were targets for those viewing research as a luxury expendable in a time of economic crisis.

In 1936 Richardson published a study of doctorate holders in America since 1862, whether foreign or native-born, and holders of domestic or overseas degrees. He identified 114 holders of foreign degrees (both native and foreign-born) compared to a total of 1,286 degree holders from United States and Canadian universities.[20] The 34 from Göttingen far surpassed any other foreign source. To this indication of impact must be added individuals like Richardson who had gone to Göttingen but not to get a degree.

Richardson's analysis of the current situation disclosed 40 foreign-born Ph.D.'s in the country as of 1930, and an estimated 20 new mathematicians arriving due to Nazi policies. He observed this actually represented a decrease in the percentage of mathematicians holding foreign Ph.D.'s in the country because fewer Americans had gone overseas since 1913. Noting that foreign Ph.D. holders—native and immigrant—tended to be more prolific in research, Richardson feared inbreeding.[21] The American increase in the award of the degree was more a matter of quantity than quality. What he did not say explicitly, but what emerges from his statistics, is the overwhelming preponderance of undergraduate teaching as a source of employment, not the conduct of research. Like the 1935 survey, Richardson predicted a shortage of mathematicians if only economics permitted hiring and upgrading. Until that occurred, even placing twenty or so leading mathematicians was a problem, considering the hazards of nationalistic and anti-Semitic reactions.

The Perils of Nationalism and Anti-Semitism

There is no doubt of the existence of nationalistic and anti-Semitic sentiments. There is considerable difficulty in precisely estimating their consequences in many specific situations. A particular hazard is the need to separate the two kinds of sentiments. Hostile comments about foreigners may serve as code words to mask anti-Semitism. Evidence exists, however, of nationalistic feelings devoid, or largely so, of any hostility to Jews. Although U.S. history, to this very decade, is one in which immigration looms as a basic feature, newcomers have always attracted a measure of antipathy.

Such reactions existed among mathematicians before Hitler and continued after the start of the migration. In 1927, for example, Richardson wrote: "With one foreigner Tamarkin in the department, we feel that it might be a considerable risk to take on another one such as Wilson. Englishmen do not adapt themselves very quickly to American ways, and generally they do not wish to do so."[22] But there were at least 39 others besides Tamarkin in the United States by 1930, if we can trust Richardson. Some must have been very self-conscious about their origins, judging by Tibor Rado's 1932 geeting to his colleagues "as a representative of those born abroad who have adopted this as their country."[23] From 1933 until 1940 Norbert Wiener kept on worrying about the need for assuaging nationalistic sentiments: "Every foreign scholar imported means an American out of a job. . . . Any appointment for more than a year would cause a feeling of resentment that would wreck our hopes of doing anything whatsoever."[24] Veblen voiced similar fears.[25] In Wiener's case it surely stemmed from concern about placing young American-Jewish mathematicians in a tight job market.

In 1934 A. B. Coble of the University of Illinois said, despite hostile questions from a state legislator, that he would hire a foreigner if better than any native prospect.[26] Wiener's reaction to such hostility was a 1934 proposal to raise new money to provide research posts not competitive with regular posts.[27] The Berkeley economist Carl Landauer disagreed with Wiener, asserting that university administrators were giving preference to Germans over Anglo-Saxons. Rather than concentrating the refugees in graduate courses, Landauer wanted them integrated into undergraduate teaching.[28]

In the same year G. A. Bliss of the University of Chicago turned down a refugee: "I must confess also that if we could secure a new man, I should want to try to get a strong American. It is pathetic to see the good young American men, who have received their Ph.D. degrees in recent years, so inadequately placed in many cases."[29] In 1941 a dean at Yale, writing about a mathematician, Einar Hille, said: "No foreigner should be chairman of a department where undergraduate work is involved. . . . One of the criticisms of these foreign importations is that they are not suited to undergraduate work or do not wish to do it. Hence they take the most desirable positions away from our American product."[30] Although educated at the University of Stockholm, Hille was born in the United States.

As to anti-Semitism, it was ubiquitous, in at least mild forms, in the genteel world of American academia before World War II.[31] To cite a few examples, in 1931 the mathematician H. E. Slaught of the University of Chicago, writing about a mathematical astronomer said: "He is one of the few men of Jewish decent [*sic*] who does not get on your nerves and really behaves like a gentile to a satisfactory degree."[32] In seeking to fill vacancies, administrators sometimes bluntly excluded Jews or asked, as in one case, for "preferably a protestant."[33] Coble stated that Illinois played it safe on appointments, "a policy with which I am not wholly in agreement." He explained that this arose because the graduate work was conducted by men paid through the administration of the undergraduate colleges, noting that "leads to selections of a rather uniform type."[34]

But Illinois and a number of other departments already had Jewish members, typically one. This produced a problem for some when presented with the option of hiring a second. As a dean at Kentucky wrote in 1935, "You know that you have to be careful about getting too many Jews together."[35] Or, as the chairman at Indiana noted in 1938, "But there is a question of two Jewish men in the same department, and a somewhat small one."[36] Wiener encountered this problem with the possible placement of one of his students at M.I.T. In a conversation in 1935 Karl T. Compton noted the "tactical danger of having too large a proportion of the mathematical staff from the Jewish race, emphasizing that this arises not from our own prejudice in the matter, but because of a recognized general situation which might react unfavorably against the staff and the Department unless properly handled." After agreeing that no one should fail to receive fair consideration because of race, Compton continued: "Other things being approximately equal, it is legitimate to consider the matter of race in case the appointment of an additional member of the Jewish race would increase the proportion of such men in the Department far beyond the proportion of population."[37] By the standards of his day, Compton was an enlightened administrator, but he responded to and perhaps adapted to the conventions of his milieu.

Nor was anti-Semitism wholly absent from the inner workings of the American Mathematical Society. In 1934, the Society elected its first Jewish president, Solomon Lefschetz (1884–1972) of Princeton University, Veblen's successor to the Fine Professorship. That prospect apparently presented a problem earlier for one of the elder statesmen of the mathematical community, Professor G. D. Birkhoff of Harvard University (1884–1944), a close friend of Richardson. Birkhoff and Veblen were probably the two most eminent of the senior American mathematicians of that day. From 1935 to 1939, Birkhoff was Dean of the Faculty of Arts and Sciences at Harvard.

Lefschetz, a great topologist, was born in Russia and educated in France as an engineer. After coming to the United States in 1905, he lost both hands in an industrial accident. He then received a Ph.D. in mathematics from Clark University and taught at universities in the Midwest. In 1924 Veblen brought him to Princeton. Eventually, the two men would break.

Richardson foresaw troubles ahead, yet managed to survive the two years' incumbency with little apparent damage. Birkhoff's opposition to the possibility had interesting overtones: "I have a feeling that Lefschetz will be likely to be less pleasant even than he had been, in that from now on he will try to work strongly and positively for his own race. They are exceedingly confident of their own power and influence in the good old USA. The real hope in our mathematical situation is that we will be able to fair to our own kind." And Birkhoff went on to say: "He will get very cocky, very racial and use the Annals [*Annals of Mathematics*] as a good deal of racial perquisite. The racial interests will get deeper as Einstein's and all of them do."

In the same letter Birkhoff also expressed distress that the two-year presidency of the AMS, usually awarded on the basis of research eminence, would probably go to individuals (incidentally, all non-Jewish) all of whom, apparently, had different ideas from those he was espousing. He wondered how to arrange that the presidency be given for service to the society, rather than for eminence in research.[38] Despite Birkhoff's strong feelings and despite Richardson's apprehensions, when the proper moment came, it was Birkhoff who reported the nomination of Lefschetz for the presidency. To Marston Morse, writing to Veblen, the result was better than selecting a weaker man "regardless of politics."[39] Birkhoff's views were apparently fairly well known during his lifetime. In 1936 Norbert Wiener's student, the subject of the Compton memo, then at the Institute for Advanced Study, wrote his teacher a letter in which he closed: "P.S. Einstein has been saying around here that Birkhoff is one of the world's greatest academic anti-semites."[40]

Reactions to Refugees and to Nazism

Even before Hitler came to power, there were signs of future American sentiments. In 1932 the *Bulletin of the American Mathematical Society* criticized the dismissal of the Italian mathematical physicist Vito Volterra for refusing to take the oath required by the Fascist government as violating "correct principles of academic tenure."[41] Perhaps that rather mild criticism influenced the early reaction to the Nazi program, but increasingly many mathematicians foresaw deeper and more serious aspects of the occurrences in Germany. None, to my knowledge, perceived Hitler's Final Solution to the Jewish problem.

In May 1933 Veblen went to the Rockefeller Foundation about Nazi moves and became a member of the Emergency Committee on its founding. From then until the end of the war, he and his colleague Hermann Weyl ran an informal placement bureau for displaced mathematicians. In Veblen's papers in the Library of Congress are lists of names with headings such as scholarship, personality, adaptability, and teaching ability. When information about a person was incomplete in the United States, Veblen wrote to European colleagues.

In April and May 1933 Richardson at Brown University saw an opportunity for America and for his university. In a memorandum of 23 May 1933 to the Brown University Graduate Council giving notes of a luncheon discussion of the German-Jewish situation, he wrote: "In 1900 we were flocking to Germany but now more come here than go the other way." Richardson agreed with the Foundation, Veblen, and the Emergency Committee about the peril of bringing in a considerable number of mathematicians with so many young people unemployed— the "danger of causing friction and even of fanning the flames of Anti-Semitism in this country." To ensure and control a proper distribution, perhaps one to three leading mathematicians in each participating university, the Society had to take a leading role. Provided funds were available, Brown could cooperate to the extent of taking two to four mathematicians.[42] By early July the Council of the Society authorized its president, A. B. Coble of Illinois, to establish a committee of three to cooperate with the Emergency Committee. Naturally, Veblen was one of the three.[43] On 15 July 1933 Richardson could write the president of the American Jewish Congress that "our organization views with dismay and almost incredulity the developments in Germany."[44] There is no doubt of the genuineness of Richardson's personal aversion to the news coming out of the Third Reich.

Despite early successes and the strong backing of men like Veblen and Richardson, the placement program had a built-in peril. It assumed the universities (or most of them) would absorb the first wave of refugee mathematicians into their regular staffs at the expiration of the two-year grants from the Emergency Committee and the Rockefeller Foundation, roughly in 1935.[45] That depended, in large measure, upon improvements in the economic climate. In turn, such improvements could provide a degree of security against the perceived dangers of nationalism and anti-Semitism. As 1935 approached and the unemployment of mathematicians continued, pessimism developed. Writing to the Danish mathematician Harald Bohr on 5 April 1935, Richardson gloomily predicted that half of those presently supported (in all fields) were absorbable by the universities. Most of the rest had possibilities for temporary placement pending later absorbtion. Beyond that—and this presumably meant others in Europe hoping to come over—Richardson, with a few exceptions, could only see the possibility of providing unemployment relief, not aiding scholarship. With seventy-five American mathematical Ph.D.'s out of work, even the small number taken on by the Committee or employed directly approached an upper limit.[46]

But before going on to discuss the unfolding reactions of the American mathematicians to the Nazis and their victims, let us consider three instances of successful placement by 1935. The fate of the Göttingen group became an immediate concern to the Rockefeller Foundation and individuals like Veblen. Bryn Mawr College provided a post for Emmy Noether (1882–1935). For all his ambitious plans, Richardson had to content himself with the youngest man in the group, Hans Lewy (1904–   ). Despite sour comments from at least one outside observer,[47] Richardson was very pleased with his new colleague.

Richard Courant presented a more difficult problem. For one thing, Richardson had clashed with him in Göttingen before World War I. Although hostile to Courant's coming, Richardson pledged not to interfere with efforts to place him. As late as 1936, Richardson was grumbling about Courant;[48] the two men would clash at the start of World War II when each had ambitions to launch an applied mathematics program. Courant also had a reputation as a promoter which both helped and hindered his placement.

Veblen first thought of the mathematics department in Berkeley, then undergoing a reorganization. Before Hitler's ascension, he had recommended American mathematicians; by May he was pushing Courant and other displaced Germans.[49] California had selected Griffith C.

Evans of Rice Institute as the new departmental chairman. Evans strongly opposed Courant, asserting:

> To say that there are too many foreigners in American universities is not chauvinism, but merely, that the careers of promising students in America are being cut off at the top. I do not see how this can be anything but an unfavorable situation in which to develop intellectual life. A generation ago we were in need of direct stimulation and there was plenty of room; now we could well interchange.[50]

To this Veblen replied with a succinct statement of his position:

> I think I would differ from you only in attaching a little more weight to the importance of placing a few first-class foreigners in positions where they will stimulate our activity. I am inclined to think that doing so will in fact increase the number of positions that are available to the better grade of American Ph.D.s, even though it may decrease the total number of positions. . . . [A]lmost any method of strengthening the local scientific group will make it easier to place our scientifically strong products.[51]

In a later letter to Richardson, Evans elaborated his position:

> It seems to me that at the present time our own young men should be the first consideration, given the fact that Europe would not reciprocate in appointing Americans in their universities. Of course, they would say "But look at the difference!" I doubt if there is much, myself, allowing for the difference in teaching programs.[52]

Even before the exchange with Evans, Veblen had moved to place Courant in New York University, arguing that it presented an opportunity for Courant's entrepreneurial skills. By 1936 Veblen was pointedly noting Courant's good works in a region "unnecessarily arid," including providing openings for American mathematicians.[53] Even before World War II, Evans and Courant were on cordial terms; after the war Berkeley made an unsuccessful attempt to move Courant's entire group to its campus.[54]

To return to Hans Lewy, he was one of the victims of the 1935 financial situation; Brown could not keep him. Richardson wrote across the country to many departments on his behalf, the previously given references to Coble, Bliss, and Kentucky being examples of some responses. But Evans hired him for Berkeley on a regular appointment without benefit of subsidy from the Emergency Committee or the Rockefeller Foundation.[55] Evans's nationalistic sentiments were not necessarily anti-Semitic. While at Rice, for example, he wrote in 1932 about a young Jewish mathematician: "But emphatically, there should not be a prejudice against him, discounting ability on the ground that he is a Jew."[56]

These actions involving Lewy and Courant occurred while the mathematicians displayed an increasing sensitivity to the implications of Nazism both at home and abroad. At the time of the Veblen–Evans exchange, for example, John R. Kline, the chairman of the mathematics department at Pennsylvania, wrote Richardson about the planned Gibbs Lecture Einstein would give in December 1934 at the meeting of the American Association for the Advancement of Science in Pittsburgh. He advised against publicity which "might involve Einstein in some unpleasantness should the Nazi sympathizers in Pittsburgh attempt to pack the meeting."[57] In fact, no incident occurred.

In the spring of 1934 American and British mathematicians were incensed by Ludwig Bieberbach's article ascribing different forms of mathematics to racial characteristics. G. H. Hardy wrote a scornful response in *Nature;* Oswald Veblen sent off a deftly scathing letter to Bieberbach.[58] In the 1934 summer joint meeting of the Society and the Association at Williams College, Arnold Dresden of Swarthmore, as president of the latter (himself an immigrant from the Netherlands), declared: "the conviction has been growing recently that no country is safe from the distress that has fallen upon Germany the past year." He and E. R. Hedrick, of UCLA, a former president representing AMS, called in defense for renewed adherence to the intellectual standards of mathematics.[59]

In 1937, the Society had an opportunity to face up to what was happening in Nazi Germany. An invitation came to attend the bicentennial of the founding of Göttingen University. Like other Americans, the previous year's celebration at Heidelberg had offended Richardson. Writing to Birkhoff, he declared himself against "science as a national tool rather than as an end to itself."[60] Richardson decided that so important a matter had to be laid before all of the present and past officers and members of the Council, almost 90. He sent out the invitation accompanied by a memorandum written by his colleague, Raymond Clare Archibald, declaring that the invitation was not from the Göttingen known to all from the old days. The university now was a different body and, like the 1936 celebration at Heidelberg, its bicentennial would provide an opportunity for Nazi propaganda in violation of the universality of science. Archibald suggested simply sending a letter complimenting Göttingen on its past and hoping that the future would be similar.

The returns were overwhelmingly against participation. Simply say no, wrote Eric Temple Bell of the California Institute of Technology. Professor C. A. Noble at Berkeley remembered his happiness at Göttingen (1893–1896, 1900–1901), his indebtedness to his professors "and to Germany of those days." Nevertheless he concurred with Archibald, as did 54 others. Only ten recommended participation for reasons such as maintaining solidarity with colleagues. One of the ten,

W. A. Wilson of Yale, thought the Jews "are largely responsible for their troubles."[61]

Another occasion implicitly to face the consequences of Nazism occurred in the 1938 celebration of the Society's semicentennial. Its planners foresaw a festive occasion for a small strong, tightly knit group. Archibald, who was also a historian of mathematics, would write a history of the Society, and a dozen or so papers would survey the development of fields of mathematics in the United States. Archibald thought the American emphasis could "open [us] to charges of provincialism." Veblen wanted to avoid a historical review of mathematical subjects, suggesting simply a survey of current knowledge. Others, like Lefschetz, objected to having "ancients" speak, simply wanting the best in each speciality. The compromise, suggested by Kline, called for contributions from outstanding mathematicians who were to be free to give or not give historical contributions, not even necessarily concerned with U.S. contributions. He also suggested excluding recent arrivals in the country as authors, even if they were the best persons for a topic. By 1937 Hedrick noted the apparent exclusion of the foreign-born and those trained abroad. (Although one speaker, E. T. Bell, was born overseas, all were products of U.S. universities.) That did not arouse controversy. Kline's other suggestion did; he wanted Birkhoff to have a prominent place in the program.[62]

Birkhoff presented a historical survey of mathematics in the United States during the past 50 years, giving praise and criticism. He then brought up the question of the foreign-born mathematicians. He felt they had an advantage. In getting research positions, they did less teaching than the native-born; they lessened the number of positions for American mathematicians who were "forced to become hewers of wood and drawers of water. I believe we have reached the point of saturation. We must definitely avoid the danger." Starting out with praise, Birkhoff then listed all of the people who had come in the last 20 years. Included in the list were such colleagues of his as Alfred North Whitehead and others who were neither German nor Jewish. Despite the nature of Birkhoff's list, many ascribed his views to anti-Semitism.[63]

The speech and its printed version elicted strong responses. Abraham Flexner of the Institute for Advanced Study eloquently argued against the presumed bad effects.[64] Lefschetz, who was listed, complained that he had been in America for 33 years; all of his mathematical work had occurred in the United States. Writing to his old friend, Richardson reported that people, "not all Jews," looking at the semicentennial volumes expressed marked disapproval of the sentiments. Perhaps reflecting feelings about his own immigrant origins, Richardson gave two omitted names, one a Briton whose residence in the United States was nearly as long as his own.[65]

If Birkhoff's public pronouncement offended many,[66] other private incidents in the life of the Society just before and after 1938 evinced sensitivity to the presence of the foreign-born. In all probability anti-Semitism was absent, or nearly so, in these cases. For example, in 1936 Veblen proposed opening the Society's series of *Colloquium Publications* to non-U.S. mathematicians. Whatever Veblen intended, the proposal was taken to refer especially to those now resident in the United States. The varied responses inclined to go along. A. B. Coble approved for those who "have identified themselves for a long period with our American program." Kline would publish high-caliber work, rejecting any provincial or narrow attitude.[67] As usual, Evans had one of the more interesting replies: "so long as the authors are expected to be permanently or semi-permanently a part of this American scene, even if the best work under conditions that are denied to the rest, and do not seem at present to be contributing to the solution of problems confronting American universities and American mathematicians."[68] Evans was mistaken in at least two instances. Lewy, right after arriving at Berkeley, wrote Richardson expressing concern about high school training in America.[69] In Courant's earliest efforts to form a center at New York University, he proposed including efforts at improving high school instruction.[70] Both men were obviously in the tradition of Felix Klein.

Concern over the foreign-born reappeared again in 1939 during the deliberations for the award of the Cole Prize for algebra. Evans, now president of the Society, wrote to Richardson, reporting that the chairman of the prize committee, Eric Temple Bell, was unhappy with the presence in his group of a recent arrival, Emil Artin. Bell strongly believed Artin would not support an American, contrary to the intentions of the donors, but "will vote prize to some strange bird of passage." Richardson's response pointed out that the Society's policy had explicitly moved away from limiting prizes to the American-born. In fact, that policy originated in 1921 when Richardson amended the terms of the Bôcher Prize specifically to include any "resident of the United States and Canada" without regard to citizenship. The crisis on the prize committee vanished when its members, with no apparent acrimony, agreed on the merits of the work of a Jewish American-born mathematician, A. Adrian Albert.[71]

The Founding of *Mathematical Reviews*

Of all the reactions to Nazism of the American mathematicians, by far the most significant was the decision in 1939 to found *Mathematical Reviews*. Into the decision entered explicit judgments about events overseas, about choices of intellectual and national ideologies, and about relationships with refugees in the United States and colleagues overseas. Reviewing and abstracting media have played roles in the life of science since their appearance in the seventeenth century. Beyond the obvious roles in dissemination and validation of new knowledge, such publications often reflect predominance or even hegemony of national research or linguistic communities. By World War I the Germans had developed an extensive network of reviewing and abstracting media clearly reflecting their extensive research activities, as well as a penchant for organization and thoroughness. Some scientists in the Allied powers reacted by moving to form alternate sources of scientific information.

Among American mathematicians the desire for a critical abstracting journal was part of the general drive to develop a well-rounded mathematical community not in a state of colonial-like dependence on Europe. In 1922 H. E. Slaught included it in his proposal for increasing funding and activities of mathematics.[72] Oswald Veblen also tried for an abstracting journal during his presidency of the Society. The Depression halted efforts in this direction. Influencing Veblen was dissatisfaction with both the German *Jahrbuch über die Fortschritte der Mathematik* and the Dutch *Revue semestrielle des publications mathématiques,* particularly concerning the time lag in publishing the abstract.

The situation changed materially in 1931 when the Berlin firm of Springer launched the *Zentralblatt für Mathematik und ihre Grenzgebiete*. It was satisfyingly prompt, critical and complete in coverage. Its editor, Otto Neugebauer, a member of Courant's group at Göttingen, was not Jewish but was politically suspect to the Nazis. Neugebauer fled to Denmark in 1934. Reputedly, he later said of his exile: "I did not have the honor of having a Jewish grandmother." As early as August 1933 Veblen proposed bringing Neugebauer to America to continue the *Zentralblatt*. Richardson disagreed because of financial conditions.[73]

Some mathematicians, perhaps Veblen, perceived a European neglect of, and indifference to, American contributions, particularly among the Germans. It was, one suspects, especially galling in view of the recent strenuous efforts to expand such contributions. Writing to Veblen in 1935 about comments in the *Zentralblatt,* T. Y. Thomas, then at Princeton University, referred to "the usual European attitude towards American work which was exhibited in a very tactless manner."[74] A mathematician at Wisconsin, Rudolph E. Langer, in 1936 commented to Birkhoff: "When Europeans give their recognition to an American, there can be no doubt that it is deserved."[75] As late as 1942, I have encountered a reference to a hypothetical "European Citation Verein."[76]

A deterring effect was simple doubt of the intellectual capacity of the

American mathematical community for this task. In a different context Evans wrote in 1936 about helping the *Zentralblatt* and the *Fortschritte:* "Or it may be desirable to have an American agency take over the entire task. I doubt however if there are a sufficient number of Americans of the required scholarship to perform the task. It must be remembered that while on the Continent there is a considerable amount of ambitious scholarship even in the secondary instruction, there is in this country, due to our methods of selecting teachers, a dearth of it both in secondary schools and in colleges."[77]

Perhaps more important than doubts like Evans's was Richardson's growing conviction of the strength of mathematics in the United States. A few months after Evans wrote, Richardson with obvious relish quoted G. H. Hardy's words to the Society at the Harvard Tercentenary: "now America could produce three mathematicians of rank to every one that could be produced by any other country." He added: "With the influx of distressed German scholars and others, mathematics has probably forged ahead relatively more than other sciences in the last dozen years."[78]

In late 1938 the *Zentralblatt* removed the Italian mathematician Tullio Levi-Civita from its board for racial reasons, presumably under pressure from the regime. Neugebauer resigned, as did many foreigners on the advisory board such as G. H. Hardy, Oswald Veblen, and Harald Bohr. On 27 November 1938, Richardson reported the resignations to Evans, as president-elect of the Society, adding news of the barring of Russians as collaborators and as referees. Not surprisingly, refugee mathematicians were also excluded from the review process.[79]

An indication of the reaction to the new policy of the *Zentralblatt* is in the correspondence of J. L. Synge, then at the University of Toronto. On 9 December 1938 he wrote Richardson: "I do not believe in a policy of appeasement. I regard the directors of the Zentralblatt as having betrayed a confidence placed in them. . . . Of course, if there is in the Society any considerable number of members who approve Nazi policy, the situation is more delicate." On 19 January 1939 Synge wrote Ferdinand Springer severing his connections with the *Zentralblatt*. He protested the removal of Levi-Civita. The bar against emigré reviewers of German papers was an uncalled for violation of scientific internationalism: "the prohibition introduced by the publishers of the Zentralblatt appears to me insulting to a body of mathematicians for whose academic eminence and personal integrity I have a high regard."[80]

Having long prepared for this moment, Veblen drafted a statement urging the founding of an American journal. In it he wrote, after referring to his efforts fifteen years previously, that the United States was then "not yet strong enough to carry the load without much of a strain on its creative elements. Since then the number of productive mathe-

maticians in our country has increased much more rapidly than antici-
pated and has also been supplemented by an influx of scholars who
found it difficult or impossible to continue their work in Europe. As a
result the mathematical center of gravity of the world is definitely in
America." Veblen called for a new journal passing judgment on new
theories, one "based on the traditional decencies of scientific and human intercourse."[81]

Although outwardly neutral, Richardson's position was now favorable to the proposal. He moved to get Neugebauer a place on the
Brown faculty so that the journal would function in Providence. Richardson's actions caused tensions with Veblen, eventually leading to a
clash after the periodical was founded.[82] Early in December 1938, Richardson organized a committee of the Society to consider what should be
done and also prepared for what he knew should be a spirited discussion at the Christmas meeting in Williamsburg, Virginia. Writing on
the question to his opposite number in the MAA, W. D. Cairns, Richardson penned a paragraph whose meaning was instantly clear to the
Society's insiders who were sent copies: "We must avoid any reference
to political, religious, or racial questions. We must under no circumstance put ourselves in a position of appearing to kill the Zentralblatt.
We must study the question objectively and make up our minds as to
what is best to be done for mathematics." What Richardson meant in
his letter was that the question of Nazi racial policies should not figure
explicitly in the committee's composition or its stated conclusions.[83]

E. R. Hedrick, newly installed as chancellor of the University of California at Los Angeles, responded to the letter to Cairns: "I believe that
the time has passed when we need to consider the tender feelings of
people in Germany connected with the Zentralblatt and I doubt
whether we need worry about the matters mentioned in the last paragraph of your letter. . . . I would not wish to say anything political in
criticism of the actions taken by the Germans, simply because I do not
see that it would do any good, but I do not believe we ought to hesitate
about any action we [care] to take on evening accounts." President
Evans, who could not come to Williamsburg, telegraphed Richardson to
have one "Hebrew" on the committee, suggesting Lefschetz instead of
Marston Morse.[84]

There were lively sessions at Williamsburg. In addition to the abstracting journal issue, a largely unsuccessful move was afoot to de-
mocratize the management of the Society. Because nearly three
hundred members had petitioned the Society to launch an abstracting
journal, the matter was discussed in a lively unusual open council
meeting attended by more than one hundred people. Richardson wrote
to Veblen about this session: "The political, religious, and racial questions which were involved were bound to come to the surface, although
people were requested to keep the discussion on a purely objective

basis." To Evans, a few days later, Richardson defended the exclusion of Lefschetz: "We don't want to make it appear that this is in any way a Jewish protest. (You can understand that there were a great many cross currents with regard to this and other questions at the meeting.)."[85] But Richardson made sure that the committee was headed by C. R. Adams, a member of his department at Brown.

The Council directed the committee to consider three necessary conditions for the launching of an abstracting journal: (1) financial assurance for the first five years of publication; (2) international cooperation; and (3) confirmation that the *Zentralblatt* was not likely "to make its reviews unimpeachable." The first condition reflected considerable hesitation, if not anxiety, about an undertaking large enough to imperil the solvency of the Society. Veblen handled that problem by getting a $65,000 grant from the Carnegie Corporation. The second condition generated a spate of letters to organizations and individuals in many nations asking for cooperation and pledging adherence to scientific internationalism.

The existing *Zentralblatt,* warmly regarded by many American mathematicians, provided the thorniest problem for the committee. Richardson observed: "I cannot now see what assurances they [Springer] could give that would be satisfactory to me." He had recently read in *Science* that German medical abstracting journals now omitted reviews of articles by Jews. As he wrote to Hardy before the matter was decided, Richardson wanted an international journal "independent of the whims of a dictatorship."[86] The Society even considered the possibility of purchasing the *Zentralblatt* from Springer. To the American criticism Ferdinand Springer replied: "I am determined to continue the Zentralblatt at the prevailing level as a non-party international abstract journal—at the moment I see this is not possible, my interest in the enterprise ceases." To the purchase offer, Springer later asserted he did not regard the journal as a commercial venture. More importantly, he offered to dispatch an emissary to America, F. K. Schmidt of Jena.[87] The Society deferred decision until Schmidt met with the committee.

Not only did considerations of equity require the delay; within the Society sentiment existed against the proposed journal because of a desire to aid Springer and German mathematical colleagues, as well as a reluctance to take any action splitting the international mathematical community. The department at Wisconsin, for example, went on record along these lines. One can doubt the depth of such sentiments. Mark Ingraham of Wisconsin, an officer of the Society, tepidly agreed with his colleagues but added a postscript to a letter: "Since writing this I have had a wave of indignation against the Nazis and feel more inclined to go ahead."[88]

Still another argument, derived from the fear of splitting the international community, called for a single abstract journal rather than two or more. That struck a responsive chord, perhaps in resonance with fiscal anxieties. Eventually, that point simply faded away. By May, Lefschetz could write that physics flourished with two parallel abstract journals (*Physics Abstracts* and *Physikalische Berichte*). He added that the printing and editorial jobs should go to Americans as "our learned world is supported by American funds."[89]

In March and April an unexpected event further tipped the balance of opinion. Marshall Stone, then at Harvard, received a letter from Helmut Hasse of Göttingen. Hasse justified the splitting of the refugees from other possible referees. "Looking at the situation from a practical point of view, one must admit that there is a state of war between the Germans and the Jews." He failed to understand why the Americans withdrew their collaboration with the *Zentralblatt* and referred to "Neugebauer's pro-Jewish policy." C. R. Adams circulated the Hasse letter to his committee in preparation for the coming meeting at Durham, North Carolina, noting: "Mr. Veblen insists that there is a war by the Germans against *civilization*."[90] The letter made a strongly unfavorable impression on the members of the committee and the Society's council.[91] In his 2 May reply to Hasse, Stone said the decision on the journal was "not likely to be taken for the purpose of passing judgment on the past history of the Zentralblatt . . . [but] primarily on the desire to assume for the future of mathematical abstracting a responsibility commensurate with America's great and growing mathematical importance. . . . As for the Americans who withdrew, I feel that they merely acted with true loyalty to their own national traditions and ideals."[92]

Nor did Schmidt's presentation in May convince many American mathematicians. Springer offered a compromise arrangement with two separate editorial boards: one for the United States, Britain and its Commonwealth, and the Soviet Union; the other for Germany and nearby countries. To avoid any imputation of racial motivations, Springer now asked that papers by German authors not be reviewed by German emigrants, whether Jew or Gentile. Writing to A. B. Coble about the meeting, C. R. Adams said that Schmidt "states that the German idea is that mathematics, like everything else, exists in a real world in which political considerations play a part: and that, like everything else, mathematics must expect to be affected in some measure by political considerations."[93] Springer's proposal was perceived as a gross affront to the ideal of scientific internationalism. One mathematician, T. C. Fry of Bell Telephone Laboratories, commented that Springer gave no assurance against a future ban against refereeing of Aryans' papers by non-Aryans who were not German refugees.[94]

Unexpectedly, Schmidt found allies among some members of the Harvard mathematics department. Fearing a German boycott of the planned 1940 international congress in Cambridge, Massachusetts, William C. Graustein argued against the proposed journal: "We would, I feel, be denying a principle for which we have long fought—that of the emancipation of science from international politics."[95] It was an ironic obverse of Schmidt's position. Although Graustein sat in on the committee's last session and addressed an open letter to it and to Council members, the effect was minimal. Richardson regarded Graustin's letter as playing into the hands of the Germans, who were acting on Harvard advice.[96] Adams told the committee on 17 May that Springer's moves were simply designed to confuse issues.[97]

Not everyone in Harvard (let alone MIT) agreed with Graustein and Birkhoff. Marshall Stone, as a member of the AMS Council, convened a meeting of Cambridge, Massachusetts, mathematicians on 18 May 1939, presided over by Saunders MacLane. The meeting unanimously repudiated Graustein's letter and supported the proposed journal with only one abstention.[98] Norbert Weiner, hard at work in arranging part of the 1940 Congress, drafted a letter to Graustein on 19 May threatening to resign with a public denunciation if Graustein persisted in his efforts.[99] Richardson, who had taken control of the process from his ill colleague, C. R. Adams, determined to push the decision through.[100] On 22 May Birkhoff proposed a compromise, a publication of bibliographic entries without any analysis.[101] On 25 May Richardson wrote his old friend that the Council of the Society had voted 22 for the new journal, 5 against, and 4 uncertain.[102] Oswald Veblen was named chairman of the committee to launch and to supervise the newest publication of the Society.[103]

The Coming of War and the Elevation of Applied Mathematics

While the leadership of the Society, the disciplinary establishment, grappled with the impact of the foreign-born already in their ranks, others continued to come across the ocean in the closing years of the 1930s. As Hitler's pressures expanded, still more became potential migrants. The unemployment situation remained discouraging. Richardson thought America had done all it could to absorb refugees. Approvingly, in 1938 he cited to W. D. Cairns the policy of the Emergency Committee (and the Rockefeller Foundation): "I think the principle laid down there [that is, in the Committee], namely that humanitarian considerations must be laid aside, should be followed and what can be done should be for those of high scientific merit. We want to save the scholar for the sake of scholarship."[104] After his retirement from the secretaryship, Richardson explained in 1941: "I have been compelled to consider these cases of appointments from the scientific and monetary point of view. If I should think of the humanitarian aspects I would get bogged down very quickly."[105]

Not everyone adopted that stance. In the 1938 letter to Cairns quoted above, on a "hopeless case," Richardson also wrote: "You might write to Veblen. That would seem the only possibility." Like his colleague Hermann Weyl, who ran a German Mathematicians' Relief Fund during the period, Veblen no longer restricted his efforts to the eminent. Nor were they alone. John R. Kline wrote to Courant in 1938 about an Austrian who "is not an outstanding mathematician and it may be difficult to do anything for him but still the need is extremely desperate and human feeling makes us wish to do anything that is at all possible."[106]

Veblen's actions rested on more than humanitarian considerations. Almost at the very time Richardson saw a saturated job market, Veblen asserted to Karl Menger, "our power of assimilation in this country is not yet exhausted."[107] Certainly, the university authorities did not agree with Veblen. Under the leadership of President Conant of Harvard, many joined in a drive to raise an endowment for aiding refugees as the existing funds were so limited. Many university authorities were anxious to shoulder the Emergency Committee aside in order to reaffirm control over faculty selections. Without new funds, Conant and his allies saw little hope for other refugees.[108]

What Veblen had in mind becomes clear from the refugees files that he and Weyl maintained. Not only were they aiding the noneminent, but also the two men had long stopped limiting placements to institutions with research capabilities. Veblen was now placing refugees in any willing four-year college or even in junior colleges. In these moves Weyl and Veblen had the cooperation of Harlow Shapley, head of the Harvard College Observatory. The matter was sensitive; it was, after all, the upgrading called for in the 1935 article on the job market and in Richardson's 1936 piece.

If teaching posts at these lesser institutions were conspicuously filled by refugees, then the existence of a substantial number of unemployed native-born mathematicians might lead to the feared nationalistic and anti-Semitic backlash. To avoid a clash, Veblen needed the agreement of Birkhoff, whose influence stemmed both from intellectual prominence and the role of Harvard as the leading undergraduate source of American mathematicians. On 24 May 1939, Shapley wrote Weyl: "When Veblen and Birkhoff were in my office the other day, it was agreed that the distribution of these first-rate and second-rate men among smaller American institutions would in the long run be very advantageous, providing at the same time we defended not too feebly the inherent rights of our own graduate students."[109]

The meeting occurred just as the Society decided to found an abstracting journal. The Society made another decision in September 1939. It formed a War Preparedness Committee, chaired by Veblen's colleague, Marston Morse. Evans and Richardson at first wanted to bar participation of individuals with German names. Morse smoothly avoided that position by insisting "it is important that such German names as we have represent the best possible choices."[110] The Society now moved, modestly but unequivocally, toward the time in World War II when mathematicians were in short supply and refugee skills could help to free the Old World from the Nazi blight.

Nazi policies contributed to an upgrading of the status of applied mathematics in the United States. To men like E. B. Wilson, a friend of Richardson from his student days at Yale, pure mathematicians had blocked applied mathematics by the control of the department at Harvard and of the section in the National Academy of Sciences.[111] To Courant in 1927, advising the Rockefeller Foundation, U.S. mathematicians were too abstract, specifically pointing to the work in topology. In 1941 he ascribed this tendency as a reaction against superficial utilitarianism.[112]

Other mathematicians were rather proud of the emphasis. In his contribution on algebra to the semicentennial, Eric Temple Bell wrote:

> It may be said at once that American algebra, contrary to what some social theorists might anticipate, has not been distinctly different from algebra anywhere else during the fifty-year period. According to a popular theory, American algebraists should have shown a preference for the immediately practical, say refinements in the numerical solution of equations occurring in engineering, or perfections of vector analysis useful in physics. But they did not. The same topics . . . were fashionable here when they were elsewhere, and no algebraist seems to have been greatly distressed because he could see no application of his work to science and engineering. If anything, algebra in America showed a tendency to abstractness earlier than elsewhere.[113]

Veblen made a related point earlier, in 1929, in opposing a proposal for AMS to found a journal of applied mathematics: "the second reason [after financial problems] is that I do not believe that there is, properly speaking, such a thing as applied mathematics. There is a British illusion to that effect. But there is such a thing as physics in which mathematics is freely used as a tool. There is also engineering, chemistry, economics, etc., in which mathematics play a similar role, but the interest of all these sciences are distinct from each other and from mathematics."[114] From this position, Veblen's priority for pure research over both teaching and applications arose naturally. He assumed that good teaching and effective applications flowed best from abstract theory.

Norbert Wiener, Richard Courant, and R. G. D. Richardson probably agreed with Veblen's position except for one crucial point. Wiener once referred to the importance "of a physical attitude."[115] Courant came from a tradition which assumed that the other sciences presented raw data for the labors of mathematicians, as in his work on Plateau's problem. One suspects that Veblen and others preferred to see mathematics simply as abstract games with symbols with considerations of reality and utility totally absent.

Richardson saw an opportunity to aid industry and engineering in the coming of the exiles. In 1941 he launched a summer school at Brown which later developed into an institute of applied mechanics. Richardson described the United States as the world leader in pure mathematics and high in a few "applied" fields like statistics and mathematical physics. He proposed to correct the lag in applied mechanics, particularly fluid dynamics and elasticity. The faculty at Brown in this area was overwhelmingly composed of German refugees.[116]

At the same time Richard Courant tried to launch a more ambitious institute for basic and applied sciences at New York University. Richardson's move killed Courant's 1941 initiative. According to Courant's biographer, Richardson was motivated, in part, by his antipathy to the foreigner Courant.[117] Resiliently, Courant obtained war contracts and later launched what is now the Courant Institute of the Mathematical Sciences, a name reflecting a viewpoint congenial to the founder's Göttingen roots and thus different from Veblen's position. The existence of a strong prewar belief in pure mathematics helps explain why World War II led not only to an efflorescence of applied mathematics but also to a great growth of abstract research.[118]

Conclusion

Even before the United States joined the conflict, writings appeared appraising the meaning of the transfer of cultural skills across the ocean. To the chemist C. A. Browne in *Science,* it was a rerun of history, recalling the Göttingen seven and the Forty-Eighters: "That Germany should now repeat on a vastly greater scale the tyrannical follies of a century ago seems too incredible for belief." Making no mention of anti-Semitism, he noted how well earlier German migrants had assimilated, enriching the country.[119] Like Browne, other writers stressed both Nazi folly and American precedent. In 1942 Arnold Dresden of Swarthmore gave a listing of refugee mathematicians, opening and closing with an account of Joseph Priestley's 1794 arrival and welcome by the American Philosophical Society. The recent migrants became another example of America's traditional role as a haven for the oppressed of Europe.[120] And in 1943, replying to a letter from Weyl about a German refugee mathematician in Chile, Griffith C. Evans expressed interest, noting: "Certainly at Berkeley we are proud of Lewy, Tarski, Wolf and Neyman."[121] It was quite a contrast to his 1934 exchange with Veblen.

The growth of the American mathematical community clearly helped absorb the individuals in the migration. At the same time, this growth, coinciding with the Depression, created a problem for the refugees. Insecurity about economic conditions reinforced insecurity about status in the world community of mathematicians. Despite these persisting feelings, the statistics disclose a different situation. In the decade of the 1920s, American universities granted a total of 351 doctorates in mathematics. The number rose to 780 in the following decade. Some universities had spectacular expansions in the same period. Princeton went from 14 to 40 in doctorates granted; MIT from 5 to 32; Michigan from 11 to 59; and Wisconsin from 8 to 32.[122] Although published figures on mathematics alone do not exist, the total number of teachers in higher education in the period 1930–1940 rose approximately 77 percent.[123] The job market problem was largely one of accommodating the increase in home-grown Ph.D.'s. Despite Depression conditions, somehow many mathematicians, including refugees, did get posts, if not always matching aspirations. As in the case of physics in the United States, the Depression period was one of growth for mathematics in terms of both quality and quantity.[124]

At the same time American mathematicians and physicists in 1933/34 were noticeably sensitive about the perils of nationalism and anti-Semitism. Afterward, in his defense against charges of Communism, Veblen described his efforts to place refugees as seriously encountering "the opposition of American anti-Semites."[125] Both perils existed, and the latter certainly affected young Americans born into the wrong faith entering learned professions in the interwar years. In retrospect, the fears expressed in 1933 and 1934 were overstated. Recent historical studies have pointed to the New Deal as the period in which the concept of a pluralistic society took root and even as a source for much of the post–World War II expansion of civil rights.[126] Foreign observers in the early 1930s reacted differently from most of the Americans. From a comparative prospect, they noted the absence of the virulent, pathological nationalism and anti-Semitism of many European countries.[127] Certainly, the American university with all its peculiarities appears innocently idyllic in contrast to the German university at the end the Weimar era. The Nazi era had the ironic effect of ending this idyll by disclosing the dangers of anti-Semitism.

Like all significant historical events, the reception of the emigré mathematicians was filled with ambiguities and contradictions reflecting the complexity of the situation. Like Richardson, many Americans were influenced both by altruism and cool calculations of national and institutional advantage. It is always gratifying to do well while doing good. Precisely disentangling the two motives is impossible. Veblen clearly convinced himself that his course favored both the refugees and the future of mathematics in the United States. Others were not so certain.

Judgments of motives are hazardous. Birkhoff does nominate Lefschetz and aids individual refugees despite his stated position. Evans refuses Courant in strong nationalistic terms and then hires Lewy. Evans and Richardson stood up for internationalism, as they saw it, in the founding of *Mathematical Reviews* but later wanted to bar the War Preparedness Committee not only to refugees but also to Americans of German ancestry. Instead of implacable historical forces or clear-cut social processes, the sources disclose troubled, inconsistent humans struggling to do right, however defined, in the face of opposing or indifferent trends. The mathematicians were just like everyone else in this respect.

Time and place clearly mattered—what Schmidt meant by mathematicians being unable to escape from the realities of the world. By the 1930s the American mathematical community had a leadership firmly committed to the primacy of pure mathematics and a general membership earning their livelihood by teaching elementary branches of the subject to nonmathematicians. Since applied mathematics, as we know it today, existed on a most modest scale, undergraduate teaching was seen as the utilitarian function justifying support of pure mathematics. Veblen and others in the interwar years attempted to broaden the utilitarian grounds for support with limited success. Birkhoff's concern for sharing the teaching load was far more typical than Veblen's absolute priority for research. Independent of any prejudices, the structural features of higher education and of the mathematical community might have frustrated the successful reception of the refugees. Even in 1939 Veblen needed Birkhoff's cooperation.

Outcomes did not flow simply from true or supposed realities of the world. Schmidt's position implied a social determinism beyond human interference. At every point alternatives existed, many quite reasonable by the standards of the day. What actually happened was partly a triumph of two ideals deliberately or tacitly accepted by the leaders of the mathematical community.

One ideal was nationalistic. Despite the immigration quotas and the economic crisis, there were Americans in the 1930s who wanted to believe in the reality of the traditional view of the United States as a refuge from tyranny. From the deliberations in the Rockefeller Foundation to Dresden's article of 1942, there was a desire to live up to that ideal.

Whatever its relation to a complex and obscure reality, the ideal of the universality of science clearly appealed to the hearts and minds of the mathematicians. The commitment to pure research in the most abstract of all fields undoubtedly reinforced the will to believe in this ideal.

Faced with Nazism abroad and its victims in the United States, the leaders of the mathematical community persisted in voting, in effect, for a world in which these two ideals were to prevail. The decision to establish the War Preparedness Committee in September 1939 was the culmination of a process of politicization. As pure mathematicians and as U.S. citizens, the leading mathematicians genuinely believed in disciplinary avoidance of both domestic and foreign politics. As far as surviving sources permit judgment, they were largely of moderate or even conservative political leanings. Leftist views were very scarce. As an activist, Oswald Veblen (the self-described old-fashioned liberal) was atypical. Like his more famous uncle, a hint of prairie populism clung to the man.

As events occurred in Europe and as the small number of refugee mathematicians settled into U.S. posts, the community of mathematicians progressively lost its belief in political neutrality and, more and more, assumed an explicitly hostile stance to Nazi Germany. Increasingly, that country was seen as a menace both to national ideals and to the ideals of science.

Notes

1. For the British effort, see Walter Adams, "The Refugee Scholars of the 1930's," *The Political Quarterly* 39 (1968):7–14; Norman Bentwich, *The Rescue and Achievement of Refugee Scholars* . . . (The Hague, 1954); and A. J. Sherman, *Island Refuge* . . . (Berkeley, 1973). The best general study for the United States is Maurice R. Davie, *Refugees in America* . . . (New York, 1947). Pertinent to this study is Donald Fleming and Bernard Bailyn (eds.), *The Intellectual Migration: Europe and America, 1930–1960* (Cambridge, Mass., 1969).

2. For the former, see Stephan Duggen and Betty Drury, *The Rescue of Science and Learning* . . . (New York, 1952). For the Foundation, see Thomas B. Appleget, "The Foundation's Experience with Refugee Scholars," 5 March 1946, in RG1/Ser. 200/Box 47/Folder 545a, RFA.

This and many later notes cite manuscript sources. Here is the list of abbreviations used to indicate them:

AMS   Archives, American Mathematical Society, Lehigh University, Bethlehem, Pa.
BHA   George David Birkhoff Papers, Harvard University Archives
CIMS   Archives, Courant Institute of Mathematical Sciences, New York University

GCE   Griffith C. Evans Papers, Bancroft Library, University of
        California at Berkeley
NW   Norbert Wiener Papers, Archives of the Massachusetts Institute
        of Technology
OV   Oswald Veblen Papers, Manuscript Division, Library of Congress
RBA   R. G. D. Richardson Papers, Archives, Brown University,
        Providence, R.I.
RFA   Archives, Rockefeller Foundation, Rockefeller Archive Center,
        North Tarrytown, N.Y. (cited above in this note)

3. Norman Bentwich, *The Refugees from Germany, April 1933 to December 1935* (London, 1936), p. 174.

4. The 1939 count is from Arnold Dresden, "The Migration of Mathematics," *American Mathematical Monthly* 49 (1942):415–429. The listing in Dresden is both incomplete chronologically and rather peculiar in some specifics. The range given here is an extrapolation. None of the sources really attempts to identify individuals who were not yet visible as members of a professional community at the time of migration, even though some entered into the field in the United States. An interesting source is Max Pinl and Lux Furtmüller, "Mathematicians under Hitler," Leo Baeck Institute, *Yearbook XVIII* (London, 1973), a revision of Pinl's series in *Jahresbericht der Deutschen Mathematiker-Vereinigung* under the title "Kollegen in einer dunklen Zeit," 71 (1969):167–228; 72 (1971):165–189; 73 (1972):153–208; and 75 (1974):166–208.

5. Based on Appleget, "The Foundation's Experience," and on Duggan and Drury, *Rescue of Science*.

6. As evidenced by many documents in 2/717/91/725 and 726, RFA, and elsewhere in the same collection.

7. In 2/717/92/731, RFA.

8. In Sherman, *Island Refuge,* p. 259.

9. See Adams, "Refugee Scholars." For contemporary comments see R. A. Lambert to A. Gregg, 20 May 1933 (2/717/91/726); Lambert memo of 2 October 1933 (2/717/92/729); Lambert diary entry, 2 July 1934 (2/717/109/840)—all in RFA. In his reminiscences, the physicist Hans Bethe recalled the distinction between Britain and the United States: "In America, people made me feel at once that I was going to be an American." Jeremy Bernstein, "Master of the Trade," *The New Yorker* (3 December 1979):100. Bethe noted about England: "it was clear that there I was a foreigner and would remain a foreigner."

10. In his autobiography, *I Am a Mathematician . . .* (Cambridge, Mass., 1964), p. 175, Norbert Wiener names Veblen and John R. Kline. The latter shows up in sources known to me but not as significantly as does Richardson. I have not located Kline's papers which may say more on his role. Kline succeeded Richardson as secretary of the American Mathematical Society in 1941.

11. Veblen to Richardson, 6 May 1935, old file, AMS. These records consist of a "new file" (an attempt to reorganize which went only part way) and the "old file" (largely an alphabetical correspondence file).

12. Veblen to H. L. Stimson, 8 December 1943; to Col. Leslie Simon, 30 September 1946; to Simon, 5 February 1947; and item 11 in Box 21, Summary of Defense, all in OV.

13. Richardson to Max Mason, 20 February 1929, 1.1/200/125/154, RFA.

Veblen's papers contain charts, graphs, and tables about the improvement in mathematics in this period.

14. Richardson to G. D. Birkhoff, 23 February 1932, BHA; Richardson to O. D. Kellog, 1 March 1932, RBA; minute of Board of Trustees, AMS, 2 January 1932, Box 20, OV.

15. The specific spur to forming MAA was the AMS refusal to take over the *American Mathematical Monthly*. In 1938 R. C. Archibald of Brown, who favored a research emphasis, took issue with Richardson's *ex post facto* characterization of MAA as "child of AMS," stressing the distinction. Richardson appealed to his contemporaries for reassurance on the closeness of the two. See Richardson to E. R. Hedrick and T. S. Finke, 15 October 1938, and Hedrick's reply of 21 October 1938, in old file, AMS.

16. See Veblen to J. W. Alexander, 1 May 1923, OV.

17. Davie, *Refugees in America,* p. 307. How this influenced social relations is shown by two anecdotes in S. M. Ulam, *Adventures of a Mathematician* (New York, 1976), pp. 90, 119.

18. E. J. Moulton, "The Unemployment Situation for Ph.D.s in Mathematics," *American Mathematical Monthly* 42 (1935):143–144. See also K. P. Williams and Elizabeth Rutherford, "An Analysis of Undergraduate Schools Attended by Mathematicians," *School and Society* 38 (1933):513–516; and "Report on the Training and Utilization of Advanced Students of Mathematics," prepared for MAA, *American Mathematical Monthly* 42 (1935):263–277.

19. Ibid. 41 (1934):612–613; R. L. Wilder to Richardson, 16 May 1935, RBA.

20. R. G. D. Richardson "The Ph.D. Degree and Mathematical Research," *American Mathematical Monthly* 43 (1936):199–215.

21. A similar conclusion was voiced by T. C. Fry to Richardson, 12 July 1935, in Semicentennial Correspondence, AMS.

22. Richardson to Birkhoff, 17 May 1927, RBA.

23. *American Mathematical Monthly* 39 (1932):126.

24. Wiener to Otto Szasz, 13 August 1933, NW.

25. For example, Veblen to B. L. van der Waerden, 18 December 1933, OV.

26. Cable to Richardson, 30 June 1934, old file, AMS.

27. See *Jewish Advocate* for 14 December 1934, 1:4.

28. Landauer to Wiener, 7 January 1935, NW.

29. Bliss to Richardson, 10 April 1935, in Lewy file, RBA.

30. Charles H. Warren to Richardson, 28 November 1941, RBA.

31. See, for example, the treatment of exclusionary practices in Marcia G. Synnott, *The Half-Opened Door: Discrimination and Admissions at Harvard, Yale, and Princeton, 1900–1970* (Westport, Conn.: Greenwood Press, 1978).

32. Slaught to Richardson, 23 January 1931, RBA.

33. D. Buchanan to Birkhoff, 18 August 1937; W. M. Smith to Birkhoff, 9 February 1937, BHA.

34. Coble to Richardson, 30 March 1935, old file, AMS.

35. Paul P. Boyd to Richardson, 18 March 1935, Lewy file, RBA. But the president of the university thought otherwise, expressing a desire for more men with European training.

36. K. P. Williams to Veblen, 6 May 1938, OV. Also see S. A. Mitchell to Veblen, 14 December 1935, OV.

37. K. T. Compton, Memorandum of a conversation with Norbert Wiener, 13 May 1935, NW. See Wiener's comments in *I Am a Mathematician,* pp. 180, 211, relating to this incident.

38. Birkhoff to Richardson, 18 May 1934, RBA.

39. Morse to Veblen, 12 September 1934, OV.

40. N. Levinson to Wiener, 1 October 1936, NW. For Wiener's view, see *I Am a Mathematician,* pp. 27–28.

41. AMS *Bulletin* (2) 38 (1932):337.

42. In RBA. T. B. Appleget of the Rockefeller Foundation was a Brown graduate, and RFA has a number of relevant letters between him and Richardson in the spring and summer of 1933.

43. Richardson to Veblen, 3 July 1933, OV; Richardson to committee, 25 July 1933, old file, AMS.

44. Richardson to B. S. Deutsch, 15 July 1934, RBA.

45. J. R. Kline to Veblen, 23 November 1933, OV.

46. Richardson to Bohr, 5 April 1935, RBA.

47. In his diary entry for 24 June 1934, Warren Weaver reported the critical comments of the physicist F. K. Richtmyer of Cornell, then head of the Division of Physical Sciences of the National Research Council. Richtmyer objected to Lewy's favored treatment while so many were unemployed. 2/717/109/839, RFA.

48. Constance Reid, *Courant in Göttingen and New York* . . . (New York, 1976), p. 227; Richardson to H. Bohr, 26 December 1933, OV; Richardson to Birkhoff, 21 July 1936, BHA.

49. Veblen to J. H. Hildebrand, 23 January and 9 May 1933, OV.

50. Evans to Veblen, 16 January 1934, OV. As part of the upgrading, two American mathematicians were dismissed and four others placed on notice. Perhaps this influenced Evans. C. A. Noble to Evans, 23 October 1933, Carton I, GCE.

51. Veblen to Evans, 23 January 1934, OV.

52. Evans to Richardson, 18 April 1934, Displaced German Scholars file, RBA.

53. Veblen to Chancellor Chase, 13 November 1933; to E. B. Wilson, 13 April 1936, OV.

54. G. C. Evans to Courant, 11 April 1939, Master Index file; Courant to G. C. Evans, 8 December 1949, General file—both in CIMS.

55. Evans to Richardson, 1 May 1935, Lewy file, RBA.

56. Evans to Birkhoff, 13 April 1932, BHA.

57. Extract of John R. Kline to Richardson, 18 January 1934, in Richardson folder, OV.

58. G. H. Hardy, "The J-type and the S-type Among Mathematicians," *Nature* 134 (1934):250; also in Hardy's *Works,* vol. 7 (1979), pp. 610–611. Veblen to Bieberbach, 19 May 1934, RBA. Ludwig Bieberbach, "Personlichkeitsstruktur und mathematisches Schaffen," *Forschungun und Fortschritte* 10 (1934): 235–237.

59. *American Mathematical Monthly* 41 (1934):433.

60. 21 July 1936, BHA.

61. Taken from the Göttingen Celebration file in AMS. Noble's letter is dated 10 April 1937, E. T. Bell's 9 April, and W. A. Wilson's 8 April.

62. Based on the Semicentennial Celebration folder in AMS, which is largely from 1935. See also E. R. Hedrick to L. P. Eisenhart, 30 January 1937, AMS.

63. In vol. 2 of AMS, *Semicentennial Publications* (New York, 1938), pp. 276–277. Reid, *Courant,* pp. 211–213.

64. 30 September 1938, BHA.

65. 20 September 1938, BHA.

66. There are later echoes of this incident in the literature. It was referred to indirectly in Marston Morse's obituary: "Birkhoff was at the same time internationally minded and pro-American." Being detached from the world, his social and political views gave rise to misunderstandings (see Birkhoff, *Collected works,* vol. 1 (1950), p. xxiv). Veblen's necrology of his deceased colleague specifically discussed the address: "It may be added that a sort of religious devotion to American mathematics as a 'cause' was characteristic of a good many of his predecessors and contemporaries. It undoubtedly helped the growth of the science during this period. By now mathematics is perhaps strong enough to be less nationalistic. The American mathematical community has at least been healthy enough to absorb a pretty substantial number of European mathematicians without serious complaint" (ibid., p. xxff).

Garrett Birkhoff recalled the incident when he discussed the problem from the standpoint of a young American mathematician (see J. D. Tarwater, J. T. White, and J. D. Miller [eds.], *Men and Institutions in American Mathematics* [Lubbock, Tex., 1976], p. 66). For a recent comment by an immigrant, see Ulam, *Adventures,* p. 101.

67. In Richardson file, February 1936, OV.

68. Evans to Richardson, 30 March 1936, new file, AMS.

69. 6 January 1936, RBA.

70. Courant to Weaver, 6 December 1936, in Master Index file, CIMS. Both Courant and Lewy were indicating an interest in the nonresearch tasks Evans and others feared were neglected by research-oriented refugees.

71. Evans folder, new file, AMS, especially Evans to Richardson, 13 April 1939, the Richardson's reply of 1 May. In the Special Funds folder, old file, AMS, is E. B. Vleck to Richardson, 18 November 1921. Richardson is clearly looking out for Canadians, although his letter of 1 May 1939 indicates some discomfort on how the policy developed.

72. H. E. Slaught, "Subsidy Funds for Mathematical Projects," *Science* [n.s.] 55 (1922):1–3.

73. Veblen to Richardson, 4 August 1933; Richardson to Veblen, 9 August 1933; Veblen to Richardson, 12 August 1933—all in Displaced Scholars file, RBA.

74. T. Y. Thomas to Veblen, 13 July 1935, OV.

75. Rudolph E. Langer to Birkhoff, 22 October 1936, BHA.

76. J. D. Tamarkin to C. N. Moore, 17 August 1942, RBA.

77. Evans to Richardson, 1 July 1936, Evans correspondence, new file, AMS.

78. Richardson to James McKean Cattell, 5 October 1936, new file, AMS.

79. H. Bohr to Veblen, 11 March 1938; Veblen to J. H. C. Whitehead, 22 November 1938, OV. Richardson to Evans, 27 November 1939, *Mathematical Reviews* file, new file, AMS.

80.　See his penciled draft of "The abstract journal problem: historical background," Box 17, OV.

81.　Synge to Richardson, 9 December 1938, and Synge to Ferdinand Springer, 19 January 1939, both in Box 17, OV.

82.　Richardson to Birkhoff, 18 January 1939, BHA, which also discloses opposition to Neugebauer's coming in Richardson's department. The later clash brought out explicitly the differences between the research-oriented Veblen and the Society-oriented Richardson: "Veblen is not interested in helping build up the Society. He thinks it is being run by a group of mediocrities and has even suggested that there be a new organization of a limited number of people who are actually doing high grade research . . . [and is a] bit comtemptuous of the ordinary way of looking at things" (Richardson to Mark Ingraham, 24 January 1940, RBA).

83.　4 December 1938, OV. Evans to Richardson, telegraph, n.d. [December 1938?], in *Mathematical Reviews* file, new file, AMS. Evans did tell Richardson to go ahead if the Lefschetz suggestion was not acceptable.

84.　Hedrick to Richardson, 9 December 1938, *Mathematical Reviews* file, new file, AMS.

85.　Richardson to Veblen, 31 December 1938, Box 17, OV; Richardson to Evans, 4 January 1938, Evans file, new file, AMS. See also Richardson to Committee, 9 January 1939, Box 17, OV.

86.　Richardson to Veblen, 11 January 1939, Box 17, OV. Richardson to Hardy, 8 April 1939, in Richardson file, OV, which also contains other comments about these events.

87.　The quotation is from Springer to Veblen, 12 January 1939, which was in response to Veblen's letter of 5 December 1938, Box 17, OV. The refusal to sell is in Springer to C. R. Adams, 24 April 1939, filed under F. K. Schmidt in OV.

88.　Rudolph E. Langer to C. R. Adams, 21 February 1939, and Mark Ingraham to Adams, 21 February 1939, in Box 17, OV.

89.　Lefschetz to Richardson, 18 May 1939, OV.

90.　C. R. Adams to Committee, 11 April 1939, enclosing translated extract of Hasse to Stone of 15 March 1939 which Stone received on 29 March, in Box 17, OV. See S. L. Segal, "Helmut Hasse in 1934," *Historia Mathematica* 7 (1980):46–56.

91.　For example, see the letter to Hardy cited in n. 86.

92.　Copy in Box 17, OV.

93.　See Adams to Coble, 3 and 8 May 1939; also Schmidt to Richardson, 15 May 1939—all in Box 17, OV.

94.　T. C. Fry to Adams, 19 May 1939, BHA.

An oral tradition exists of anti-Semitism in the *Fortschritte*. Whether true or not, it did not figure specifically in the events in the United States.

95.　Graustein to Adams, 11 May 1939, Box 17, OV.

96.　Richardson to Veblen, 13 May 1939, Box 17, OV; Richardson to Evans, 28 June 1939, Evans file, new file, AMS.

97.　Adams to Committee, 17 May 1939, Box 17, OV.

98.　Stone to AMS Committee, 24 May 1939, enclosing minutes of meeting, Carton XIV, GCE.

99.　Draft, Wiener to Graustein, 19 May 1939, NW. It is uncertain whether the letter was sent.

100.   Richardson to Evans, 25 May 1939, enclosing his letter of 24 May to AMS Committee, Carton XIV, GCE.

101.   Birkhoff to Committee, 22 May 1939, Box 17, OV.

102.   25 May 1939, in *Mathematical Reviews* file, new file, AMS. In his obituary of Birkhoff, Veblen noted that he loyally worked for the new publication "after main issues decided against his judgment." See Birkhoff's *Collected Papers,* vol. 1 (1950), p. xxi.

103.   In 1947/48, when F. K. Schmidt and others decided to revive a German abstracting journal for mathematics, both Veblen and Courant were bothered by what Courant described as the "aggressive German nationalistic attitude." (Schmidt to Courant, 1 December 1947; Courant to Schmidt, 14 January 1948; Courant to Veblen, 3 February 1948—all in Master Index file, CIMS. Veblen to F. K. Schmidt, 25 February 1948, OV.) In 1938/39, Courant stayed out of the discussions because of his relations with the Springer firm.

104.   23 May 1938; in Cairns file, new file, AMS.

105.   Richardson to Weyl, 25 June 1941, RBA.

106.   12 December 1938, General file, CIMS.

107.   22 July 1938, OV.

108.   See Duggan and Drury, *Rescue of Science,* pp. 96–101; and the comments in David C. Thomson, "The United States and the Academic Exiles," *Queens Quarterly* (Summer 1939):212–225. This proposal continued the theme of protecting young American scholars while clearly reflecting an impatience with a process in which a nonacademic body exerted pressure.

109.   In Box 29, OV.

110.   Morse to Evans, Carton XIV, GCE.

111.   See E. B. Wilson to H. Shapley, 5 November 1925 and 21 March 1928, Shapley Papers, Harvard Archives. I am indebted to Karl Hufbauer for these references. For Wilson's views on applied mathematics, see his letter to Richardson, 11 June 1941, in RBA.

112.   See A. Trowbridge to W. Rose, 27 May 1927, in International Education Board, 1/1/110, RFA, and related documents in same location; also "the little black book" with its ratings of mathematicians in 1/10/142. Courant's 1941 comment is in his draft memorandum on "A National Institute for Advanced Instruction in Basic and Applied Science," p. 2, copy in Richardson Papers, RBA.

113.   AMS, *Semicentennial Publications* (New York, 1939), vol. 2, p. 1.

114.   Veblen to Richardson, 5 April 1929, in the R. L. Moore 1931–1935 folder, old file, AMS.

115.   Wiener, *I Am a Mathematician,* p. 34.

116.   For Richardson's views, see Richardson to Birkhoff, 10 April 1937, Birkhoff Papers, BHA; Richardson to Courant, 18 March 1941, with enclosed proposal, Master Index file, CIMS.

117.   Reid, *Courant,* p. 228f; Courant to T. Saville, 19 March 1941, in Richardson folder, Master Index file, CIMS.

118.   For some indications of the development of applied mathematics in the United States in World War II, see the recollections of M. Rees in "The Mathematical Sciences and World War II," *American Mathematical Monthly* 87 (1980):607–621.

119.   C. A. Browne, "The Role of Refugees in the History of American Science," *Science* [n.s.] 91 (1940):203–208.

120. Dresden, "The Migration of Mathematics."

121. Feb. 11, 1943, in Frucht file, Box 30, OV. The addition of a Czech and two Polish mathematicians indicates the widening effect of Nazism.

122. Lindsay Harmon and Herbert Soldz, *Doctorate Production in United States Universities* . . . (Washington, D.C., 1963), from tables in Appendix 1.

123. U.S. Bureau of the Census, *Historical Statistics of the United States* (Washington, D.C., 1957), p. 210.

124. Spencer R. Weart, "The Physics Business in America, 1919–1940: A Statistical Reconnaissance," in N. Reingold (ed.), *The Sciences in the American Context: New Perspectives* (Washington, D.C., 1979), pp. 295–358.

125. Duggan and Drury, *Rescue of Science,* p. 68; Veblen's Summary of Defense, item 11, Box 21, OV.

126. See, for example, the recent article by Richard Weiss, "Ethnicity and Reform: Minorities and the Ambience of the Depression Years," *Journal of American History* 66 (1979):566–585.

127. In RFA, 2/717/92/730, is a note of a conversation of 6 November 1933 between A. V. Hill and R. A. Lambert of the Foundation on this point. See Lambert's earlier statement of 2 October 1933 in 2/717/92/729.

13. Vannevar Bush's New Deal for Research; or, The Triumph of the Old Order

For many years I had a sense of uneasiness about explanations of how the World War II experience transformed the role of research and development in the United States. There were, of course, first-rate examples of historical research to which we are all indebted, but all too often assertions of a fundamental change of status had very little in the way of supporting analysis.

One source of my uneasiness was the obvious impact of previously existing policies, organizations, and folkways of the research community. I perceived degrees of continuity with the past, not an abrupt schism. Given the past history, for example, it was no surprise to me that a member of the Army's Corps of Engineers headed the Manhattan Project. Less obvious as a source of uneasiness was the wholly understandable and necessary emphasis on a number of rather spectacular big events and themes. The coming of nuclear warfare and its impact on international security is one example. To my way of thinking, this resulted in a literature giving considerable attention to the atomic scientists movement and the hydrogen bomb decision, often at the expense of less spectacular long-term trends of great, if not greater, significance.

Academics properly noted and applauded the troubled gestation of the National Science Foundation and then the later appearance of the National Endowments for the Arts and the Humanities. I joined in the applause without voicing my sense that long-term trends in the U.S. government would have resulted in such support in one form or another. All God's children had a claim on federal largesse, or so it seemed, before the Reagan presidency. I never believed that the fate of learning or of the nation hinged on whether or not scholars could get a particular kind of support. A case can be argued that the period between the two World Wars in the United States was a golden age for the growth of science, the arts, and various kinds of learning. The federal presence then was not so visible or so essential.

What was lacking in the literature then (a change is now taking place) was a statistical sense leading to a critical scrutiny of where the action really was occurring. The most obvious was the large defense research and development operations *not* focused on ultimate weapons of destruction. Given the size of the expenditures, the defense research and development has had consequences we are still assessing. Something similar can be said about the rise of a health sector in our economy and its vast research support by the National Institutes of Health. By contrast, the relative sparsity of federal aid for industry in general and for the environment until relatively recently has contributed to a different set of outcomes. I also feel a need for more studies of the private sector, particularly industry and the foundations. The federal patron is far from the whole story.

In the literature Vannevar Bush appears prominently but cryptically. The organizer of great wartime programs of research and development writes a report on postwar science policy, *Science—The Endless Frontier*, that ultimately leads to the National Science Foundation. A few years before writing this article I began an intensive hunt through various bodies of unpublished personal papers and institutional records. Bush just fascinated me as a person and as a historical challenge. Always allowing a few honorable exceptions, the existing writings did not make sense of the man I found in the sources.

What follows is an attempt to sketch what made Bush tick in order to explain two things. First, what did he intend by the recommendations in *Science—The Endless Frontier*? Second, why was his program largely disregarded? I do not see Bush as the progenitor of a new post–World War II order but as a Moses who pointed out the way to a promised land, remained behind while others moved into the new domain, and realized to his dismay that the new promised land did not match his vision. This is a sketch of Bush as a loser, as a believer in certain old-fashioned virtues others chose to ignore. This is a sketch of a very impressive human being in whose losing we may have all sustained wounds.

A version of this article appeared in *Historical Studies in the Physical and Biological Sciences* in 1987.

Introduction

In 1970 two old friends corresponded about the past. James B. Conant, president emeritus of Harvard, had just published a brief piece, "An Old Man Looks Back: Science and the Federal Government, 1945–1950." Written during the campus disturbances of the Vietnam years, it concluded with an "I told you so." Not adopting Vannevar Bush's proposals for postwar science resulted in the troubling inordinate role of the military in the nation's research and development.[1]

Bush's response to Conant's essay differed significantly from the tone of his recently published autobiography, *Pieces of the Action*, with its generally rosy picture of the United States in the opening pages and in other passages. Bush praises how far the United States has come while expressing reservations about this, that, and the other aspect of the contemporary scene. In the book he even mistakenly implies with satisfaction that the National Academy of Sciences (NAS) provides (by right?) the names of the members of the President's Science Advisory Committee.[2] Privately to Conant, Bush wrote almost an expression of dismay at the current scene. The two men apparently agreed that the course of the Office of Naval Research (ONR) and of the National Science Foundation (NSF) frustrated their hopes for the nation. To Conant, Bush recalled, not surprisingly, the unpleasantness of "those few years after the war." What is surprising is his almost vehement insistence that what he had done in the past was because of the National Academy of Sciences.[3]

What follows is a gloss on the text of that exchange between Conant and Bush, or, if you wish, an essay in revisionism. This is an attempt to explain what Bush intended and why that was frustrated. While not denying the existence of Senator Harley N. Kilgore and his postwar plans, this essay tries to place that motivation of Bush's within a broader framework.[4] Nor is there much respect here for the significance of the argument between President Truman and others over whether or not the director of the proposed National Science Foundation be a presidential appointee or appointed by the National Science Board.[5] With the passage of time that issue seems like much sound and fury over relatively little. Nor am I inclined to go along with the unctuous judgment of NSF's first director, Allan T. Waterman, that his agency largely fulfilled the expectations of Bush's 1945 report *Science—The Endless Frontier*.[6]

Bush's image today derives from the great report, the autobiography, an oral tradition of the exciting days of the Office of Scientific Research and Development (OSRD) in World War II, and small, growing body of historical writing. Still lacking are reliable studies of many important topics amply documented by available massive archival sources and personal papers. The underlying thesis here is that Vannevar Bush was a creative anachronism, someone with many viewpoints not wholly representative of his era but who attempted to adapt his viewpoints to the exigencies of the moment and the perceived shape of the future. In *Science—The Endless Frontier* (SEF), the very conservative Vannevar Bush proposed a new deal for research consonant with some strands of the historic New Deal. The National Academy of Sciences and the armed services had key roles in the development of his views and in the frustration of his hopes. Quite a number of individuals and groups—even those ostensibly favoring SEF's proposals—misunderstood Bush's rather idiosyncratic position. Consciously or not, they were choosing a different future than Vannevar Bush.

The Mind of Vannevar Bush

In the absence of many needed studies, Vannevar Bush is seen almost in a vacuum, as though the entire array of national research and development policies and programs hinged on whether the proposed National Science Board did or did not appoint the director of NSF. What made him tick? What was in the mind of an electrical engineer and university administrator best known for a pioneer analog computer who just before World War II became the head of a private nonuniversity organization, the Carnegie Institution of Washington (CIW), devoted to basic research with no obvious relationship to electrical engineering or the development of computing devices?

Not everyone liked Vannevar Bush, the man. Opposition to his views produced dislike of the personality, criticism of his excessive self-confidence and "elitist" stances. Even such critics were impressed by the many evidences of intellectual and administrative abilities. But there is ample evidence of Bush's ability to establish warm friendships, relationships furthering the impact of his views. In retrospect Bush comes through as shrewd, witty, and ebullient, fully capable of charming politicians, industrialists, and scientists who might have reservations about particular positions. He was far more of a realistic compromiser than his opponents inside and outside the government realized. In those trying years after World War II, it slowly dawned on him that others were not as ready to compromise on their strongly held beliefs and fervently sought goals.

With the easy, perhaps risky, certainty of hindsight, we can see that Bush had an extraordinary mixture of skills and experiences required for the leadership of the Office of Scientific Research and Development. He was an outstanding electrical engineer of his period. Much of what OSRD accomplished was engineering not science, as Bush later correctly asserted, but engineering with rather distinctive attributes by the standards of the day. Nor did it hurt to have served as Vice President and Dean of Engineering of the Massachusetts Institute of Technology before coming to Washington. That experience was valuable in at least three ways:

First, the role of higher education was the apparent great novelty of the wartime experience, and consequently higher education's role was a major issue in postwar science policy. Familiarity with the ways of the professoriat and of educational administrators eased the conversion to wartime conditions and influenced Bush's thinking as late as the early years of Eisenhower's presidency.

Second, and more specifically, Bush's experiences as an educational administrator was one of the influences on his decisions as to how to organize first the National Defense Research Committee (NDRC) and then OSRD. That organizational preference colored his thinking both in regard to the organization of NSF and of the Defense Department. This was not merely a preference for the comfortably familiar but was linked to strongly held policy views.

Third, Bush's MIT experiences prepared him specifically and by analogy for a number of issues encountered during the war and afterward. The unit administering industrial grants and contracts at MIT before World War II added the increased government relations to its responsibilities after 1940. A separate unit for government grants and contracts came into existence at MIT well over a year after the end of World War II. The relation with the federal patron was seen as another instance of the problem of somehow dealing with an external source of support so as to benefit from the funds without injuring or compromising whatever was defined as the essential characteristic of the university. The university orientation was one of the differences producing strains between Bush and Frank B. Jewett, the president of the National Academy of Sciences and a vice president of the American Telephone and Telegraph Company.

The two had many viewpoints in common, particularly an antipathy to the New Deal. When the two men referred to the "Chief" in their wartime correspondence, they meant Herbert Hoover, not Franklin Delano Roosevelt.[7] Although Bush's autobiography has passages supporting what we now refer to as the "safety net" aspects of the New Deal program, he remained strongly against the welfare state and its centralizing tendencies. During the war years and immediately afterward, Bush took strong exception to what he conceived as dangerous trends counter to traditional American values.[8] In his draft of the report to the President on OSRD during the war, for example, Bush noted the antilabor nature of his postwar proposals in the area of national defense. Bush did not deny the point: "if the shoe fits, let it be put on." He went on to call for no artificial limits to opportunity "to become rich—say it, become rich." Not that Harry Truman, for example, thought otherwise about opportunity, or that Truman could not cite, as Bush did in his draft, the shirt sleeves to shirt sleeves myth of the rise and fall of wealthy families. Individuals like Truman who had been part of the New Deal simply placed issues in a different framework from that of conservatives like Bush and Jewett.[9]

Bush and Jewett had in common a vehement antipathy to government agencies. It was more than an understandable extrapolation from personal experiences giving primacy to the private sector. The reaction was emotional, a deep-seated aversion to organs engaged, in their view, in subverting the American Way. When Harold L. Ickes, the Secretary of the Interior, made an unsuccessful move in 1941 to have the regular agencies play a role in the wartime mobilization of research and development, Jewett commented to Bush that it was against the best interests of the country "to have entrusted the mobilization to the whims of politically appointed and temporary departmental officials."[10] Such officials represented the detested New Deal politics and not the true interests of the nation, let alone science and technology. Implicitly, Bush and Jewett had reservations about the legitimacy of an elected government and its properly appointed administrators.

SEF was a deliberate move against government science, as Kevles previously noted.[11] In the post-SEF grants and contracts age, viewing the federal establishment as an alternative to universities and industries and, therefore, a threat to them may seem strange. But that was the perception of many in the 1940s.[12] Beyond ideology, many believed the federal agencies were largely incapable of top-notch science.[13] Given the subsequent awarding of Nobel Prizes for research performed in the National Institutes of Health, the Naval Research Laboratory, and the Veterans Administration, perhaps the perceived threatening alternatives were not wholly impossible. That aside, the opposition to SEF, covert and otherwise, of the federal establishment helped frustrate Bush's aims.

During the 1940s Bush and Jewett had differing adjustments to the inescapable existence of the federal government. Although Jewett would be the only scientist to speak against expansion of the federal role and for private support in the Senate hearings triggered by SEF, his adjustment was more in harmony with what became the post–World War II order than Bush's.[14] For the period 1940–1950, when the Korean War changed the situation, Vannevar Bush was unsuccessfully fashioning a new political economy for research and development in the United States.

Earlier, during the period of SEF's gestation, Bush worked on and published a quite different statement of his world view. For an engineer, Bush had a surprising zest for the literary. That penchant resulted in the publication in January 1945 of a short, widely reproduced piece, "The Builders."[15] It is a metaphor of the process by which the structure of organized knowledge is built. Bush's metaphor is a rather trite one, clearly the building of a cathedral. The language is fervent and even striving for the poetic. To Bush, the process is highly unorganized even though the edifice is as though "predestined by the laws of logic and the nature of human reasoning." The materials are varied and so are the kinds of participating individuals. Parts of the edifice are in use during construction because applications are taking place;

other parts are merely admired for aesthetic qualities. As to the men, they include those of rare vision, "master workmen," and others of lesser status. Beyond those concerned with the raising of the edifice are individuals who give the structure meaning (philosophers one presumes) and those who "labor to make the utility of the structure real" (engineers and businessmen?). Nor does Bush omit mention of the people bringing food and drink to the workers. There is even a reference to those "who sing to them, and place flowers on the little walls that have grown with the years." "The Builders" is a strongly felt, if gushy, expression of a vision of society whose components are somehow articulating meaningfully although not by any laws of logic and reasoning, let alone any conscious planning by government. It is an organic metaphor useful for a conservative's argument against improper planning. Somehow the edifice grows; somehow the various groups perform their roles.

During the 1940s Bush was translating the image of his essay into concepts, structures, and procedures for the proper organization and management of research and development in the United States. That involved fashioning a definition of planning consonant with his conservative viewpoint. Bush synthesized his views and experiences in three separate, yet related, areas: first, the meanings with which he invested such terms as "science," "engineering," and "research;" second, notions of professionalism; and, third, organizational preferences meshing with the first two.

A principal problem confronting Bush was public confusion over terms like "science," "research," and "engineering," at least according to his views. Throughout the war and for many years afterward, he tried to clarify their meanings to colleagues and to the public with only modest success. To much of the public, even professionals who should have known better, the triumphs of OSRD and of the Manhattan Project represented "science." In the autobiography, Bush gave strange anecdotes about how engineering lost out. The military regarded engineers as salesmen; the British had a snobbish class attitude toward engineers. Or he noted that the physicists in the Radiation Laboratory at MIT actually became engineers. As early as 1946, he publicly avowed that the contribution of sciences to the wartime research and development program was "indirect."[16]

Later commentators on science policies would argue about the term "science" and its relation to "engineering" or to "technology," producing various expressions showing differing degrees of theory and/or of utility in research and development activities—like "mission-related basic research." Some historians of technology lambasted Bush for creating an environment in which science reaped the laurels for the achievements of technology. That is, the accusation was that Bush demeaned engineering and technology to the "mere" status of applied science.[17] *Pieces of the Action* makes it quite clear that Bush was not exalting science and cared very much about engineering.

What confused contemporaries as well as later writers was the distinction in Bush's use of "research" from both "science" and "engineering." "Research" was the broadest of the three. In a 1946 letter to K. T. Compton, Bush defined it as "the ordered pursuit of new knowledge." He went on to observe:

> Within the definition, therefore, would come the ordered pursuit of new knowledge regarding the behavior of a chip on the tool of a lathe, as well as the knowledge of the constitution of the stars. Moreover, the method of search is not specified, and the pursuit of new knowledge by empirical methods is included if it is ordered, just as fully as is research which employs all of the resources of the scientific method. Finally, there is no specification as to whether the knowledge is or is not of direct utility.[18]

With this definition of research in mind, certain rather obvious points followed in Bush's world view. Research was performed by both scientists and engineers; the various scientific and engineering specialties corresponded both to social groupings and to bodies of knowledge. Bush made a distinction between "applied science" and engineering, the latter aiming for effective use within an economic nexus. Although OSRD operated outside of the normal economic nexus, Bush believed time constraints during the conflict performed a function similar to economic considerations in peacetime. "Science" tried simply to expand knowledge, as of the constitution of the stars.

"Applied science," in Bush's worldview, engaged in research outside of the economic nexus but with the intention of converting pure knowledge to a form suitable for application—at which point, one presumes, engineers practicing engineering (not necessarily doing research) took over. Or, as Bush stated in a public address after the war: "The function of the engineer . . . is the application of science in an economic manner for the benefit of mankind."[19] Engineers contributed to science by their research. Their disciplines encompassed knowledge created by their professional predecessors, as well as by individuals from other disciplines like physics, chemistry, and mathematics. Bush undoubtedly would have rejected any suggestion that his work as an electrical engineer was not in some meaningful sense an application, among other things, of Maxwell's equations. Engineers also had roles as professionals not ordinarily exercised by scientists. Engineering fields included knowledge peculiar to their subject matters and social functions not part of the scientific disciplines. The science–engineering distinction was one instance of the more general theory–practice relationship looming large in national science policy questions from World War II to the present.

Vannevar Bush was more than an outstanding engineer. He was significantly atypical of his profession at that period. Bush had a rare doctorate—a joint one from Harvard and MIT. The autobiography recalls with warmth the experience of being the only enrolled student in a class at Tufts in non-Euclidean geometry.[20] As noted earlier, between the wars Bush's greatest achievement was the invention of the differential analyzer, an important pioneer analog computer.[21] From that developed a concern for promoting developments in a wide range of information processing activities. His mathematical bent influenced aspects of his position.

In Bush's definition of terms like "science," "engineering," and "research," mathematics counted greatly. As early as a 1938 letter to Jewett, he explained how mathematics influenced his policy at MIT. In addition to his own proposed center for mechanical analysis (i.e., a computer facility), Bush supported the moves for expansion of "pure or formal" analysis in the Mathematics Department. More important, he felt, was to place and develop "men of mathematical power" in all of the science and engineering departments.[22] The spread of sophisticated mathematics into so many fields provided a kind of underlying unity to significant research and a form of identity to the various disciplines. Although Bush toward the end of his life spoke strongly of the role of theory (presumably mathematical in form), he could still insist on the "necessity of getting a barnyard grasp of things."[23]

That attitude and reference to the Wright brothers during the fashioning of SEF reflected not only Bush's practical engineering interests and his personal delight in inventive tinkering but his openness to the unexpected.[24] Bush was not looking to the "bright inventor in a garage" for technological advances. As he later noted, the situation was different now because of the advance of fundamental science, mathematical theory, and controlled experimentation. He asserted in 1946 that science's contribution to technology was "indirect" prior to World War II and then brought to fruition during the war only by collaboration with engineering, technology, government, and finance. But he could and did cite operations research, as well as new devices and instruments beyond the routine capabilities of American industry, as contributions of science. In this situation, Bush saw a need for the involvement of both scientists and engineers at all stages of both research and development (R&D) and of other relevant social processes.[25]

Quite in keeping with the spirit of "The Builders" was how Bush construed professionalism and the resulting obligations of those within a profession. Even before the war he publicly asserted that a profession "takes charge by right of superior specialized knowledge." Toward the end of his life he could write of the small, intelligent minority who ruled in the "last generation" and whose kind would also rule in the future.[26] Elitism is the term often used to characterize such views, but in Vannevar Bush's mind this was not a matter of retaining power in favored hands nor of crude cronyism.

Professions did not simply take charge, they were "ministers to the people." (Bush's father was a clergyman.) Bush stressed the altruism and idealism of the professional man. As he once stated: "The hallmark of a profession is ministry to the people . . . disinterested authoritative employment of special or superior knowledge for the counseling and the safeguarding of others." Bush went on to call for service, rather than merely pecuniary return, as the primary motive for professions. A profession had an obligation to all the groups involved in the building of the edifice. By extension, the concept of ministry was applied to Bush's experiences and expectations with industry: "the task of management here [the modern corporation] becomes fiduciary in essence, closely akin to trusteeship with obligation to four not three groups—government, employees, owners, consumers—all properly weighted." Bush was well aware that reality did not always match his words.[27]

Bush cannot be fathomed without keeping in mind another aspect of his career—the concern with issues and practical problems of management and of organizational structure. At one time or another he belonged to an impressive number of boards, committees, councils, and the like. In public utterances and private correspondence recur comments on the need for a head of an organization to have a good board, the roles of members of collegial bodies, and their responsibility to those represented and to the general public. All this accords well with the ministerial concept of a profession but was linked in Bush's mind to another concern.

In 1946 Don K. Price wrote Talcott Parsons expressing surprise at Bush's opposition to the administration-supported Kilgore Bill for a National Science Foundation. According to Price, the bill followed the example of OSRD in having a single, accountable administrator. Bush had insisted on having an unnecessary structure of councils and committees in OSRD and in imitating those arrangements in his proposed NSF.[28] Price did not understand how Bush perceived and ran OSRD. To Bush, the committees and the councils were the heart of the matter, not any single leader, himself or some future director of NSF.

Price might have noted Bush's academic background. Collegial governance is highly prized and sometimes actually practiced in the university world, even with strong leaders like K. T. Compton and Vannevar Bush at MIT. And CIW, with its autonomous, high-quality research, self-governing units, represented "small science." In his consulting as an engineer and in his administration of MIT and CIW, Bush followed a pattern applied to OSRD and to his postwar proposals for NSF and for defense research. He carefully defined sectors and delimited the boundaries between the sectors whether they were grandiose entities (like government, science, engineering, and finance) or miniscule (like individual actors in a research laboratory). Having classified the entities concerned, Bush then analyzed their interactions. Wherever possible, at MIT, CIW, OSRD, or in the Department of Defense, he concerned himself with the proper relation between the entities and the efficient flow of ideas and of men between them.

Even before the outbreak of war Bush became much interested in various kinds of "scientific aid" to learning, eventually specifically to the hardware and theories of information science and technology. That concern was because he saw information science as more than just a pragmatic tool for advancing research. It was a natural consequence of a belief that management of complex organizations required oversight of the flow of data and ideas. Later, drawing upon the work of his associate and successor at CIW, Caryl P. Haskins, Bush cited the need for a pyramidal structure for efficiency.[29] In OSRD and in the Defense Department Bush made much of the need to have disputes settled at or close to the working level where the technical competence existed. It was an R&D equivalent of the ideal of a republic of learning, of a belief in a kind of grass roots democracy of those in the ministry. Bush's role in OSRD and in the Research and Development Board (RDB) was judicial, settling differences before they became irreconcilable.[30] To Bush, New Deal centralization demeaned the local groups on which creativity depended and from which true democracy derived. He was quite sincere in his views and enough of a realist to know at least some of the problems they engendered.

During World War II, the National Advisory Committee for Aeronautics (NACA) and NAS–NRC provided the immediate models for OSRD and Bush's NSF proposals. NACA was a committee running an agency; the director, was responsible to the committee. Below this top structure was an elaborate array of technical committees with vaguely defined, but real, influence in policymaking and program oversight. NACA and its subordinate technical committees provided vehicles for ministerial representation.[31]

NAS–NRC was even more significant for Bush's views and his actions. The honorific Academy represented something important to Bush's worldview but something not always made clear in public statements. Collegial rule, not single powerful administrators, prevailed in the Academy. But it was the National Research Council's structure that closely approximated what came into existence first as NDRC and then as OSRD. In fact, in medicine and metallurgy NRC units acted as the operating arms for OSRD.

When Bush had authority over the postwar military R&D program (first with the Joint Research and Development Board, then RDB), he established an elaborate committee structure quite analogous in spirit and form to those of OSRD, NRC, and NACA. Through Bush's influence, when the National Science Foundation came into existence, each operating division originally had a statutory committee in addition to the panels awarding grants. In both RDB and in NSF these committees were not merely advisory but intended originally to play real roles in policy formulation and in program review. Bush thought of them as analogous ideally to boards of directors and trustees of corporations, foundations, and universities. A concept so markedly different from the prevailing norm in American political and administrative theory and practices puzzled and even infuriated some politicians and federal administrators.[32]

The NAS was important to Vannevar Bush during the war and in the years immediately afterward for two reasons. It was, first, by definition the institutional locus of key elements of his ministry and, second, the Academy with its associated National Research Council illustrated the correct organizational form for managing R&D. From 1939 until sometime after Truman assumed the presidency, Bush was in effect acting as a consultant to the Office of the President. Perceiving that the virtues of NAS–NRC were not adequate to the needs of the wartime emergency, Bush fashioned a related yet distinctly different alternative (NDRC, later OSRD) whose management he assumed.

In the formulation of SEF, Bush, acting again as a consultant to the presidency, proposed how to follow his preferred pattern in peacetime while evading both the NAS–NRC and the emergency powers conferred upon the government during wartime. In his autobiography, Bush proudly noted his prewar consulting deliberately was not limited to his specialty. Bush liked and was very good at trouble shooting.[33] Up into 1945, he moved with assurance to extraordinary achievements. The unhappy postwar dénouement indicates a gap in his otherwise shrewd field of vision. The presidential consultant had somehow failed in his analysis. The very successes of OSRD arising from his great talents helped frustrate Bush's plans for a new postwar order for research and development.

The Forbidden Fruits of the OSRD

To the Office of Scientific Research and Development Vannevar Bush brought his well-developed, partly articulated set of values and concepts. The distinctions, the boundaries, suffused his administration even though the emergency necessitated waiving peacetime constraints. Not surprisingly, the achievements of OSRD became part of the justification for both the SEF program and the proposals of his opponents after the war. Understanding Vannevar Bush and his aims, however, requires clarification of what was not permitted to OSRD, and how and why some boundaries were stretched.

Not disturbing the old order was a widely understood policy in World War II Washington. New Deal reforms were deferred in order to concentrate on winning the war. Not disturbing the old order was literally impossible given the vast expenditure of funds, the movements of millions of citizens within and without the country, and the sheer extent of physical construction resulting from the national mobilization. Bush certainly agreed with the sentiment. Not disturbing the old order for Bush specifically meant protecting the NAS and NACA. But while Bush consistently minimized actions disrupting the older order, quite a number of organizations and individuals were carefully positioning themselves for what had to be a new era. Like Bush preparing SEF, they were choosing their futures.

Bush, for example, knew that Princeton University's Physics Department had closed down its atomic research program after attaining limited results. Many of its best scientists moved to the Metallurgical Laboratory at the University of Chicago, including H. D. Smyth, the Department's chairman. When Smyth heard during the war that Chicago was negotiating with Enrico Fermi to switch from Columbia, he denounced it as "unethical." Although A. H. Compton wrote a careful defense of Chicago's actions, Bush remained suspicious of the fairness of the motives and actions of Chicago's president, Robert Hutchins, even after hostilities had ceased. To Smyth and to Bush, Hutchins was improperly using the emergency to attain changes in the existing order.[34]

As a matter of both principle and personal inclination, Bush largely concentrated his talents and his agency's efforts upon the combat needs of the armed services. Bush's restraint, his scrupulous reluctance officially to transgress his narrow definition of the OSRD mission, had unexpected consequences for his postwar proposals. Unofficially, he responded to presidential requests for advice on scientific and technical matters outside of the realm of military R&D, even in the very areas in which be banned OSRD participation.[35] Early in the war Bush declared he had no postwar planning mandate and formally avoided that function until the arrival of the Oscar Cox–inspired letter leading to SEF.[36] In 1944, wholly consistent with his general position, Bush excluded OSRD participation in the planning for the postwar rehabilitation of Europe.[37] Without the action of Cox, would Bush have ventured so far from his preferred scope? After Harry Truman took office as President in 1945, Bush was seen as a valuable specialist on defense research whose general views of the role of the federal government in the sciences were out of step with the politics of the new administration. Bush became increasingly frustrated and angry as Truman rejected the continuation of the casual easiness of Bush's former unofficial role as the science advisor to the President.

Within the large domain of national defense during a global war, Bush was adamant about staying out of civilian defense production problems. Bush had no difficulty in opposing any such extension of OSRD as not related to the actual conduct of warfare. The War Production Board established a small unit for technological advice and called upon industry for aid.[38] In marked contrast to Bush, Frank Jewett was more than willing to have the National Research Council act in a wide range of civilian engineering problems. The vice president of AT&T and president of NAS–NRC had no scruples about aiding both the armed forces and civilian industry. To Jewett, such aid was a clear consequence of both the organic act of the Academy during the Civil War and of President Wilson's executive orders in World War I.[39]

Despite the great scope of the wartime mobilization, many pure and applied fields were excluded from OSRD, or, at best, were only marginally involved. Very early in the war the geologists moved unsuccessfully to carve out a larger role in the scientific mobilization. The social scientists had a somewhat greater success. OSRD did have an Applied Psychology Panel. In the Social Science Research Council, nevertheless, concern existed about Bush's intentions for their disciplines toward the end of the war. Investigators in biology and in agriculture could not crack Bush's resolve to limit OSRD to research and development focused on the actual conduct of military operations. He consistently turned down proposals and informal feelers. OSRD was not to serve as a vehicle for the expansion of government.[40] Among the excluded were those who would join in the questioning of Bush's postwar proposals. In 1944 the geneticist L. C. Dunn of Columbia worked with Senator Kilgore's staff on legislation for the new era.

Medicine was a significant exception to the relative exclusion of the biological and social sciences from OSRD. In 1944 an Insect Control Committee was formed at the request of the Committee on Medical Research (CMR), close to conventional government biology and agriculture. In later years Bush justified the presence of the CMR on the ground of the field's obvious importance to the conduct of the war. He did not mention that both the Army and the Navy had medical arms engaging in R&D. Instead, Bush gives an account of how he intervened informally to prevent embarrassment to the administration, which was about to give mobilization authority to individuals and groups recently found guilty of antitrust violations.[41]

Bush may be accurate as far as he went, but his account is incomplete. In the early days of the war a committee existed concerned not only with military medical research and development but also with the host of health-related social questions. One can describe this as a body parallel to but less influential than the two other committees, NDRC and NACA. The administration decided to split the two functions. The

R&D operations were under the medical sciences division of NRC headed by Lewis R. Weed of the Johns Hopkins Medical School. Because of tensions between Weed and Surgeon General Thomas Parran of the U.S. Public Health Service, CMR became one of the constituent units of OSRD but headed by A. N. Richards of the University of Pennsylvania Medical School. Whatever the intentions in the Executive Office of the President, the effect was to sever the scientific and the technical from social policies and programs in accordance with Bush's basic stance.[42]

CMR, like metallurgy, operated out of the National Academy through Weed's division of NRC. Despite Bush's position and despite the general conservative stance of much of the medical profession (and such leaders as Weed and Richards), CMR displayed an interest in questions outside of its formal, narrow operations-support-oriented mission. Despite the formal split early in the war, CMR retained an awareness of broader social programs and issues bearing on the physical and mental condition of members of the armed services.[43] Bush did not perceive the medical community as a serious problem to his postwar plans for two reasons. By and large, his relations with its leadership was harmonious. More important, Parran's organization, the U.S. Public Health Service with its National Institutes of Health, simply did not loom large in Bush's view of the structures of federal power. That was a serious blind spot.

Even within his carefully circumscribed scope for OSRD, Bush prescribed tactics and policies minimizing the impact on the old order. The use of the contract as the administrative instrument for carrying out the OSRD program was linked to the desire to use existing entities. Using existing entities minimized the time and the costs required for establishing new ones. Contracts with existing organizations minimized future disruptions occasioned by newly launched R&D installations. Even with the organizational structures of the old order, contracts had a virtue. By definition, contracts were arrangements of limited duration and scale.

Given the size and nature of the war effort, avoiding the establishment of new laboratories was impossible. The existing network of facilities in industry, government, and the universities could only handle so much without drastic changes in scope, as well as in scale. Wartime shortages of trained personnel made some existing organizations cautious about defense obligations. Bush's contractual strategy minimized the permanency of any new facilities occurring in the existing institutional structure. Where defense needs were so large and so novel as to strain the existing structure, Bush had to establish new operational entities. The principal examples were Section T of OSRD, responsible for the work on the proximity fuse; the Radiation Laboratory at MIT;

and the nuclear program launched within OSRD and transferred to the Army's Corps of Engineers.

Section T was headed by Bush's colleague at CIW, Merle A. Tuve, who had participated in the early stages of the atomic program and played an important role in discussion of policies for military research and development toward the end of the war. Very early, Section T was detached from the regular division structure of NDRC and placed directly under Bush. The Section's operating arm was the Applied Physics Laboratory administered under a Navy contract by the Johns Hopkins University, an arrangement still in effect.

Early in the war Jewett sniped at the MIT Radiation Laboratory. Like Section T, the Rad Lab was draining people from the universities, especially younger scientists and engineers seeking draft deferments. Perhaps with an eye to real and potential competition with the Bell Telephone Laboratories, Jewett saw the Rad Lab as a harbinger of a new situation.[44] Jewett was, of course, correct. The Rad Lab as such no longer exists, but its descendants, lineal and otherwise, are still at MIT or nearby.

Although the "uranium problem," as it was called in the early days, started out on a modest scale, Bush and his associates realized not only its potential for growth but also that the effort would get into functional areas OSRD wanted to avoid. Bush remained an important policymaker during the period when the Manhattan Project of the Corps of Engineers administered the program. Bush was well aware of the perils faced by Major General Leslie R. Groves, its leader, in the near impossible task of minimizing disturbances of the *status quo* following the successful detonations of the atomic bombs in 1945.[45]

Less spectacular and perhaps more consequential in the long run were the problems arising from the path taken first by NDRC and then OSRD in regard to the design, procurement, and introduction of devices, systems, and procedures into the armed services. Officially, Bush's agency dealt with that part of R&D anterior to design and field testing, let alone the placing and supervision of production contracts. Nor was Bush anxious to get into the touchy matter of the introduction of novelties into the armed services, necessarily involving matters of doctrine, strategy, tactics, and even the training of personnel. By and large, he was successful. The War and Navy Departments picked up the results of the work of OSRD and carried the process forward. On paper, the transfer of the uranium problem to the Manhattan Engineering District was the exemplary case. Bush's principled reluctance, however, could be breeched.

Emergency pressures influenced Bush's actions involving Section T, the Rad Lab at MIT, and the entire matter of the "few quick." An aggressive, able administrator, Tuve saw his mission as extending across the entire spectrum from research to the question of correct combat utilization. A basic research scientist both before and after the war, he was under no illusions about the preponderant nature of Section T's work: "It takes a war to make an industrialist out of a physicist."[46] Contracting with industry was done by the Navy, which maintained a special administrative unit at the laboratory. As one of the great successes of the wartime program, the Section T experiences greatly influenced Bush's later position on defense research and development.

The Radiation Laboratory apparently did not come into being with the same charter as Tuve's organization, possibly because its much larger domain was not so obviously at the point reached by the proximity fuse early in the war. As R&D proceeded, particular devices emerged to the point where the Rad Lab edged into design and even into the areas of procurement and combat introduction. The OSRD staff knew they were going over a real, if invisible, line, and not all were pleased. The Rad Lab actions bore a resemblance to the "few quick" in their origins but evolved into an approximation of the Section T pattern.[47]

By "few quick" Bush meant the practice of rushing into existence a small number of full-scale examples of newly developed devices for evaluation. OSRD's civilian status and its hierarchical placement directly reporting to the President enabled Bush to extend his mandate. As late as the early 1950s, he urged the "few quick" practice, under civilian aegis, on the Defense Department to the great dismay of the military. It was blatant poaching on their preserves.

Bush had justified the practice as the understandable extension in a wartime emergency of normal R&D. But Bush was not shy about also arguing for the "few quick" because of the slow, cumbersome nature of the procurement procedures of the Army and the Navy. A more interesting Bush argument in defense of the "few quick" stressed that the armed services and civilian industry were geared to mass production. Not only were the products of mass production relatively cruder than the "high-tech" devices emanating from OSRD, but they inevitably represented an early freezing of design, making major innovations difficult. To this Bush added that his agency's products—by implication all products for combat—required a higher level of quality control than normal in civilian mass production and in military procurement.[48]

Another wartime OSRD move into supposedly forbidden areas involved introduction of new developments into use in the armed services. Again, as in the "few quick," Bush justified his moves (such as the creation of the Office of Field Services headed by Karl T. Compton) by the pressure of events during the emergency. Although the activities of the Office of Field Services and similar earlier moves always stressed cooperation with the Army and the Navy, OSRD clearly was impinging on sacred territory. Waterman, who succeeded Compton at OFS, muted the accounts of civil–military frictions in the published official history of the organization.[49]

While Bush went from triumph to triumph, moving carefully from within to without his self-proclaimed boundary, Frank B. Jewett had an ambiguous position. Not only was he president of the Academy and a vice president of AT&T, but at the same time he had a key role in NDRC and its successor, OSRD. Jewett was often very critical of Bush's creation. To him, it was an amateur operation with great potential for upsetting the existing order.[50]

Jewett also had an ambiguous role in the matter of what was in and what was out of OSRD. As president of NAS–NRC, Jewett forwarded to Vannevar Bush, the director of OSRD, favorable recommendations for proposals in biology, agriculture, and the social sciences. As an OSRD official, he concurred with Bush's rejections. Then, as president of NAS–NRC, Jewett backed Ross G. Harrison, the chairman of NRC, in attempts to carry out rejected proposals within the structure of the Council using funds from private sources or other government agencies.[51]

By its charter, NAS was subject to calls for advice from any federal organization. By traditions arising from not well-articulated beliefs, NAS–NRC covered any nonhumanistic field whether pure or applied. By the outbreak of World War II, the organization comfortably accommodated requests from the private sector. To Jewett, NAS–NRC was a proven instrument for the harmonious cooperation of government and the private sector; for the fruitful interaction of the abstruse and the practical; and for the maintenance of a healthy balance between a swelling officialdom and the traditional rights of the individual. Jewett's policies, curiously reinforced by the trends in the larger society, contributed to the Academy's emergence after the war with an enhanced role as adviser and spokesman, enhanced both as the willing ally of the federal administration and sometimes as a source of open and covert dissent. Jewett, the dissenter from the wartime policies of Bush, is the functional ancestor of the academicians who spoke out against the military during the Vietnam War.[52]

Jewett emerges in retrospect as a consistent, principled conservative whose points often were disturbing to opponents because they resonated with views held by individuals as far apart ideologically as Vannevar Bush and Harlow Shapley. As the head of the Academy, Jewett's call for the independence of science from politics fitted the beliefs of many in the research community. As the head of Bell Telephone Laboratories, he spoke as a representative of a "mature industry" whose experiences influenced Bush and others in their views of how research produced practical benefits. Jewett could not be shrugged off.

Jewett saw no reason why NRC could not function as *the* research organization during the war and even after the war. After all, that was how NRC originated in World War I. Nor was he as concerned as Bush with the administrative problems and the opposition of the Bureau of the Budget and its Director, Harold D. Smith. Jewett could and did point to medicine and metallurgy, among other areas, as instances in which NRC aided the government. There were legal strategems available for overcoming legal obstacles. Toward the end of the war, he pushed for a Research Board for National Security (RBNS) in NAS–NRC to coordinate defense-related research and development.[53] Earlier, in 1943, Jewett had seen his way to his future: NAS–NRC as a voluntary peacetime OSRD–NDRC–CMR after transfer by "another executive order," with Vannevar Bush as president of the National Academy of Sciences.[54] Bush did not choose that future. During the war Bush bore Jewett's criticisms with good humor, yet they must have rankled, the latter's position being so close to his own; he was merely more pragmatic he noted soon after peace came. The two were in agreement on one point. Bush and Jewett both wanted OSRD dismantled as quickly as possible and opposed any postwar plan predicated on forming a postwar civilian analogue. A peacetime OSRD was a forbidden fruit to men with their conservative views.[55]

Science the Endless Frontier:
Between Vision and Reality

On September 30, 1944, Rear Admiral Julius Augustus Furer, the Navy's Coordinator of Research and Development, noted in his diary a visit from the physicist I. I. Rabi, of the Radiation Laboratory at MIT. Rabi reported great consternation at the Rad Lab over Bush's announced policy of terminating OSRD following the momentarily expected defeat of Nazi Germany. Termination perturbed the Rad Lab staff, many of whom worried about their uncertain employment prospects in peacetime.

If Bush did not relent, Furer wrote that the Navy's Bureau of Ships was ready to continue the life of the laboratory by a contract with MIT "for an indefinite period."[56] Under pressure, Bush put off the start of termination to the end of hostilities with Japan. Termination of OSRD was very much on Bush's mind when the President of the United States asked him to prepare a report.

Despite its great success, OSRD had to go. Born of wartime necessity, the agency was an unclean thing, a spectacular example of how the government could manipulate the private sector for public purposes. Even the laudable purpose of defeating the Axis and a benign administration like Bush's could not excuse the violation of preferred boundaries and norms. Senator Kilgore's legislative ambitions simply illustrated the more general peril.

Somehow, the National Academy of Sciences had to play a greater role in the new postwar order for research and development. In wartime NAS had to step aside for OSRD. A true federal operating entity was called for. After peace came, perhaps a greater role for NAS was possible within the framework of a peacetime organization. Proper, of course, meant being acceptable to someone like Bush, who had supported the Liberty League's crusade for Alf Landon in the presidential election of 1936, as well as to the party in control of the White House.

To help him respond to the letter from President Roosevelt, Bush appointed committees to report on each of the four questions posed to him.[57] Isaiah Bowman, the president of Johns Hopkins University, headed the key committee on the third question, how to aid the research activities of public and private institutions. In February 1945, during the gestation of SEF, Bowman, in his capacity as an official of NAS, had occasion to correspond with Bush. In his responses, Bush touched on OSRD and the postwar situation. On February 6, he wrote to turn down a nomination to the NAS Council:

I also think it highly desirable that individuals who have occupied prominent positions during the war in connection with scientific matters should stand aside when it comes to the post-war period. This would apply particularly to such individuals as Conant and myself. There is a feeling, quite naturally, that some of us ran away with the ball, and I know there is a sentiment at some places in the Academy that many of the aspects of OSRD's work should have been carried by the Academy itself, or that OSRD should have been a subordinate body under the Academy or something of the sort.[58]

On February 19, Bush wrote about the load of work about to descend on NAS:

It is true that I am creating some of this worry myself. I have purposely been steering matters in the direction of the Academy. A part of this is due to the feeling on my part that the existence of OSRD during the war inevitably stood in the light of the Academy in its service to government to some extent, although we avoided this having an adverse effect on the Academy's fortunes by various steps. Part of it also however is due to the fact that there are important things to be done in this country at the end of the war which the Academy should do, because of its constitution and position.[59]

After publication of SEF, Bowman formed the leading private group lobbying for Bush's proposals, an organization later supplanted by the more inclusive Inter-Society Committee pushing for what became the National Science Foundation in 1950.[60]

SEF is the second of a sequence of three influential official reports on

304 The Perils of Maturity

national research policy. Its predecessor, the three-volume *Research: A National Resource*, emanated in 1938 from the Science Committee of the National Resources Committee (later, National Resources Planning Board).[61] Bush was aware of NRPB. The head of NRPB, Frederick Delano, the President's uncle, introduced Bush to Harry Hopkins in 1939, setting in train the events leading to NDRC. SEF's successor, *Science and Public Policy* (5 vols., 1947), ostensibly came from a board including Bush, who bluntly told President Truman afterward he had nothing to do with the final product.[62] As influential state documents, the three are unsurpassed to this date. Great as SEF's impact was, the other two reports limited the effect of Bush's thrust.

They differed from SEF in two major respects. First, their emphases were on governmental programs—actions of the federal establishment—relatively more than on somehow aiding or stimulating private bodies. Bowman's committee and Bush's final text largely disregarded governmental bureaus. In his public response to Senator Kilgore's proposals during the war, Bush noted how the private universities had stimulated the state universities in research. Initiatives in science and technology preferably came from outside the government.[63] A second difference is that both *Research: A National Resource* and *Science and Public Policy* are largely the products of social scientists who implicitly regard public policies as instruments of their disciplines. Unlike SEF, both have a tacit identification of sound social science and enlightened public policy. Theory to Bush's opponents in the social sciences tended to the instrumental. Theory in the natural sciences to Bush ideally existed independent of social purposes. The "master workmen" did not "labor to make the utility of the structure real."

Bush needed a proper administrative mechanism for planning and executing his postwar program. That meant keeping "out of politics," specifically a degree of insulation from the regular scrutiny of the Congress, the Office of the President and its Bureau of the Budget, as well as exemption in whole or in part from the regulations of the Civil Service Commission and the General Accounting Office. In the years after SEF, Bush repeated again and again that the President had the right of appointment to the proposed National Science Board and that appropriations of the new agency would go through the usual legislative channels. To counter charges of NSF's isolation from the concerns of the rest of government, on several occasions he contemplated adopting the NACA pattern where private citizens served with representatives of federal agencies on the governing body.[64]

Bush, Jewett, and the like-minded were convinced that the regular government agencies were inherently flawed. More than efficiency and corruption was involved. "Politics" meant the intrusion of the extraneous—in Bush's worldview violating the intrinsic nature of specific programs or institutions by favoritism to individuals, by pandering to economic, regional, and other interest groups, and simply by concern for irrelevant issues. Bush's ministry decided what was relevant and, by a system of checks and balances, kept "politics" out of their deliberations. Bush's medical advisory committee took the extreme step of proposing an autonomous endowed foundation for medical research. At one point Bowman's committee considered recommending the formation of a government corporation, a familiar administrative device for minimizing the normal political process.[65]

What Bush did not propose (nor countenance afterward) was what Don K. Price later dubbed "government by contract." Those arrangements could and did violate Bush's sense of the proper, differing roles of industry, government, the universities, and the like. Government by contract, to meet Bush's standards, would have required each federal agency to behave in the manner projected for NSF, that is, to reflect governance by qualified individuals imbued with a sense of ministry. In differing ways, Truman and Bush both doubted that possibility.

Perhaps the closest model to what Bush aimed for in the governance of the NSF was provided by the efforts of Herbert Hoover and the early New Deal, particularly in the short-lived National Recovery Administration (NRA) of 1933–1936. Moral suasion combined with the muscle of the state to yield a voluntary merging of interests, public and private. NRC reflected the same sense of the desirability and possibility of getting individuals of good will to submerge their special interests in order to achieve a common good. (From July 1936 to January 1, 1940—both before and after the formation of NDRC—Bush had served as the chairman of the Division of Engineering and Industrial Research of NRC.) While the language used by Hoover and by Bush stressed individualism, the effect could represent a swamping of the unrepresented in a corporate consensus. Not every denomination had its ministerial spokesmen.

Very much in the minds of Bush and many of his contemporaries was the proposed educational program of his report, perhaps more than the aid to research activities. Wartime services of trained scientists and engineers combined with the drop in college enrollments to create an awareness of a national need. The effects of the Depression on undergraduates was still fresh in the minds of Bush and others from science and engineering departments in higher education. To them, increasing the numbers of scientists and engineers was a good in itself as fostering a desired culture. Henry Allen Moe of the John Simon Guggenheim Foundation headed Bush's advisory committee on education which recommended allocating undergraduate scholarships on a geographic basis but awarding graduate support strictly on merit.[66]

The educational program had a purpose besides simply the increase of specialized personnel. It was to correct a perceived flaw in the national scheme of things. To Bush and many others, the existing arrangements blatantly violated the desired meritocratic nature of the R&D community. Entry too obviously depended on socioeconomic origins. The effects of the Depression on young people seeking education strengthened the existing feelings. New Deal moneys from agencies like the Works Progress Administration (WPA) and the National Youth Administration (NYA) flowed into many campuses, even with conservative-minded administrators. Although the proposals by Bush for educational aid were well received and relatively noncontroversial, when the NSF came into existence the agency never fully followed through.

In part, events overtook both Bush and NSF. The GI Bill would aid many who otherwise could not have attended college. Women and blacks were barely mentioned in 1945, let alone singled out as possible sources for specialized "manpower." The GIs were one of the rare exceptions to Bush's avoidance of nonmmilitary entanglements. He furnished advice to planners of programs to provide books and courses to help enlisted personnel prepare for higher education. When he informed the President, Truman approved, perhaps remembering his own reconversion problems after World War I. In the postwar years, Vannevar Bush approvingly cited Jefferson on governmental aid to education in support of his own proposals.[67]

What generated passions from 1945 to 1950 was how to carry out the recommendations falling within the scope of Bowman's committee. Although presenting the texts of all the committees, *the* report was Bush's own set of recommendations not quite matching the conclusions of his committees. As part of the maintenance of a healthy research community in the nation, he saw his proposed foundation as *the* federal entity supporting good quality research, whether pure or not. Initially (and even after NSF came into existence) Bush did not see the new foundation primarily, let alone solely, as a means of funneling federal dollars to the universities for the support of pure research.

As the new foundation was to be *the* federal entity supporting research in general, medical research and military research were specifically included in its scope as vital national functions requiring nonconventional approaches and solutions. Bush believed in the possibility and the desirability of splitting off the best research from other scientific and technical programs. His new foundation would not only support the best but would also target those advances perceived as promising for national needs, whether in defense, medicine, or civilian technology. Bush did not propose peer review for project grants but, rather, the use of contracts with the grant reserved for general support purposes. The targeting was like the policy of Warren Weaver at the Rockefeller Foundation; the use of the contract arose from OSRD and was followed by the Office of Naval Research's program right after the end of World War II.

As to civilian technology, the report to the President is rather brief and vague—quite understandable in light of political realities. Aiding industry had less popular attractiveness than fighting disease and strengthening national defense. Deferred for another occasion was the expression of Bush's views on the reform of the patent system. Bush did write in SEF about the need to shorten the interval between research and its technological application.

Related to the reticence about civilian technology was the mild call for the new foundation to have an undefined role in the coordination of research programs within federal agencies. Most likely a residue from Bush's informal wartime presidential advisory role, coordination probably meant steering the results of research into use, clearly within the scope of various federal agencies. Bush later soft-pedaled the theme of coordination, as it attracted near universal opposition from the existing agencies and their allies within and without the Congress. Implicit in the SEF program and its coordinating function was a threat to the existing and contemplated intellectual and administrative expansions of the agencies, limiting, for example, their support and conduct of mission-related basic research and even of sophisticated applied research. They did not want another entity standing between themselves and the President, commenting to budget examiners, and looking at their programs from an alien framework. Lumping medical and military research with the more general basic investigations was Bush's contribution, his particular vision. Whether intended or not, that made his proposed foundation look like a peacetime version of OSRD.

Strictly speaking, none of the circle of influential scientists around Bush favored an OSRD for civilian science in the postwar years. But both Bush and Jewett were aware of strong feelings in at least some quarters for an OSRD-like body for national security. Jerome Hunsaker, Bush's successor at NACA, proposed his agency as the model. But Karl T. Compton and Merle A. Tuve each had ideas for agencies of the OSRD type for national security. An Executive Order would establish Compton's. Tuve's proposal was more elaborate, requiring Congressional action.

Jewett proposed reliance on the NAS–NRC, leading to the curious, brief incident of the Research Board for National Security (RBNS). His reasoning resembled Bush's in certain respects but placed far greater emphasis on the role of the armed services:

(1) The Army and Navy have a statutory responsibility to the nation for its defense which they can neither avoid nor delegate.
(2) Any organization which failed to recognize this statutory responsibility would be inimical to the interests of the nation and highly dangerous to its safety. It would be an irresponsible intrusion of civilians which the Congress neither should or would sanction or tolerate.

Bush's conclusions were markedly different. Jewett's Research Board was to be simply a temporary arrangement pending the formation of the new foundation. Perhaps Jewett and others hoped RBNS would somehow survive, even functioning for Bush's new agency as CMR had served OSRD. Perhaps Bush had similar thoughts.[68] More than anything else, Bush's national security proposals guaranteed the attenuation and demise of his postwar plans. Even before SEF appeared, a far different outcome was in the making.

Dénouement

At first Bush had much optimism about the prospects of his proposals under President Truman. In a memorandum on a conference of June 14, 1945, with the new chief executive, Bush noted:

We first discussed my report in response to President Roosevelt's request. The President has read this and told me he likes it. He is now releasing it. . . .

We then had an interesting discussion in regard to the difference between governmental study and planning for the future and what is known as "planned economy." The President regretted that there was no place such as the old National Resources Planning body for the proper type of forward looking study.

So far so good, from Bush's point of view. But the larger portion of the memorandum related to defense research and the temporary role of the Research Board for National Security under the National Academy. Here Bush still conveyed an upbeat tone but there are overtones suggesting difficulties:

he [Truman] evidently therefore agrees fully with the point of view that OSRD should not do post-war research. He stated that he thought the operations ought to be in governmental hands rather than in the Academy, which should primarily be advisory, and I told him I agreed with this and that I felt the Academy did also. . . . He would, however, like to have the matter worked out, even although it is temporary, so that it is in general accord with the principle that government operations should be under the usual governmental controls. I told him that I thought the solution that I proposed . . . provided for this, making RBNS advisory to the Services and the Services placing the contracts. I also said I thought it was essential to handle the matter in such a way that the scientists would not feel that they had been treated roughly after the long discussion that there had been on the matter.

The memorandum concluded almost with a sigh of relief. "It is interesting that Mr. Smith's name was not mentioned by either of us during the conversation."[69] Harold D. Smith, Director of the Bureau of the

Budget under Roosevelt, held the post in the early years of the Truman administration. Having had earlier encounters, Bush knew that Smith was a formidable opponent.

In his autobiography, Bush rated Truman a great President but one with few and rather ineffective channels for information and advice. Bush noted: "Later he relied heavily on me, for a while, for information on scientific and technical matters. We had an interesting relationship, while it lasted."[70] The deterioration of that relationship had many causes. Bush suspected that one cause was that Vice President Truman, unlike his predecessor in that office, Henry A. Wallace, was not briefed officially about the atomic bomb by Bush until after Roosevelt's death.[71]

Quite different in background and outlook, the Director of the Bureau of the Budget in retrospect resembles Bush in ability and in verve; they were two of a kind. A Midwesterner, Smith had served as the state of Michigan's budget chief before coming to Washington. Smith viewed himself as in the social sciences. American experience had demonstrated the "scientific" truth that organizations required individual heads, not governing boards, committees, and councils. (Only the later Atomic Energy Commission violated that precept.) Not only was that organizational form "democratic" in the sense of establishing responsibility and accountability for policymaking and program execution, Smith regarded it as essential for economy and efficiency. Responsibility and accountability loomed large in his mind, rather than dispersal of decisions and actions to private hands not necessarily committed to the public good. In 1942 he noted with approval that Donald Nelson, the head of the War Production Board, had severed his connections with Sears, Roebuck. Bush maintained his position in the Carnegie Institution of Washington.[72]

Responding to the moves of Bush, Smith kept referring to the need for balance in the postwar research world. That meant two things: the social sciences needed support to counter the excessive concern of the physical scientists with warfare, and the regular government research and development agencies needed strengthening and coordination. Bush wanted entanglements with neither the social sciences nor the regular agencies. Smith saw Bush's proposals as attempts to get special privileges for the physical sciences at the expense of proper democratic principles and procedures. By early 1945 Smith had devised a strategy and tactics that eventually nullified Bush's plans.

Research coordination, both military and civilian, became a concern of Smith's at least by the end of the summer of 1944, particularly as he increasingly feared a one-sided approach, rather than a broad-gauged study, from Bush. He also pondered the possibility of a different study of national science policy. In January of 1945, Smith was talking about perpetuating OSRD as an independent agency concerned both with the physical sciences and the social sciences, rejecting the suggestion of the Social Science Research Council (SSRC) that his Budget Bureau coordinate federal research.[73] By January–February of 1945, word was spreading about the recommendations of the committees advising Bush, with talk of the involvement with the military and even, incorrectly, of a war on cancer. The former particularly concerned Smith, who went to see the ailing President Roosevelt on research in the government. In his diary entry on the meeting, Smith noted that Roosevelt did not remember asking Bush to do a study. Smith recorded telling Roosevelt about OSRD having more "leaf-raking and boon-doggling" than in the WPA. In a few days the President would sign a letter against RBNS ultimately dooming Bush's program.[74]

Smith, like later advisers to Truman, was most conscious of national defense as a significant constitutional responsibility of a President. In no way could national defense policy formulation and administration be delegated to private individuals and organizations. Smith noted two such moves involving the National Academy of Sciences. OSRD proposed that NAS have the task of coordinating the release and dissemination of the scientific and technological data generated by wartime research. In Roosevelt's letter, that was the first query posed to Bush. Civilian and military agencies complained to Smith about the arrangement. One day before the Roosevelt letter, Smith wrote Conant of his opposition, asserting that OSRD would stay in existence to do the job. Indicative of the importance of that matter to Smith, the release program eventually was under a cabinet-level board. The question of patent rights to wartime inventions was very much in the air at that time. Unlike the armed services, OSRD followed the strict federal policy of reserving rights to inventions done under federal subsidy. Smith's reaction probably stemmed from concern that the release would weaken claims to patents.[75]

Smith would have no part of the proposal for RBNS and won the backing of Roosevelt and of Truman. In his opposition, Smith had an ally: the Navy, acting through Lewis Strauss, was opposed to what was seen as essentially an arrangement between Bush, Jewett, and Secretary of War Stimson.[76] The split followed the earlier division in the Committee on Post-War Research (the Wilson Committee).[77] By April and May, Secretary of War Stimson and Bush, aware of Smith's hostility, tried without success to deflect the Director of the Budget Bureau. In May, Smith cut off funds to RBNS and directed that a Joint Army–Navy Board should advise OSRD. (Bush already had such bodies in place.) He also decided to brake crippling budget cuts so that OSRD could continue to function.[78]

A new President now sat in the White House. Smith discovered that Truman knew about the atom bomb project but not of its title, the Manhattan District Project. Truman shared Smith's antipathy to NAS and his belief in the necessity of keeping vital administrative functions in responsible governmental hands. By June Truman directed the halt to the liquidation of OSRD particularly so that defense R&D could function properly within the government. Against Bush's protest that OSRD was an emergency agency only and not suitable for a permanent role, Truman held firm as late as August 22, 1945. By now, the administration wanted to continue OSRD until the passage of the proposed NSF legislation covering the spectrum of governmental interests, not simply military R&D.[79]

Smith's position was succinctly expressed in two memos at the end of the month. In an August 27 meeting with Bush and others, Smith declared: "that he could not agree with [Under Secretary of War Robert P.] Patterson that scientists are a special breed with special consciences and special intelligence. He [i.e., Smith] said he disliked any such approach to governmental organization for research because it vitiates sound handling of the problem. He approved of the statement made by a representative of the Department of Commerce who had stressed the importance of social research. The Director indicated further that he disliked seeing the physical scientists meshed with the military, for he is concerned with getting balance in the whole research field."[80] A memorandum to Truman the next day asserted: "I share his view [referring to an SSRC officer] that the physical scientists need their approach softened by the impact of social science. . . . I am concerned that in our effort to resolve bureaucratic conflicts in the field of research, we may come out with an unbalanced research program as between the various agencies of the Government and as between the physical sciences and the social sciences, with the social sciences being overshadowed by the impact of the atomic bomb." It is a credit to Bush's forcefulness that he pushed through the liquidation of OSRD in the face of the opposition of Truman and Smith. It is an intriguing "what if" of history to speculate what might have happened if Bush had compromised with his strongly held views and kept OSRD alive pending passage of postwar science policy legislation. But Bush could not countenance a metamorphosis of OSRD into the new foundation.[81]

Instead, Bush increasingly lost contact with the Truman administration. New people supplanted Bush in his unofficial role as science adviser to the President. Reacting to Bush's aggressive moves, in March 1945 Smith brought in Don K. Price, then a young officer in the Coast Guard.[82] On June 1, 1945, a future Chief Justice of the United States, Fred M. Vinson, advised Truman to have a broader review of the recommendations of *Science—The Endless Frontier.*[83] James R. Newman of the Office of War Mobilization and Reconversion (OWMR) became involved in science policy planning.[84] By October Smith pointedly reminded Bush of Truman's position on the NSF. By November Bush bitterly protested to OWMR his secondary role in the science policy field.[85] Although Bush would move to cement relations with Smith's successor, James Webb, by and large he was viewed as conniving against the Truman administration. At the very least, one can describe Bush's defensive rationalizations of his actions as disingenuous.[86]

What concerned Bush most in his dispute with the administration were not abstract questions of constitutional prerogatives nor of professional autonomy. Political intrusion into research represented a greater danger, leading to dubious, if well intentioned, research programs and to the use of R&D funds for extraneous social and economic purposes. Easy money would tempt the universities and the best researchers to engage in inferior, costly investigations and to take on roles properly exercised by industrial corporations and by government.

Two postwar examples illustrate his position. Oscar Cox had an important role in the early history of NDRC and in the events leading to SEF. A Wall Street attorney, Cox came to the Treasury Department in Roosevelt's second term. He drafted the legislation resulting in Lend-Lease and served with distinction in the wartime mobilization. When Cox later returned to private practice, he acted as Bush's personal attorney. Shortly after peace came, he wrote Bush proposing a billion dollar war on disease, obviously influenced by the World War II successes of OSRD. His draft bill, like the Lend-Lease legislation, bestowed great powers in relatively brief terms. Politely, Bush demurred from a war not solidly based on existing science. Nor was he swayed by Cox's point that the bill authorized other worthy activities besides research. In 1972 Bush similarly privately opposed Nixon's war on cancer.[87]

What Bush may not have known in 1944 was Cox's linkage of the proposal for what became SEF with regional development—geographic frontiers and "frontiers of the mind." In 1961 when he privately opposed the NASA space program, Bush's criticism was in terms of what was a proper R&D program. Bush's opposition would have been even more adamant had he known that the NASA Administrator viewed the space program as a deliberate vehicle for regional development. The Administrator was James Webb, Smith's successor at the Bureau of the Budget, who recommended the Truman veto of the first NSF legislation.[88]

From Bush's standpoint, the mixed signals from his various ministerial communities compounded his serious problems with the Truman administration over the SEF program. The most obvious opposition came from the social scientists. At an April 1946 meeting of the SSRC's Committee on the Federal Government and Research, the problem facing the social sciences was correctly summed up: "It is difficult to gauge the extent of the current opposition, but it seems to rest upon confusion between social science research, social philosophy, and social action."[89] The Committee itself had more prosaic origins at the start of 1945. Before the end of 1944 SSRC had good sources for Bush's views of the miniscule role for the social sciences in the future he planned. Besides Harold Smith, there was the official historian of OSRD, James P. Baxter III, a member of the Committee, who had access to policy planning documents within the OSRD headquarters.

At first the SSRC Committee took a position that would have delighted Bush and Jewett had they known of it. The members reaffirmed the "traditional" opposition to governmental subsidization of research as a threat to scholarly integrity. That soon gave way under the prodding of the Council staff; recognition of the value of some federal research, notably in the Department of Agriculture; and simple dismay at exclusion of the social sciences from Bush's proposals in the report to the President. By their meeting of April 14, 1945, if not sooner, the Committee resolved to fight for federal support. By the end of July, SSRC decided to prepare their own report justifying support. Particularly disturbing was the exclusion from the scholarship and fellowship programs and the perceived hostile viewpoints of the physical and biological scientists.[90]

By the fall of 1945, when the Kilgore and Magnuson bills were competing for primacy in defining federal civilian science policy, the exclusion of the social sciences had produced a serious split in the ranks of Bush's supporters. Isaiah Bowman, president of Johns Hopkins University, organized a citizens committee for Bush's program. Henry Allen Moe of the Guggenheim Foundation refused to sign the statement issued by the committee, specifically criticizing the exclusion of the social sciences.[91] By this date and until the very passage of the NSF legislation in 1950, keeping the social sciences out was a firmly held position of the political right. The SSRC strongly resented the willingness of many of their colleagues in the natural sciences to take the path of least resistance.

To some physical scientists, the social sciences represented a threat more serious than the opposition of conservatives. As the federal government's programs expanded slowly and unevenly after the end of the war, Bush and others noted the growth of private foundation support for the social sciences. Two disturbing assumptions accompanied this growth at the expense of the natural sciences, or so it seemed to them: first, that the federal government would provide for research in the physical sciences; second, that the physical sciences were linked with warfare, that is, death and destruction, a viewpoint powerfully reinforced by John Hersey's *Hiroshima,* which appeared in the *New Yorker* in 1946. With dismay, Bush, a Trustee, watched the Carnegie Corporation propose a major expansion of its efforts in the social sciences. Eventually, the Corporation largely ended its support of the physical and biological sciences. At the end of 1946, Bush reacted strongly:

There is a tendency, all too prevalent these days, to regard the natural sciences as providing the implements for evil men, and the social sciences as a counter to this situation. Actually, of course, both groups of sciences are on exactly the same basis; their products provide tools for good and evil. . . . We would not stop our progress in the social sciences because better understanding of mass psychology, for example, is one of the most powerful tools of a dictator, any more than we would stop progress in the natural sciences because of an atomic bomb. We pursue both because of the faith that the acquisition of knowledge, while dangerous, is worth the risk. We either proceed down the path of conscious effort to control the course of evolution or we quit the game.[92]

To Bush, the growth of the social sciences represented a threat to the independence of science from government because of the potential drying up of private support. To the physicist C. C. Lauritsen, advising the fledgling Ford Foundation, assurance of support from that source was a great relief in view of the overall thrust of the new philanthropy.[93]

Bush never opposed the social sciences as such, arguing for their development along the lines of the sciences and engineering. As late as 1955 he wrote John Gardner of the Carnegie Corporation expressing his concern over the policy favoring social sciences. The natural sciences needed foundation money to set a standard against the project system with its premium on salesmanship, compromise, and safe ventures. To the best of my knowledge, only one foundation in that period, the Sloan Foundation, responded implicitly to Bush's words. Not coincidentally, Warren Weaver had moved to Sloan from the Rockefeller Foundation.[94]

Far more embarrassing to Bush was the hostile reaction of many biologists, including a number prominent in the affairs of the National Academy. Not as spectacular nor as well known as the opposition of many atomic physicists to the May–Johnson Bill (reluctantly supported by Bush), the opposition of the biologists shattered any pretense that Bush was *the* spokesman for the research community. What aroused the opposition was the original language of the Magnuson Bill placing biology in a subordinate position to medicine, in accordance with the emphasis of Bush's report. The contrast with the physical sciences was simply too much. Among the leaders in the attack on Bush was Robert Chambers of New York University, who headed the NRC Committee on Biology and Agriculture toward the end of the war. In the fall of 1945, when efforts were underway to fashion acceptable legislation for NSF, Ross Granville Harrison wrote, "If called to testify, I shall certainly try to make it clear that the biological sciences should be created on a parity with the physical sciences and not be made merely the handmaiden of medicine." By the end of 1945, W. J. Robbins protested to Bowman's committee working for Bush's program about the demeaning treatment of biology. At the April 1946 meeting of the Academy, the University of Chicago physiologist A. J. Carlson delivered what Bush described as a "violent speech" in opposition. At Woods Hole that summer, Chambers publicly attacked Bush.[95] Although suitable wording entered subsequent drafts of the NSF legislation, there were now two professional groups who sympathized with much proposed by Bush but doubted his concern for and understanding of their needs.

Harrison, Chambers, Carlson, Dunn, and others well remembered Bush's disinterest in their efforts to aid the war effort. But Jewett in 1944 had urged the Committee on Post-War Research (the Wilson Committee) that the armed services not exclude any field of science, specifically referring to health and agriculture. Jewett doubted the validity of a research-supporting body administratively divorced from user organizations like the Army, the Navy, and the Public Health Service. More important, Jewett saw little point in the narrow scope produced by Bush's scruples. The pattern he favored worked very well in AT&T.[96] A substantial body of scientists, widely scattered on the ideological spectrum, implicitly leaned to arrangements markedly deviating from Bush's stance.

With the public spotlight on atomic affairs, research for the armed forces, and the stubborn refusal of SSRC to accept exclusion, Bush and other key participants in policy debates immediately after the war often overlooked and discounted the potential impact of biomedicine. Quietly, the Public Health Service fashioned a powerful research establishment at its National Institutes of Health in the years after the war.[97] Medicine did interest Bush; it is a curious minor theme in his public and private utterances. Early in World War II, he published a discussion of medicine in his essay, "The Case for Biological Engineering." It is odd that Bush characterized the physician as more an applied scientist and "certainly not an engineer." After the war he criticized aspects of medical education (a son became a physician). By 1955 Bush corresponded with John Gardner about the need to integrate medicine with the community. Nor is it inappropriate to note Bush's service as chairman of the board at Merck after leaving CIW.[98]

In Roosevelt's letter asking for Bush's views on postwar policy, medicine was the second question posed. In *Science—The Endless Frontier,* medical research became *the* topic of the first substantive chapter, as though Bush knew the subject had fewer pitfalls than support of military research and of research in the universities. His medical advisory committee originally proposed a completely separate independent, endowed foundation for medicine. With some reluctance, they agreed to Bush's proposal. There is evidence of great hostility among the medical community of that day to a federal expansion into health areas and, particularly, a strong antipathy to the Public Health Service (PHS) and its head, the Surgeon General, Dr. Thomas Parran.[99]

Bush's insistence on the liquidation of OSRD resulted in the transfer of outstanding CMR contracts to PHS, which earlier had succeeded in amending existing legislation authorizing an expansion of the programs of its constituent bureau, the National Institutes of Health (NIH). While this was occurring, Bush—in marked contrast to Jewett's position—preferred excluding medicine from the scope of his Joint Research and Development Board, the predecessor of the Research and Development Board of the Department of Defense.[100]

At the same time, medicine played a strange role in the legislative history of the NSF. While the March of Dimes (the Polio Foundation) adamantly opposed an expansion of the federal role in either NSF or NIH, other health foundations (particularly the Cancer Foundation under the aegis of John Teeter), placed language in the proposed legislation vetoed by Truman calling for commissions to study and to recommend programs for particular diseases. The implication is that the new NSF would then act through NAS and the existing health foundations—that is, minimizing the disruption of the older order. Like Bush, these foundations did not favor the expansion of the old governmental order represented by PHS and its National Institutes of Health.[101] From 1947 to the 1950 enactment of the NSF legislation, a federal administrator like E. U. Condon at the National Bureau of Standards—and others—would point to NIH especially and to ONR as successful examples countering the need for the agency envisioned by Bush.[102]

Within the defense establishment, the creation and rise of ONR was not the principal threat to Bush's position. Rather, what bothered Bush most and taxed his energies and imagination for many years was the entire mind-set of the services. Many in the services perceived Bush's proposals as a threat to the integrity of their mission and an assault on their professional competence. Even before the end of the war Bush contemplated writing a book on the problem of military research in a democracy and eventually produced *Modern Arms and Free Man* (1949), a best-seller. Bush did not pull many punches. On the situation between the two wars he wrote:

> Civilians felt that this was a subject for attention only for military men; and military men decidedly thought so too. Military laboratories were dominated by officers who made it utterly clear that scientists or engineers employed in these laboratories were of a lower caste of society. When contracts were issued the conditions and objectives were rigidly controlled by officers whose understanding of science was rudimentary, to say the least. To them, an engineer was primarily a salesman, and he was treated accordingly.[103]

Bush clearly thought poorly of a continuation of this particular old order:

There is nothing more deadly than control of the activities of scientists and engineers by men who do not really understand, but think that they do or must at least give others that impression, and the worst control of all is by individuals who have long been immersed in a particular subject and have made it static.[104]

Bush did not keep his attitudes and his proposals secret during the war when he recommended to a Congressional committee that the armed services confine themselves to improvement of existing weaponry.[105]

Perhaps even more disquieting to officers of the Army, Navy, and Air Force was how Bush's concern for R&D broadened to an intense contemplation of the entire question of defense organization, military planning, and even conduct of operations. Bush carefully gave credit to the military; he often asserted his high regard for the officer class in what he considered their proper sphere and took great pains to affirm his support for various defense activities. But there was no way of evading Bush's basic lack of respect for all kinds of sacred cows of the military. He spoke out. For example, in an address of March 30, 1946, before the Joint Air Defense Conference, Bush succinctly summed up his opinion of the way national security decisions were reached: "This kind of organization would not be tolerated a week in a manufacturing concern producing bobby pins."[106] The military had reasons for concern. Bush's views still counted.

From 1944 until the early years of the Eisenhower administration, defense organization, policy, and procedures evolved accompanied by tensions and anxieties among both military and civilian participants. First, demobilization had to proceed in a manner satisfying popular demand for a return to the old order of peacetime, while somehow new arrangements came into being suitable to the changed world situation. Second, that meant, to most participants, a unified defense establishment with peacetime selective service. Unification engendered an enormous amount of controversy, echoes of which still persist. Third, the new national security arrangements had to accommodate the strange new scientific and technological world disclosed by OSRD, particularly its spin-offs in nuclear weaponry, missiles, and electronics. Fourth, the new developments kept alive the memory of Pearl Harbor and raised the specter of even worse catastrophes. For that reason, the widening rift with the Soviet Union, the loss of the monopoly of atomic weaponry, and the victory of the Chinese Communists intensified concerns over organization, strategy, and technological advances. The shock of actual combat in Korean War, above all, convinced many scientists and engineers, including Bush, of the need to contribute to strengthening the defense effort and to improving the organization and administration of military R&D.[107] For many, peacetime never came. While aspects of Bush's program were co-opted, the results were quite far from his vision.

Bush underestimated the determination and resiliency of the officer class; effective linkages developed between industry and the armed services (Eisenhower's "military-industrial complex"). In the university world there was no shortage of individuals interested in serving their nation while gathering in contracts and consulting fees, not to mention the power and prestige of advising on great issues of life and death. In retrospect, Bush was not quite the menace he seemed to the military in the late 1940s when he cooperated with the Budget Bureau to cut and reallocate R&D funds for the armed services. His words in 1951 had only limited consequences then and immediately afterward:

we must avoid domination of our scientific affairs by the military. The military mind is likely to be a single track mind. War and preparation for war are far too important to be entrusted to generals. There is much of this sort of thinking in the public mind today; in the light of some, but by no means all, of the Korean experience there is real reason why the public should be questioning and distrustful of the military mind. When the Founding Fathers placed our military organization underneath our political civilian system they knew what they were doing, and we had better keep it that way.[108]

Unwittingly, Bush's success at OSRD had created an obstacle to the acceptance of his views. He could write in 1949 in the same letter about both the unreasoning fear of new military weapons and the "appalling defense establishment." In 1948, to cite one of many examples, Bush noted to Karl T. Compton the inclination of some of the military to downplay new developments by stressing the essentiality of preparedness of conventional forces.[109] Unreasoning fear produced an atmosphere in which large sums were given for poorly thought out overlapping crash programs, not the "unitary" programs Bush favored at RDB, "orderly competition within a well ordered framework." Unreasoning fear fostered an atmosphere conducive to coercion against dissenters. Unreasoning fear gave the military a great advantage in policy arguments with civilians.

Ironically, Bush strongly favored maintenance and upgrading of conventional forces. Bush had a well-developed antipathy to the idea of "push-button" war and strong doubts about the practical value of missiles and the H-bomb in the real world, as distinct from the paper world of military planners. Having studied the analysis of strategic bombing in World War II, he had little tolerance for the arguments of the Air Force: "the enthusiasts of air power are often not stopped by such a minor obstacle as an obstinate fact."[110] Bush was quite skeptical as to the development of the intercontinental ballistic missile (ICBM):

It [the ICBM] would never stand the test of cost analysis. If we employed it in quantity, we would be economically exhausted long before the enemy. It makes little difference in war whether one's people, facilities, and materials are destroyed or whether they are employed in making devices that are then destroyed without advantage.[111]

From the very early days of the S-1 project, the progenitor of the Manhattan District, until his last years, Bush regarded nuclear weapons as an unfortunate, necessary evil, stressing the grim prospects of even worse chemical, biological, and radiological horrors.[112] He encouraged the development of systems of defense and, like some other scientists, later promoted the use of smaller tactical weapons as an alternative to the strategic bombing doctrines of the Air Force. To no avail, Bush tried to stop the first U.S. H-bomb test and to halt a U.S.– Soviet H-bomb arms race.[113] He did not think it would bring real security; Bush clearly saw that weapon as a great threat to the old order he so greatly cherished. Although Bush's reputation guaranteed an audience for his words, increasingly the players in the science policy games in the United States in the years 1945–1960 lifted bits and pieces from his suggestions as it suited their needs, ignoring the rest of the message with impunity.[114]

Nowhere was that more evident to Bush than in the effects on the universities and in the course followed by NSF. Bush's dismay in 1970 was not new. Shortly after the launching of the NSF, Bush had a conversation with Warren Weaver of the Rockefeller Foundation, who had chaired ONR's scientific advisory committee. Both agreed that NSF, as newly constituted, represented a defeat for their hopes and aspirations for science in the nation.[115] ONR had acted, they believed, as the surrogate for the proposed NSF, but the Navy refused to transfer funds to the new agency as Bush and Weaver expected.

Bush and Conant were not considering a problem created by the Vietnam War, nor one made acute by that conflict. A basic structural flow in the American scheme of things had developed, they believed, as a consequence of the events foiling Bush's proposals in SEF. Not simply research but the universities had suffered. NSF itself played only a marginal role in the criticisms of Conant, Bush, and Weaver. What it did not do loomed far larger than what its activities actually encompassed.

Not that Bush liked the Waterman-directed agency's activities. Far from it. As early as 1951 he commented on the ambiguous effects on the quality of research of increased accessibility to government funds. Shortly afterward, Bush complained about the project system supported by grants, noting the necessarily high costs of administration and the premium placed on "salesmanship."[116] By 1953/54, when

Waterman incautiously placed him on an NSF committee on government–university relations, Bush strenuously challenged aspects of both policy and procedure in the agency. The entire agency was too obviously skewed to the benefit of the universities, with inadequate concern for both the needs of the sciences and the needs of the nation.[117]

In a 1953 letter to his former colleague at MIT, J. A. Stratton, a fellow member of the committee, Bush summed up his position:

It would be fatal if the general attitude of the public was that this was a committee made up largely of beneficiaries under a plan explaining to the public how the plan could be revised to be of more benefit to the beneficiaries. We need to delve much deeper than that.

The fundamental question, I think, is whether federal support of research in universities is aimed at subsidy of the universities, or whether it is merely aimed at broadening the base of research in the country.

Bush went on to call Stratton's attention to the "criticism at the present time of the conduct by universities of great programs of research for the federal government separated more or less completely from regular university operations, where the university acts primarily as manager." Bush did not favor the recent suggestion of the Joint Chiefs of Staff to turn these over to industry. While he saw virtues in separating research from "the industrial, profit-making organization which may later be called on to manufacture weapons or devices in large quantity," he was troubled by the implications for the universities, interesting for a future honorary chairman of the MIT Corporation.[118]

Nor was he alone as the course of the nation's science policy increasingly called on higher education to assume managerial roles in both civilian and military R&D. Consider the experiences, for example, of Lee DuBridge, president of the California Institute of Technology (CIT) and a key lieutenant of Bush at OSRD. CIT now operates the Jet Propulsion Laboratory for the civilian National Aeronautics and Space Administration (NASA); the activity originated within the armed services. In 1947 DuBridge gingerly fielded a proposal from the Air Force to evaluate its guided missile contracts. In 1951 CIT played host to Project VISTA, an influential summer study concerned with developing a battlefield role for nuclear weapons. The Army and many scientists supported VISTA in reaction to the strategic warfare doctrines of the Air Force. By 1959 DuBridge was outraged at a Department of Defense listing showing CIT (and MIT) as one of the leading defense contractors in the nation. The Institute, he wrote the Department, was not a profit-making private corporation but an educational organization devoted to teaching and to expanding human knowledge.[119] Whatever DuBridge's qualms, functionally CIT was one of the Defense Department's top contractors.

Writing to the general public in his best-seller, Bush stressed a basic point underlying his policies in OSRD, a point consonant with his hostility to the New Deal's programs as undermining, as he saw it, desired older values. More important for national security than any organizational structure or any weapons system was the internal strength and goodness of the nation.[120] It was a conservative's vision of what other individuals of a quite different stripe would call a "people's war."

In his concern for patents, Bush acted out of a belief that technological progress came from small innovative industrial units.[121] In his qualms about government by contract and the proliferating project grant system, Bush reflected a view that science flourished best in small bodies like those of CIW or in the hands of the few of great talent in university departments. In struggling to keep decision making in the armed services as close as possible to the small creative technical groups, Vannevar Bush acted not only in accordance with what was comfortably familiar but as a creative anachronism.

His was a striving to hold back a future with blurred boundaries between public and private, between contemplation and action, between power and ideals, between national goals and self-aggrandizement. His was a worldview of discrete "local" entities, largely self-governing and self-motivated, and each requiring careful attention to their intrinsic natures. It was a conception of minimal government and of individual initiative—of personal responsibilities and of attention to the rights of others. Striving to work within the system, Bush often endangered his consistency of purpose as he tried to compromise with others.

Even before coming to Washington, even as Bush worked on the differential analyzer at MIT, his vision scarcely matched national realities. Because of that, *Science—The Endless Frontier* was more honored than literally heeded. Bush's vision threatened serious disturbances to various old orders and the enticing futures they discerned in post–World War II America. Nevertheless, the issues implicit and explicit in Bush's proposals are very much with us today. A memory, perhaps flawed, of the World War II scientific mobilization is still alive in national science policy discussions. Because of that memory Bush survives as a symbol with two principal elements. One element is as a precedent of what the national will could accomplish. When that resulted in the space program and the war on cancer, Bush took strong exception. Undoubtedly more to his liking was the other element: the possible role of knowledge in furthering the internal goodness and strength of the country.

Notes

1. *Bulletin, New York Academy of Medicine* 47 (1971):1248–1251.
2. *Pieces of the Action* (New York, 1970). The NAS reference is on p. 308.
Of interest for the argument of what follows are the passage on medicine (p. 8),
a reference to Frank B. Jewett (p. 39), a comment against the OSRD patent
policy (p. 84), and his satisfaction that the young people supporting Eugene
McCarthy in the New Hampshire primary worked within the system (p. 248).
Bush believed in the system and in the need, indeed virtue, of working within
it even when strongly against specific policies or actions.
3. Bush to Conant, Dec. 11, 1970, Bush Papers, MIT Archives. This collection dates from the Bush's departure from Washington during the Eisenhower
administration. The collection for the years in Washington, ca. 1938–1955, is
in the Manuscript Division of the Library of Congress (LC). The former is a
smaller collection of files of a retired individual; the latter, the many files
maintained in Bush's office at the Carnegie Institution of Washington (CIW):
these have documents of both a personal and professional nature, including
important items on the administration of CIW and OSRD.
4. Among the historical writings are A. Hunter Dupree, "The *Great In-
stauration* of 1940: The Organization of Scientific Research for War," in G.
Holton (ed.), *The Twentieth Century Sciences: Studies in the Biography of Ideas*
(New York, 1972), pp. 443–467; J. Merton England, "Dr. Bush Writes a Report: 'Science—The Endless Frontier'," *Science,* 191 (1976):41–47 and *A Patron
for Pure Science: The National Science Foundation's Formative Years, 1945–
57,* (Washington, D.C., 1982); Daniel J. Kevles, "FDR's Science Policy" (letter
to ed.), *Science* 183 (1974):798, 800; and "Scientists, the Military, and the Control of Postwar Defense Research: The Case of the Research Board for National
Security, 1944–1946," *Technology and Culture* 16 (1975):20–47; and "The National Science Foundation and the Debate over Postwar Research Policy . . ."
Isis 68 (1977):5–26; and *The Physicists* (New York, 1978); Robert Franklin
Maddox, "The Politics of World War II Science: Senator Harley N. Kilgore and
the Legislative Origins of the National Science Foundation," *West Virginia
History* 41 (1979):20–39; Carroll Pursell, "Science Agencies in World War II:
The OSRD and Its Challengers," in N. Reingold (ed.), *The Sciences in the
American Context: New Perspectives* (Washington, D.C., 1979), pp. 359–378.
While indebted to all, I am not in complete agreement with any one of these
writings.
5. This was the principal reason given for Truman's veto of the first NSF
bill on the grounds that the appointment by the National Science Board violated democratic principles of governance. I doubt that any method of appointment would have changed the course of the new agency. The Board or rather
one of its members, Lee DuBridge, after all, successfully prevented the appointment of Truman's first choice for Director, Frank Graham. For the veto
message and other relevant items, see J. L. Penick, Jr., et al., *The Politics of
American Science, 1939 to the Present* (Chicago, 1965).

6.   See Waterman's introduction to the 1960 reprint of *Science—The Endless Frontier,* again reprinted in 1980 by Arno Press (New York). Nor do I agree with Kevles who thinks the final NSF legislation largely a triumph for Bush ("The National Science Foundation"). As I shall show, Bush did not think so either, although he politely said so in the preface to the 1960 reprint. Bush believed in working within the system.

7.   Jewett to Bush, May 27, 1943, Bush Papers, LC.

8.   Among many possible examples are Bradley Dewey, Sr., to Bush, Sept. 3, 1944 and Bush's reply of Sept. 28, in Bush Papers, LC.

9.   The draft of October 1945 is in Box 140 of Bush Papers, LC. See p. 2f. In hindsight, the language used in the 1930s and 1940s by Jewett, Bush and other conservatives about the "revolutionary" measures of the New Deal appears excessive; Truman is now often seen as representing a moderate or conservative position in Washington and his administration—with the notable exception of civil rights—as a partial retreat from New Deal reformist thrusts. We can view Bush as maintaining a genuine civility in his avoidance of harsh language while conscious of the strong feelings among many of his friends and associates.

10.   Jewett to Bush, March 15, 1941, Bush Papers, LC.

11.   See "The National Science Foundation."

12.   See below for some consequences for SEF.

13.   At least two factors were involved. The first was the restrictive nature of civil service regulations and pay scale; the second was the perception that government bureaus tended to routine duties in well-established fields, as well as an overly applied viewpoint. See England, *A Patron for Pure Science,* pp. 90–91.

14.   The university orientation of many scientists and of some later historian produced a slighting of the role of industry.

15.   Originally in the January 1945 issue of *Technology Review* 47:162–163. By September 1944 (Box 138, Bush Papers, LC) a text of the piece existed. Bush included "The Builders" in the two anthologies of his writings: *Endless Horizons* (Washington, D.C., 1946), pp. 179–191, and *Science in Not Enough* (New York, 1967), pp. 11–13, the only writing so treated.

16.   *Pieces of the Action,* pp. 53–54, 138; "Planning in Science," p. 2, a speech of May 16, 1946, Box 129, Bush Papers, LC.

17.   See, N. Reingold, "Clio as Physicist and Machinist," *Reviews in American History* 10 (1982):264–280.

18.   Bush to Compton, Sept. 13, 1946, Bush Papers, LC.

19.   Lawrence Scientific School anniversary address, Mar. 31, 1948, Box 131, Bush Papers, LC.

20.   *Pieces of the Action,* p. 352.

21.   I have benefited from a reading of an unpublished essay of Larry Owens on Bush's work on the differential analyzer.

22.   Bush to Jewett, July 12, 1938, Bush Papers, LC.

23.   *Pieces of the Action,* pp. 262–264. See also *Endless Horizons,* pp. 27–28.

24.   The reference to the Wright brothers occurs in Bush's memorandum of Jan. 15, 1945, to C. L. Wilson, OSRD NA, Ser. 2, Box 3. From the immediate context, Bush is clearly disturbed at the apparent downgrading of engineering research and the overemphasis on pure research in the university.

25. In addition to "Planning in Science" (n. 16), see the address "The Scientist and His Government" of Feb. 22, 1946, Box 128, Bush Papers, LC.

26. *Endless Horizons,* p. 133; *Pieces of the Action,* pp. 16–17. For a hostile comment, partly based on the elitism issue, see Sam Bass Warner, Jr., *Province of Reason* (Cambridge, Mass. 1984), p. 206. I think Warner is quite wrong when he asserts, "In his [Bush's] vision the vast majority of mankind have no voice."

27. "Science, Strength, and Stability," address, June 8, 1946, p. 6 (Box 130, Bush Papers, LC); see also remarks in the Lawrence Scientific School address on engineering as a profession ("minister to the people") from which (pp. 9–10) the quotation is taken. The point I want to make is that Bush was sincere in these expressions, that he tried to live up to his ideals, and that his viewpoints have a curious, muted resonance with what we now designate as civic humanism and republican ideology.

28. Price to Parsons, Dec. 6, 1946, Correspondence files, Social Research Council (SSRC) records, New York.

29. *Pieces of the Action,* pp. 269–270; Caryl P. Haskins, *Of Societies and Men* (New York, 1951). Bush argued in "The Scientist and His Government" (p. 4) for a "voluntary collaboration of independent men," that is, a democracy, as being superior to absolutism, a point repeated in his *Modern Arms and Free Men.* Following Haskins, associated groupings (i.e., voluntarism) were contrasted with ant-like herds. These Bush identified with totalitarian centralization, language used to characterize the Soviet menace during the Cold War.

30. Expressions of the viewpoint are in "Military Organization for the United States" [1946/47], esp. p. 4, Bush Papers, LC. But note that Bush assumed that the workings of democracy did not always turn out to everyone's satisfaction, that there was a risk involved. For example, during the furor over security charges leveled against E. U. Condon, the Director of the National Bureau of Standards, he wrote Jewett that scientists only liked the democratic process when it worked for them (Mar. 15, 1948, Bush Papers, LC).

31. Alex Roland, *Model Research: The National Advisory Committee for Aeronautics, 1915–1958,* 2 vols. (Washington, D.C., 1985), is a recent and fine history of the agency.

32. For a comment on the judicial role and how Bush viewed the committees, see his letter to Jewett, Oct. 2, 1946, Bush Papers, LC. The Divisional Committees in NSF were quietly eliminated by Waterman's successor Leland Haworth. There is a need for a realistic description and analysis of the roles of various committees and boards in the governmental structure, particularly those in the National Institutes of Health, the National Endowment for the Humanities, and the National Endowment for the Arts with statutory origins and presumed authority beyond the merely advisory.

33. What one sees in Bush's pre-Washington career and even the work at CIW is a great curiosity and a talent for exploiting this curiosity. Bush was both a formidable specialist and, at the same time, a formidable generalist. Perhaps there was too much self-confidence and an occasional slighting of the abilities and interests of others with narrower angles of vision. Comments on Bush's consulting are in chap. V of *Pieces of the Action.*

34. For this incident, see folder 86, Box 98, with two A. Compton items (the S-1 records of Bush and Conant, OSRD NA); Smyth's comment is in a letter to E. P. Wigner, July 10, 1944, 5:1, Smyth Papers, Library, American Philosophi-

cal Society, which also has Compton's "Mr. Fermi, the Argonne Laboratory, and the University of Chicago" of July 28, 1944; and Bush to S. K. Allison, Oct. 10, 1945, Allison Papers, Regenstein Library, University of Chicago.

35. *Pieces of the Action,* p. 50.

36. Bush to C. W. Eliot, Oct. 8, 1942, OSRD records, National Archives (RG 227), Ser. 13, Box 70.

37. A. MacLeish to Bush, Jan. 12, 1944, with Bush note of Jan. 15: "Told him we had better leave science out." OSRD, NA Ser. 13, Box 70.

38. See Pursell, "Science Agencies in World War II."

39. R. C. Cochrane, *The National Academy of Sciences: The First Hundred Years, 1836–1963* (Washington, D.C., 1978); A. Hunter Dupree, *Science in the Federal Government* (Cambridge, Mass., 1957); Kevles, *The Physicists.*

40. F. B. Jewett to W. B. Cannon, May 31, 1944, Bush Papers, LC. In the Archives, NAS, see the file on the two wartime bodies, the Committee on War Biology and the Committee on Biological Research, in the Policy Files of Ross G. Harrison, Chairman, NRC. Material on the geologists is in the W. W. Rubey Papers, LC.

41. *Pieces of the Action,* pp. 43–48. Bush here also cites the importance of medicine and the need to get an appropriate committee to consider bovine mastitis. Toward the end of the war Bush regarded the settlement of the penicillin patent rights as an exemplary way of allocating credit to all the contributors, including basic scientists. Medicine had acquired an importance to him in late 1944 it previously lacked. See John Patrick Swann, "The Search for Synthetic Penicillin During World War II," *British Journal for the History of Science* 16 (1983):154–190. Bush's comment on penicillin in the context of patents is in his letter to John Tate, Feb. 8, 1945, Ser. 2, Box 3, OSRD NA.

42. Official file 2240, White House records, Franklin D. Roosevelt Library, Hyde Park, N.Y. The existing body was the Health and Medical Committee, originally attached to the Council on National Defense but in 1941 attached to the Federal Security Agency. In a memorandum to Harry Hopkins of March 3, 1941, Bush recommended severing the R&D functions and giving more funds for use through NAS, not adding to his jurisdiction. See also the records of the Committee on Military Medicine, ca. 1940–1945, in the Archives of the NAS, as well as various wartime files of the NRC's Division of Medical Sciences.

43. Based on the CMR records in the papers of A. N. Richards, Archives of the University of Pennsylvania, Philadelphia.

44. Jewett to Bush, Jan. 6, 1942, Bush Papers, LC. In this letter Jewett accuses Bush's organization of favoring MIT by treating physics differently than chemistry. In the case of chemistry, the contracts were allocated broadly to minimize the impact on chemistry faculties. (The Manhattan Project would aggravate that effect on physics.) Jewett thought in terms of industry primarily, not educational institutions. In a letter to O. E. Buckley, Oct. 15, 1942, Bush Papers, LC, on the R&D needs of the War Production Board, Jewett believed the board would have to follow the NDRC pattern, by which he meant going to industrial research laboratories, not the university world.

45. R. G. Hewlett and Oscar E. Anderson, Jr., *The New World 1939/1946* (University Park, Pa., 1962).

46. Tuve to R. C. Tolman, Apr. 21, 1942, OSRD, NA Ser. 8, Box 4; *Pieces of the Action,* pp. 106–112.

47. There is a real need for a fresh, independent look at the Rad Lab, one of

the great successes of the World War II R&D mobilization. To I. I. Rabi, who was there and at Los Alamos, the Rad Lab's achievements vindicated the goodness of science, unlike the tainted or ambiguous results of the Manhattan Project. Others in the heady days before the Vietnam War period stressed the defensive, nondestructive nature of radar as they promoted continental missile defense systems. There is relatively little literature on the consequences for the interrelations of the universities, the military, and industry.

48. See "A Few Quick" in Box 134 of Bush Papers, LC. Jewett saw the practice differently, as a way of getting the creators to let loose and turn over "bread board" models to design engineers (Jewett to Conant, July 6, 1942, Bush Paper, LC). See also *Endless Horizons,* p. 108; Minutes, Advisory Council, Nov. 13, 1941, OSRD NA, Ser. 13, Box 63; and on procurement problems "Analysis of the request of Lt. General Somervell . . . " 11 pp., Apr. 24, 1942, Ser. 2, Box 5, OSRD NA. In these World War II discussions are the harbingers of the issues raised by such commentators as Seymour Melman.

49. The first OSRD folder, Box 32, Waterman Papers, LC, relates to the OSRD history. P. Scherer writing to Waterman on the draft of "Combat Scientists," July 24, 1946, commented that the "draft does not give the impression of 'close coordination' and 'extraordinary and intimate relations between the military and the scientific and technical men.' " Not that warm relations did not exist, particularly at the working level, in many instances. But there was a tendency to soft-pedal the civil–military differences in the heady postwar days. The notable exception was the squabble over atomic energy legislation.

50. Jewett to Bush, Oct. 29, 1942, Bush Papers, LC: "NDRC is for this particular job in my judgement a group of amateurs doing an amateurish job in creating an organization to be run by amateurs in industrial research and development." NDRC–OSRD overstepped boundaries recognized by both men. Bush, with his university orientation, had less concern in this instance than did Jewett, coming from industry.

51. Jewett to W. B. Cannon, May 31, 1944, Bush Papers, LC. There are many examples in the Archives, NAS–NRC.

52. The overwhelming consensus in the professional communities of scientists, engineers, and medical personnel in 1945/46 was for government aid. Disputes centered on the nature of the aid and its administration. Few, apparently, followed Jewett in his opposition when he went public.

53. See Kevles on RBNS. I do not agree with Kevles that this was a lost chance to avoid the problems Bush and Conant corresponded about in 1970. The military would still have placed contracts at willing universities acting on the advice of the NAS board.

54. Jewett to Bush, May 23, 1943, Bush Papers, LC.

55. See Irwin Stewart, *Organizing Research for War* (Boston, 1948), chap. 21, for the official account of demobilization of OSRD.

56. In Manuscript Division, LC.

57. The questions were (1) how to make known, consistent with military security, "the contributions which have been made during our war effort to scientific knowledge"; (2) how "to organize a program continuing in the future the work which has been done in medicine and related sciences"; (3) what "can the Government do now and in the future to aid research activities by public and private organizations"; (4) and how to discover and develop scientific tal-

Vannevar Bush's New Deal for Research **327**

ent in American youth. *Science—The Endless Frontier,* p. 1. I have used the
1960 reprint edition. A rough draft of the Roosevelt letter is enclosed in O. Cox
to I. Stewart, Oct. 18, 1944, in Ser. 13, Box 70, OSRD NA.

58. Bush to Bowman, Feb. 6, 1945, Bowman Papers, Special Collections,
Eisenhower Library, Johns Hopkins University.

59. Bush to Bowman, Feb. 19, 1945, Bowman Papers.

60. The Inter-Society Committee evolved in 1946/47 to represent groups
and interests broader than the perceived viewpoint of NAS. After Truman's
1947 veto, the Committee headed the lobbying effort for an NSF, with Bush
playing an intermittent, limited role. The best single source known to me on
the Committee is the files on it in the Harlow Shapley Papers in the Harvard
University Archives.

61. The Board was an early victim of conservative anti–New Deal ire early
in World War II. Although best remembered for its encouragement of regional
and local planning involving the environment and natural resources, NRPB
was edging into other areas, as in the report on research.

62. Bush to James Forrestal, Sept. 10, 1947, Bush Papers, LC. The exchange with Truman is in Bush's memorandum of his conference with the
President of Sept. 24, 1947, Bush Papers, LC.

63. Bush's letter appeared in *Science* 98 (1943):571–573. To counter Senator Kilgore's proposals for a "New Deal" agency for R&D, Bush declared that
because of the war there was a temporary centralization preserving the independence of the research groups—correlation not unitary control. After the
point about state and private universities, Bush declared himself for the support of the best in order not to increase mediocrity.

64. Bush apparently assumed that the representatives would behave re-
sponsibly, like the agency officials serving on NACA, if only because of the
presence of strong individuals from the private sector. But note the comments
of Bush's colleague Paul Scherer (in Bush to Steelman, Apr. 4, 1947, Bush
Papers, LC) that the Steelman Report contained "a wealth of outright pork,"
that half of the science advisers to the NSF director were to come from the
government, and that "outright grants replace contracts."

65. Arnold Miles's memorandum to Director, Bureau of the Budget, Mar.
10, 1945, reports hearing from Wassily Leontieff that the Bowman Committee
was against involvement of NAS but supported a "National Scientific Corporation." A draft bill had been prepared by Oscar Cox's staff. File 39.27, General
Records of the Director, 1939–1946, RG 51, NA.

66. Appendix 4 of SEF has the report of Moe's committee. Education was
another area where Jewett and Bush differed despite their agreement on a
basically conservative stance. In 1943 Jewett compared the "revolutionizing"
thrust of Senator Kilgore's proposals with New Deal actions in the areas of
labor and education (Jewett to Bush, May 27, 1943, Bush Papers, LC).

67. Bush to O. Ruebhausen, Sept. 23, 1946, Bush Papers, LC.

68. For example, Bush wrote D. W. Bronk (Aug. 16, 1946, Bush Papers,
LC) that the Joint Research and Development Board would reconcile differ-
ences with the military (i.e., his judicial function) while NRC, then headed by
Bronk, would advise the government and be its official link to science. For E. J.
Cohn (to H. W. Smith, May 23, 1945, in Ser. 2, Box 3, OSRD NA), who was
against the separation of natural and medical sciences in a new agency, the

role of NAS in RBNS was a hopeful precedent. But earlier, W. H. P. Blandy, chief of the Navy's Bureau of Ordnance, wrote to Bush (Nov. 13, 1943, Ser. 13, Box 70, OSRD NA) advocating laboratories in both military and civilian organization and the need for a "feeling of mutual respect" between "rugged individuals" from within and without the armed services.

For Bush's views, see SEF, p. 27f, and "Research on Military Problems," *Endless Horizons,* pp. 82–100, for the text of his statement of Jan. 26, 1945, before the House Select Committee on Postwar Military Policy, presented during the gestation of SEF. For his view of the temporary nature of RBNS, see Bush to C. L. Wilson, Jan. 15, 1945; and for a broader statement on military research, the letters to F. G. Fassett, Jr., of Feb. 3 and 9, 1945—all in Bush Papers, LC.

The discussion of differing views is based on a table and associated documents dating from July to Sept. 1944 in Folder 23, Box 74, of the Compton–Killian Collection, Archives, MIT. The Tuve Papers, LC, contain many files on postwar military policy. In "Notes, July 27 [1944] on Army–Navy Research," Tuve called for an independent OSRD-like agency to cooperate jointly, "not just *subsidiary* collaboration" with the services (his emphasis). Tuve ruled out NAS as not broad enough and was concerned that the Wilson Committee ignored the research conducted by the armed services.

69.   Bush, "Memorandum of Conference with the President," June 14, 1945, Ser. 2, Box 2, OSRD NA.

70.   *Pieces of the Action,* p. 293.

71.   Bush to George C. Marshall, June 12, 1947, Bush Papers, LC; *Pieces of the Action,* pp. 278f, 293.

72.   Bush memo, June 15, 1943, Box 21, Bush Papers, LC, discloses that OSRD paid his transportation expenses while CIW covered other costs. Bush's status, as not quite a federal employee, was one of his justifications, in 1945/46, for his independence from the administration's position on the NSF legislation.

73.   Smith Diary, Aug. 31, 1943 (on the Budget Bureau's planning being integrated with operation and policy determination, unlike Bush's views), Aug. 31, 1944, Jan. 4 and 21, 1945, Smith Papers, Franklin D. Roosevelt Library, Hyde Park, N.Y.

74.   Smith Diary, Mar. 23, 1945; Smith memorandum to Roosevelt, Mar. 30, 1945, against RBNS and NAS, and for continuing OSRD; the letter to Bush went out on Mar. 31 (Official file 330b, Franklin D. Roosevelt Library, Hyde Park, N.Y.) On Mar. 31, 1945, Smith sent a letter to Conant about not liquidating OSRD and stressing the need for democratic controls. Minutes, Committee on Federal Government and Research, Feb. 21 and 24, 1945, SSRC Records, New York.

75.   Smith's Daily Record, entries of Feb. 27, Mar. 23, April 3, 11, 17, May 2 and 3; Smith to Conant, Mar. 30, 1945, Box 17, Smith Papers; May 5, 1945 entry, Furer Diary, LC.

76.   Smith Diary, May 4, 1945, and Daily Record entries of Apr. 18 and 30, May 26 and 29, 1945, Smith Papers. In RG 51 NA, file 39.27 of Smith General Records as Director, various memos, especially R. J. Bounds and W. H. Shapley to Smith, Mar. 5, 1945. Assistant Secretary of the Navy Struve Hensel supported RBNS, unlike Lewis Strauss.

77.   In 1944 Charles E. Wilson, the Deputy Director of the War Production

Board, headed a mixed committee of civilians and military on postwar military research. Although the majority, including Merle Tuve and the Navy representatives, favored an independent civilian agency, the minority (Jewett and the Army) came out for RBNS. Smith had another reason to be unimpressed by RBNS and NAS. The question to Smith was whether a temporary RBNS violating sound principle was necessary even as a temporary expedient.

78. Smith Diary, May 4, 11, and 21, 1945, Smith Papers.

79. Smith memorandum to President, Apr. 27, 1945, in file 39.27, General Records of Director, RG 51, NA; Smith Diary, May 21, June 8, 1945; Smith Daily Record, June 8, 13, July 20, 1945, Smith Papers, Franklin D. Roosevelt Library; Official file 52, White House Records, Truman Library, Independence, Missouri. After Roosevelt stopped funds to RBNS, the Army and the Navy tried to transfer funds from their accounts but were blocked by Smith and Truman. See James Forrestal memorandum to Truman, July 28, 1945, in Official file 330-C, Truman Library.

Two interesting comments on Smith's animus against OSRD and scientists are in Smith to C. L. Wilson, July 18, 1945, Ser. 2, Box 2, OSRD NA, about SEF really being "Endless Expenditure," to which Wilson commented in a note to Bush: "Smith's rather sardonic humor comes out. For him this is quite a decent letter." In Admiral Furer's diary, May 22, 1945, he reports Smith's complaint about the scientists not cooperating on government research: "[They] do not know even the first thing about the basic philosophy of democracy."

80. Smith Daily Record, Aug. 27, 1945, Smith Papers, Franklin D. Roosevelt Library.

81. Smith Diary, Aug. 28, 1945, Smith Papers, Franklin D. Roosevelt Library. For comments on this point, see Lyman Chalkley, *Prologue to the U.S. National Science Foundation* (privately published [New York?], 1951), p. 20.

82. Smith Diary, Mar. 19, 1945. Interspersed among the notes to his recent book *America's Unwritten Constitution . . .* (Baton Rouge, 1983), Don K. Price has an interesting autobiographical account. Although not specifically useful in the context of these remarks, Price is very revealing about his broad viewpoint.

83. Attached to Bush to S. Rosenman, June 1, 1945, Official file 53, Truman Library.

84. Smith Daily Record, June 14, Aug. 13 and 20, 1945, Smith Papers, Franklin D. Roosevelt Library.

85. Smith to Bush, Oct. 10, 1945, Official file 53, Truman Library; Bush to Jewett, Nov. 6, 1945, to John Snyder, Nov. 23, 1945, and Snyder to Bush, Nov. 30, 1945, Smith Papers. For how badly the Bush–Truman relationship deteriorated, see Truman to R. P. Patterson, July 22, 1946, President's Secretary's file 112, Truman Library, recounting Bush's refusal to appear at a ceremony Truman arranged for the award of the Medal of Merit.

86. J. Donald Kingsley memorandum to John R. Steelman, Dec. 31, 1946, and Apr. 1, 1947, Official file 192-E, Truman Library. Bush to Webb, Dec. 27, 1946, and May 13, 1947; to G. Rentschler, Jan. 28, 1947; Conant to Bush, Jan. 28, 1947; Bush to Jewett, Mar. 29, 1947; to B. M. Baruch, May 2, 1947; P. A. Scherer memorandum to Bush, June 12, 1947; Bush to H. M. Shapley, Nov. 10, 1947—all in Bush Papers, LC. The letter to Jewett notes that the Senate version of the NSF bill has the President appointing the Director, but Bush opines

correctly that this feature will be dropped. Bush states that, even so, apart from the initial appointment, the proposed agency will be completely removed from the control of the President. For Jewett's views, see his letter to C. A. Calverton, Mar. 31, 1947, Bush Papers LC. Another expression of conservative opposition to the NSF legislation is in E. B. Wilson to Jewett, June 16, 1947, in the file of the Committee on Names for NSF, NRC Policy files 1946–1949, Archives, NAS–NRC. Wilson was already embarrassed by the pressure of the scientists to grab government funds and thought many of those "liberals" were really "totalitarians."

87.   Cox memorandum to Truman, Sept. 24, 1945, and Cox to Bush, Sept. 30, 1945, Cox Papers, Franklin D. Roosevelt Library; V. Bush, "Cancer," Nov. 20, 1972, Bush Papers, MIT.

88.   Webb to V. Kennedy, May 16, 1966; memorandum to Lyndon B. Johnson, May 23, 1960; to E. F. Buryan, July 18, 1961—all in Webb Papers, Truman Library—with their avowals of support of regional development. Bush to Webb, May 18, 1961, and Apr. 11, 1963, Bush Papers, MIT. Cox to S. Rosenman, Sept. 30, with attached memorandum of conversation, Oct. 21, 1944, Cox Papers, Franklin D. Roosevelt Library. *Pieces of the Action*, pp. 54–55, reports Bush's change of mind after the moon landing.

Contrary to Bush's stance separating basic research from social policies but indicative of the acceptance of adding to knowledge as a proper federal function is Webb to J. Wiesner, May 2, 1961, Webb Papers, where that motivation is a proper fallback if the Soviets reach the moon first. For analogous sentiments on the merits of scientific research, see the diary of Webb's predecessor, T. Keith Glennon, in the Eisenhower Library, Abilene, Kans., entries of Apr. 29, May 5, and June 3, 1960 (in spite of the sarcastic comment of July 8 on supporting scientists in the style to which they would like to become accustomed).

89.   Minutes of Council, SSRC, Apr. 6–7, 1946, SSRC Records, New York.

90.   Minutes, Committee on the Federal Government and Research, Feb. 5, Apr. 14, May 31, June 25, July 18, July 28–29, Nov. 24, 1945; Minutes, Council SSRC, Apr. 6–7, 1946—all in SSRC Records, New York. Smith Daily Record, Jan. 23 and Mar. 2, 1945, Smith Papers. The SSRC commissioned Talcott Parsons to prepare the report, which was not published. I have benefited from a reading of the manuscript of Samuel Z. Klausner's study of the Parsons' report incident.

91.   H. A. Moe to H. D. Smith, Nov. 21, 1945, with attached refusal to sign Bowman statement, and D. Young to P. Herring, Nov. 8, 1945, Correspondence File; Minutes, Committee on Federal Government and Research Nov. 15, 1946, in SSRC Records. See I. Bowman to H. Shapley, Nov. 6, 1946, in Bush Papers, LC, for a strange statement on the social sciences by an erstwhile geographer.

92.   Bush, "Memorandum of 'Role of the Social Sciences in Modern Society,' " Dec. 10, 1946; Bush to D. Josephs, Sept. 12 and 19, 1946; Josephs to Bush, Sept. 18, 1946—all in Bush Papers, LC. For a different view of the Carnegie Corporation study, see Don K. Price to T. Parsons, Dec. 6, 1946, in Correspondence File, SSRC Records, New York.

93.   C. C. Lauritsen to R. Gaither, Dec. 9, and reply of Dec. 13, 1949, Box 7, C. C. Lauritsen Papers, Archives, California Institute of Technology.

94.   Bush to Gardner, Feb. 14, 1955, Bush Papers, LC.

95.   Bush to Jewett, Apr. 26, 1946, and Jewett to Bush, Aug. 7, 1946, Bush

Papers, LC. See the files on the Committee to Support the Bush Report (for the Robbins reaction) and on the NSF legislation hearing (with Harrison's letter of Oct. 6, 1945) in the NRC Central Policy Files of R. G. Harrison, 1940–1945, Archives, NAS. The hearings file is filled with interesting expressions of opinions going beyond the particulars of the legislative proposals.

96.   To greatly oversimplify, Bush had scruples inhibiting tendencies to build either tightly centralized empires or freewheeling conglomerates. If Bush was more pragmatic about compromises during the war, Jewett had a keener vision of a future bottom-line expediency.

97.   Donald C. Swain, "The Rise of a Research Empire: NIH, 1930–1950," *Science* 138 (1962):1233–1237. See also H. D. Smith to Elbert Thomas, June 14, 1945, Official file 7-B, Truman Library.

98.   "The Case for Biological Engineering," in *Scientists Face the World of 1942* (New Brunswick, N.J., 1942), pp. 33–45. For some examples of Bush expressions on medicine, see A. N. Richards to Bush, Feb. 11, and Bush's reply of Feb. 13, 1947; Bush to K. T. Compton, Jan. 25, 1951; "After Hippocrates", Apr. 11, 1951 (Box 134); Bush to Gardner, Feb. 14, 1955—all in Bush Papers, LC. Additional statements are in *Endless Horizons* and *Science Is Not Enough*.

99.   For example, John F. Fulton to Bush, Jan. 4, 1944, in Ser. 13, Box 31, OSRD NA.

100.   Bush to Jewett, Oct. 22, 1946, Bush Papers, LC. Bush acted on the advice of CMR.

101.   Bush to B. M. Baruch, May 2, 1947, and John Teeter memorandum to Bush [Aug. 29, 1947], Bush Papers, LC. H. Shapley to L. DuBridge, July 5, 1948, DuBridge Papers, California Institute of Technology. For ONR, I am indebted to chap. 3 of Harvey Sapolsky's draft history, with its account of that agency's opposition to NSF.

102.   Condon memorandum to John R. Steelman, July 28, 1947, Official file 192-E. Truman Library. Various documents in the Condon Papers, Library, American Philosophical Society, Philadelphia, disclose Condon's actions against the NSF legislation until its final passage.

103.   *Modern Arms and Free Men* (New York, 1949), p. 19.

104.   Ibid., p. 72.

105.   For Bush's statement, see Box 128 of Bush Papers, LC, which contains other documents on military organization. Bush had his testimony reprinted in *Endless Horizons* in 1946.

106.   An early example is his 1941 report on defense organization, enclosed in Smith's memorandum to Roosevelt, March 17, 1941, Official file 2240, Franklin D. Roosevelt Library. Bush believed the existing organization was fine but stressed the need "for formulation of military policy, not only in terms of present technical knowledge but also in terms of best scientific and engineering knowledge with regards to possible future developments in the methods and materials of warfare." But that cut right into the professional area of the officers of the armed services, for which see Furer Diary, Apr. 14, 1943, LC. The text of the address with the bobby pin quotation is in box 129, Bush Papers, LC. For favorable comments on military men, see *Pieces of the Action,* pp. 209, 298. These and other passages can be read as condescending. For Bush's functional analysis of the problem, see his memorandum to L. V. Berkner, Oct. 29, 1946, Bush Papers, LC.

107.   One of Bush's purposes in writing *Modern Arms and Free Men* and in

other public and private utterances was to mitigate the "unreasoning fear" engendered by modern weaponry—with its many bad consequences such as the excesses of security. He wrote on p. 2 of that book, the "technological future is far less dreadful and frightening than many of use have been led to believe," a point repeated in *Pieces of the Action,* p. 118. The Korean War marked a turning point leading to his active role in the Committee on the Present Danger. See Box 134, Bush Papers, LC; also Samuel F. Wells, Jr., "Sounding the Tocsin: NSC 68 and the Soviet Threat," *International Security* 14 (1979):116–158. Bush, however, never wavered in his doubts about the excesses of the military and the dubiousness of much modern weaponry.

Bush, "Science and War," 1951, Box 134, Bush Papers, LC; the quotation is from pp. 2–3. For his opposition to continued Navy support of fundamental research (for Reber, the early radioastronomer), see the letter to Jewett, Sept. 17, 1946; for Bush's role in shaping the defense R&D budget, see the letter to K. T. Compton of Dec. 30, 1946—both in Bush Papers, LC. For a contrary viewpoint stressing that "a considerable part of the research, development and testing of new weapons and devices must be carried on in laboratories of the armed forces," see Rear Admiral W. G. Switzer to C. C. Lauritsen, in 2.11, C. C. Lauritsen Papers, Archives, California Institute of Technology. An important statement is Bush's position presented to the Steelman Commission, Bush to John Steelman, Apr. 4, 1947, Bush Papers, LC.

109. Bush to C. Wilson, Mar. 4, 1949, and to K. T. Compton, Dec. 27, 1948, Bush Papers, LC.

110. At the time of the Oppenheimer security hearings, at which Bush openly expressed his opposition in print, he was privately also strongly criticizing the abuse of the security system for thought control (to L. Bradford, June 3, 1954). When asked in 1962, routinely, to fill out a security clearance by NSF to serve as a consultant to the agency, he firmly and indignantly refused (Bush to Waterman, Jan. 22, 1962, Bush Papers, LC). The quotation on the Air Force is from *Modern Arms and Free Men,* p. 96. The words on "orderly progress" are from a 1946 address at Wright Field, "Partnership in Military Research" (Box 130, Bush Papers, LC). For examples of Bush's attempts to sway the military leadership, see his letters to General Omar Bradley of Feb. 23, Mar. 6, and Apr. 13, 1950, all in Bush Papers, LC.

111. *Modern Arms and Free Men,* p. 85f.

112. Ibid., pp. 107, 142–149, *Pieces of the Action,* p. 13.

113. Bush to Conant, Jan. 16, 1951, and Mar. 29, 1954, Bush Papers, LC. For Bush's actions, see Robert Divine, *Blowing in the Wind: The Nuclear Test Ban Debate, 1954–1960* (New York, 1978), p. 16.

114. In 1945 Bush doubted that the Navy would support continuation of work at MIT on "electronic arithmetical machines" because the military would insist on security classification and a focus on Navy needs (Bush to S. H. Caldwell, Oct. 3, 1946, Bush Papers, LC). Bush simply underestimated the adroitness of the military. More important, to Bush, it was "anomalous" for essential support of the universities to "reside in an armed service" (*Modern Arms and Free Men,* p. 247). A fine example of how the armed services and the research community co-opted "science" with Bush's grudging approval were the "Summer Studies" which began before the Korean War and had a real measure of popularity until the Vietnam era. These Summer Studies revived the excite-

ment of the wartime days to many participants while going back to Bush's model of small groups of the ministry.

115. In Weaver Diary, Dec. 8, 1950, Rockefeller Foundation Archives, Rockefeller Archive Center, North Tarrytown, N.Y. Sapolsky's draft history of the Office of Naval Research is particularly useful since it gives a Navy viewpoint, not simply that of the civilian administrators. See his "Academic Science and the Military: The Years Since the Second World War," in N. Reingold (ed.), *The Sciences in the American Context: New Perspectives* (Washington, D.C., 1979), pp. 379–399.

116. In addition to previously cited documents, see Bush to C. Dollard, Sept. 20, 1951, and to B. M. Baruch, Dec. 3, 1952, Bush Papers, LC.

117. For materials on government–university relations, see Boxes 63 and 85, Bush Papers, LC. The records of his service on the NSF committee are in the latter. Note, for example, the minutes of the meeting of Mar. 5, 1954.

118. Bush to Stratton, Nov. 13, 1953, in Box 85, Bush Papers, LC. An earlier expression of concern for the universities in the new future emerging after World War II occurs in an exchange of personal letters with General Ira Eaker (Eaker to Bush, Apr. 12, 1946, and Bush's reply of Apr. 15 in Box 208, Tuve Papers, LC). Stressing the Air Force's appreciation of OSRD, Eaker writes Bush of the arrangement with Douglas Aircraft not for any specific product but for broad study and research on "inter-continental warfare." This is the origin of the present Rand Corporation. Bush strongly opposed going to Douglas because the defense funds would place CIW and the universities at a disadvantage in attracting able investigators.

119. As late as 1951, Bush thought NSF would advise the President and have a research mission involving defense: "problems which are not likely to be followed up directly in the Department of Defense" (to P. H. Douglas, Jan. 11, 1951, Bush Papers, LC), a perception shared by some others. In the Science Advisory Committee in the Office of Defense Mobilization, the Minutes of June 15, 1952, note the query by Congressman Wolverhampton why SAC exists since NSF supposedly does defense research (186.8, DuBridge Papers, Archives, California Institute of Technology). In the same collection, see, as examples, the L. C Craigie memorandum to DuBridge, Mar. 25, 1947; the Project VISTA files in 174.3; DuBridge's indignant reply to a Defense Department query to its contractors about retired officers on the payroll of June 5, 1959, in 175.2. In fiscal year 1958, according to that file, three universities ranked among the 100 leading contractors: MIT was fifty-second, CIT fifty-seventh, and Johns Hopkins ninety-sixth.

120. *Modern Arms and Free Men,* p. 128f.

121. It is hard to imagine Bush extolling monopoly as Jewett did at the Chicago World's Fair in the early years of the Depression. See R. Rydell, "The Fan Dance of Science . . .," *Isis* 76 (1985):525–542; the reference on p. 532 is to a public address. During the gestation of SEF, he wrote a succinct statement of his views on patents (to J. Tate, Feb. 8, 1945, OSRD, general correspondence re Reports to President) stressing the encouragement of small concerns and independent inventors (including individuals at nonprofit organizations). A later expression is in his letter to L. Cutler, Oct. 4, 1949, Bush Papers, LC.

14. Metro-Goldwyn-Mayer Meets the Atom Bomb

For years I had said that popularization of science is historically important but had never done anything in the area. This first foray resulted from the happy coincidence of receiving an invitation to a 1981 conference on popularization and discovering a file in the Manuscript Division of the Library of Congress on the production of a commercial movie about the building of the A-bomb.

Initially, I was concerned with how scientists reacted to the issue of what was the truth when encountering Metro-Goldwyn-Mayer's handiwork. Quickly that gave way to studying how Hollywood and the scientists interacted, each trying to produce a great public mass myth while somehow protecting the integrity of quite different ventures.

The interactions were unexpectedly enlightening. Take Vannevar Bush's objections in 1946 to his portrayal with President Roosevelt. At that date Bush was anguished over his deteriorating relations with President Truman. Was he concerned to stress the past soundness of his advice and his awareness of the proper role of the science adviser to the President of the United States? Or consider a point not in the paper—Lise Meitner's complaint about her shabby living conditions in Copenhagen after fleeing Germany, according to the MGM script. She really had a very nice place there, she insisted in a note in the MGM archives. Was MGM implicitly following a hallowed cliché derived from Emma Lazarus's words on the Statue of Liberty about the poor huddled masses striving to be free? Was Meitner implicitly making a point about socioeconomic status? Or consider General Groves's meticulous attention to detail. The Manhattan Project was a great triumph for a military administrator. But even in 1946, it must have become increasingly clear to Groves that he had a dubious future in the postwar Army. His third star came as he left for civilian employment. Was Groves anxious to have his role given just right for posterity?

To my mind, the high point in the film occurs during a scene of a conference between Roosevelt and Bush. Bush comes in, announces that he has something important to report and that it is top secret. At that point the President's black Scottish terrier, Fala, picks himself up from the floor and leaves the Oval Office under the approving gazes of the President of the United States and the Director of the Office of Scientific Research and Development. As a well-bred, well-trained, intelligent dog, Fala clearly understood that he lacked the proper clearance. Without a doubt, one of the great moments in the annals of Hollywood.

This paper was first published in *Expository Science: Forms and Functions of Popularization* (1985).

Late in 1945 Metro-Goldwyn-Mayer (MGM), the premier Hollywood motion picture studio, decided to produce a film on the building of the atom bomb. The result, *The Beginning or the End,* was released in the spring of 1947. To depict living, well-known individuals, MGM had to get their permission, resulting in written and oral interchanges. What follows is based largely on materials in the correspondence of J. Robert Oppenheimer, director of the Los Alamos Laboratory during World War II; Vannevar Bush, head of the Office of Scientific Research and Development (OSRD), which administered the wartime research and development effort including the atom bomb program in its early crucial stages; Leslie R. Groves, a general in the Army Corps of Engineers who headed the Manhattan Project, which took over from Bush; and Albert Einstein, whose letter to President Roosevelt in 1939 started the effort.[1] The correspondence makes possible a comparison of the original screenplay with various suggested revisions and with the final version as seen by the public. I hope to illuminate the differing viewpoints of the individuals involved in the development of the movie. The process depicted contains clues on issues important for historians and sociologists, as well as those in the mass media. Every genre of exposition—the scientific journal article, the historical monograph, the newspaper article, the novel, the popular motion picture, etc.—imposes a structure and dynamic on its subject matter. In ways sometimes subtle, sometimes gross, different messages are conveyed not always matching the intentions and needs of the creators of the exposition, their audiences, or the actual participants in the events described.

Origins of the Film

Shortly after Hiroshima, contacts between scientists in the Manhattan District and MGM spawned the suggestion for a movie on the building of the A-bomb.[2] These scientists were part of the so-called atomic scientists' movement which eventually resulted in the establishment of the Federation of Atomic Scientists and the well-known *Bulletin of the Atomic Scientists*. Doing a commercial movie was one, atypical incident in the efforts of individuals in the movement and other scientists to educate the lay public about the nature of the new weapon and its implications for both domestic and international policy. Besides the use of the periodical press and public lecturing, many individuals had personal contacts with influential persons. Talks on radio also occurred, the only other use of the newer forms of mass media. Unlike the MGM film, these talks were essentially like conventional public lectures.

Only in this instance was there an attempt at something markedly different from familiar expository forms. Even here, the awe of the new power caused MGM to approach the project with exceptional care. Sam Marx, who later produced the film, visited Oak Ridge in 1945 and then spoke to President Truman in Washington. Truman was friendly but noncommittal. To avoid his possible hostility, an MGM official assured Truman of the company's high intentions. People "sit in theatres and listen" for entertainment, "but in a film of this nature we are certainly going far deeper than ordinary entertainment." MGM foresaw "a great service to civilization if the right kind of film could be made."[3]

MGM hired a successful screenwriter, Robert Considine, to prepare the initial story with the advice of the men from the atomic scientists' movement. They were active in the early stages of formulation of the script because of their hope the film would further their viewpoint, as well as yield money for their meager treasury and even funds for scholarships.[4] By early 1946 these younger and generally little-known individuals had given up those hopes as the evolving script in their eyes glorified the military and "put foolish words" in the mouths of scientists. They did not wholly understand what was happening and expressed confidence that the leading scientists would never permit themselves to be so depicted.[5] What they did not know was that General Leslie R. Groves on the last day of 1945 had signed an agreement with MGM giving permission to depict him in return for $10,000 and the right to review the script. Only the military participants personally received cash for their permission to be depicted in the film. All of the scientists gave their releases to MGM gratis. Perhaps they assumed the film was a professional obligation not calling for cash.[6]

The Original Script

Considine's original story was elaborated and converted into a screenplay by Frank Wead. To get the consent of the prominent individuals in the plot (no longer a legal requirement), Wead's script, completed in March 1946, was submitted to many of the participants in the development of the A-bomb.[7] As in the early discussions with those in the atomic scientists movement, dramatic continuity was provided by two fictional males; Matt Cochran, a young physicist, and Jeff Nixon, a young Army Engineer Colonel on Grove's staff (later depicted on the screen by Robert Walker). Matt was given a young bride, Jeff a girl friend serving as the General's secretary.[8]

The script opens with a newsreel showing the burial of *The Beginning or the End* under a grove of redwood trees so that, no matter what, five hundred years later the truth about the A-Bomb will be known. J. Robert Oppenheimer (eventually played by Hume Cronyn) next introduces the story, which starts with Lise Meitner working in Berlin. She flees to Niels Bohr in Copenhagen when Nazis take over the laboratory. Word of the new work on uranium fission comes to America, and Albert Einstein writes a letter to President Roosevelt with the assistance of Matt Cochran (later depicted by Tom Drake). The bomb project is launched, leading to a reenactment of Enrico Fermi's first controlled chain reaction at Stagg Field at the Metallurgical Laboratory of the University of Chicago. Introducing the scene, the script advises: "Tremendous history is about to be made by unhistoric looking people." Soon the Italian navigator appears: "Dr. Fermi, scientifically detached from the world, enters. . . ." In a harsh review of the film, *Life* criticized the dramatic qualities of the "biggest event since the birth of Christ."[9] But Matt has qualms about continuing in the development of such a terrible weapon. Convinced to stay, he has to leave his bride behind. Vannevar Bush regretted these passages which he saw as "an American trait . . . to tie a serious matter and a romance together."[10]

Skipping the other important work of the OSRD period, the script enters into the domain of the Manhattan Engineering District where Groves (played by Brian Donlevy) exhorts industry to join up. In the released film, the DuPont representative then grandly waives all patent rights, an easy position for DuPont to take fictionally (if not really) as those very conservative gentlemen Vannevar Bush and Leslie Groves had no intention whatsoever that any such patents would go into private hands because of the nature of the technology. We next get a panorama of great factory structures, endless rail shipments, and cryptic production lines.

The script switches to Los Alamos—where rather little is shown, given the requirements of security. There is a dramatic account of the first A-bomb explosion at Alamogordo, quite impressively re-created in the studio. Jeff and Matt, who have appeared throughout, now go to Tinian to prepare the first two bombs for their use against Japan. In an impossible accident, Matt suffers a fatal radiation injury while setting up the bomb one evening all by himself. Hiroshima is devastated in a spectacular film sequence reaffirming Hollywood's skill at special effects. (Apparently little, if any, of the footage derived from the actual bombing.) Matt dies, and the screenplay (and the picture) ends with his now pregnant widow, Jeff, and Jeff's girl friend talking inspirationally how the new world coming will justify Matt's sacrifice. This bare outline does not do justice to the nuances of the screenplay, some of which are treated later.

Metro-Goldwyn-Mayer's Dilemma

MGM realized that their A-bomb film was an unusual, difficult project, but the full extent and nature of the problems only became known in three stages: (1) up to the preparation of the Wead screenplay in March 1946; (2) the reactions to the screenplay from April into the autumn; (3) from October 1946 (when a first complete film was shown to interested parties and sneak previewed) to January 1947. MGM received strenuous complaints in the last stage forcing extensive cutting and reshooting. By February 1947 the final print was available for release. The reactions of the scientists in stage (1) did not prepare the company for what occurred later.

As General Groves was bluntly told, MGM was not an endowed institution like Harvard but a commercial organization.[11] To make a picture salable, it was necessary to fictionalize for dramatic purposes. An MGM memo sent to Albert Einstein conceptualized the point: "It must be realized that dramatic truth is just as compelling a requirement on us as veritable truth is on a scientist."[12] "Veritable truth" consisted of a great mass of details. To achieve dramatic truth required selection and compression—Matt Cochran to stand for all the young physicists and Jeff Nixon for all the engineer officers under Groves. "Veritable truth," presumably a complex knowable reality, is a difficult concept for philosophers, historians, sociologists, and others who have to deal with the intractability of imperfect bodies of evidence. As disclosed by their letters and memoranda, the MGM officials were sophisticated and intelligent enough to realize the problems. "Dramatic truth" was a kind of abstract of "veritable truth" capable of commercial dissemination to a mass audience.

The depiction of Fermi's great Stagg Field experiment provided two examples of the conversion of veritable truth to dramatic truth. Matt Cochran's hesitation about continuing stood for all the qualms of those who had strong feelings about dropping the bomb on enemy targets, like the scientists who later endorsed the Franck Report in the Metallurgical Laboratory. Deliberately, all such individuals appeared in the film in one fictional Stagg Field scene as men who withdrew because of scruples after the experiment succeeded. Correspondence and the script make clear that these were intended to be Quakers. The film does not explicitly identify them as such, nor did any such incident occur in "veritable truth." By hinting at Quaker pacifism, the film glides silently by strongly felt issues, blurring why more than traditional pacifism was involved.[13]

Another problem was the desire on the part of both MGM and the scientists and engineers to show the internationalism of the effort. Besides the early European scenes, the screenplay has Bohr later playing a visible, consequential role in the A-bomb project. But depicting the participation of the Canadians and the British posed problems for "dramatic truth." It increased the casting and complicated the plot. Groves was conscious of the importance of Canada as a past and future supplier of uranium. He was less adamant than the scientists about crediting British participants in the bomb project like Rudolf E. Peierls, James Chadwick, and William G. Penney. MGM, consequently, had its way in symbolizing the Commonwealth's role by having a delegation of observers at the Stagg Field experiment, a complete fiction. Besides, MGM argued, they were not against a British presence as they expected "British shillings" to provide an appreciable portion of the profits. They were even willing to contemplate making a special Commonwealth version, presumably with a greater specific role for Commonwealth scientists.[14]

Selecting from the mass of details did more than simplify a complex sequence of events. "Dramatic truth" acted as a pressure to mold the screenplay and film into a familiar narrative form with stock characters and stock situations as far as possible. Hollywood had little experience with the realities of scientific research, let alone the unique circumstances of the A-bomb project. What they had done in the past were stories of heroic physicians and inventors, triumphs of individual wills. Here they faced a great collective enterprise. Giving Leslie R. Groves as the hard-driving organizer of a great industrial enterprise was an easier challenge than displaying Bush's managerial skills in directing research and development or Oppenheimer's role on the mesa in New Mexico. Hollywood had difficulties in depicting situations of lesser moral ambiguity and tragedy than presented by the A-bomb. Success stories where virtue triumphed were more comfortable and familiar. Even without the agreement with Groves, there is little, indeed, to suggest an inclination on MGM's part to present the antimilitary viewpoint of the atomic scientists' movement so soon after the conclusion of a popular, victorious conflict. Not that it was impossible for Hollywood to produce a great, meaningful film on the A-bomb project. Given the prevailing situation in the industry, the probability was low of achieving the necessary critical and artistic insight into the story of the Manhattan Project. Perhaps more important was the need to couple such insight with the skill to succeed within the constraints of the commercial genre or somehow to transcend those limitations.

Reactions of Key Participants

J. Robert Oppenheimer and his wife told Sam Marx in no uncertain terms of their hostile reaction to the screenplay in April 1946. Marx had no problem adjusting to the criticisms of factual details and depiction of personalities. Oppenheimer's private views were, in fact, quite mild: "[the] script is not bad generally but that 'real' characters are stilted lifeless and without purpose or insight."[15] Reassuring Oppenheimer, Marx wrote that "the character of J. Robert Oppenheimer must be an extremely pleasant one with a love of mankind, humility and a pretty fair knack of cooking." The film would make plain that Oppenheimer, not Groves, was in command at the Alamogordo test. Fresh from a reading of the Smyth Report and a viewing of newsreels, Hume Cronyn added: "I gather that simplicity, warmth and a complete lack of affectation are essential to your character." Oppenheimer signed a release shortly afterward in May 1946.[16] When queried incredulously by a scientist in the atomic scientists' movement, Oppenheimer defended his signing by asserting the main points were satisfactory: "namely scientists were ordinary decent guys, that they worried like hell about the bomb, that it presents a major issue of good and evil to the people of the world." Although the screenplay was not "beautiful, wise, or deep . . . it did not lie in my power to make it so."[17] A fictional Oppenheimer consequently appeared in movie houses, an earnest scoutmaster who accidentally had a Ph.D. in theoretical physics from Göttingen.

If Oppenheimer withdrew from further meaningful involvement after May 1946, Groves and Bush continued to interact with MGM into the late fall of 1946, particularly after seeing the first film version. Groves was quite explicit about his intentions. After checking for possible security violations, he was determined that the film would not reflect any discredit on anyone engaged in the Manhattan Project.[18] In fact, of all the participants whose reactions are presently known, Groves was the most assiduous in objecting to inaccuracy of details, tone, and interpretation, much to the annoyance of MGM.

As to discredit, he was most alert to depiction of the Army in general, the Corps of Engineers in particular, and especially himself. As late as November he complained that the film falsely implied that the Army (and its chosen instrument Leslie R. Groves) was handed a complete project requiring only routine subsequent steps. Groves insisted he had signed big contracts even before Fermi's chain reaction. Robert

Walker as his aid outraged Groves. The fictional Jeff Nixon, zestfully portrayed by Walker, chased women, was disrespectful to his superiors, and was too long haired for a combat zone like Tinian. A real Corps of Engineers Colonel did not behave that way, and a real Corps of Engineers general officer would not tolerate such a person on his staff.[19] As to his own film image, the pudgy, slightly rumpled Groves had no objection to the handsome, dapper Donlevy.[20]

Groves did not get Walker's depiction of Jeff Nixon to conform and only reluctantly conceded other points. He ended up essentially conforming to MGM's plans. Perhaps he took comfort from an aide's report on the final film version: although the film regretfully overlooked the possible economic impact of atomic energy, security was maintained with no danger of arousing popular revulsion; public impact would be minimal because the film would flop.[21]

Despite strong reservations about the screenplay and the first film version, Vannevar Bush was careful to limit his responses to MGM to the depictions of OSRD and himself as its Director. Particularly outrageous to him was the original screenplay's crediting Lyman J. Briggs as having the crucial discussion with President Roosevelt on launching the A-bomb project. Briggs, the Director of the National Bureau of Standards, headed the early Uranium committee which was absorbed by OSRD before it launched the full-scale effort that became known as the S-1 project.

Once that was corrected, Bush also strongly objected to the suggested dialogue between himself and the late President. Obviously a historic moment, Bush took the matter very seriously. But so did MGM in its own way. They wanted to depict the dead, still popular President, even to showing his well-known black Scotch terrier Fala. MGM also saw the scene as introducing an essential element of dramatic tension.

Bush did not relish the dialogue about the uncertainty of the costs where he almost reluctantly concedes the project may even require as much as $2 billion. Admitting that costs did go upward, Bush did not welcome the implication of administrative uncertainty. But Bush compromised. More annoying was the text having him say that it was uncertain whether the A-bomb could fit into a plane. Nor did he like the text saying that serious doubts existed if the A-bomb could be finished for use before the end of the conflict. On the contrary, Bush insisted he knew the weapon could fit into a plane and be ready in time. As the responsible administrator of the wartime scientific mobilization, he only advised President Roosevelt to launch the project after receiving advice from some of the best scientists and engineers in the world. The released film retained in softened form the doubts about size and timing but added Bush citing the advice of the National Academy of Sciences. It was important to MGM to enhance the tension of a deadly race with Nazi Germany.[22]

On viewing the first film version of the scene with President Roosevelt, Bush became concerned about its ending as depicted by the actors: after a discussion of the possibility of a full-scale project, Roosevelt thanks Bush, saying he will consider the matter. The veritable Bush objected to how the dramatic Bush reacted as he left, displaying dissatisfaction at not getting an immediate decision by facial expression and body language. MGM thought the scene a success; the sneak preview audience applauded. Bush's spokesman made clear what was objected to. The scene implied that American science through its spokesman [i.e., Bush] "is arrogant enough to feel it should either make the decision itself or force the Commander in Chief into making it then and there."[23] The released film has a rather prosaic parting of the two men followed by the President placing a call to Winston Churchill. By December 1946, Bush could write Bernard M. Baruch that "history was not unreasonably distorted" by the film.[24]

Albert Einstein agreed to his depiction by a quite different path. Having been early told by the atomic scientists' movement that the film reflected the military view, he twice declined permission to MGM. But the film company persisted, offering to change objectionable details. They very much wanted Einstein's appearance. Einstein apparently let Leo Szilard handle further negotiations. Szilard telegraphed him on July 27, 1946, to sign up, which Einstein did that very same day.[25] When Groves viewed the first film version, he correctly complained that an early scene purporting to show an experiment at Columbia University with Enrico Fermi, Eugene Wigner, and Leo Szilard simply slowed the action. The scene's presence, MGM explained, was insisted on by Einstein.[26]

Not all scientists agreed to their depiction. James B. Conant, the president of Harvard and a key administrator in the A-bomb project, only agreed to being shown at the Trinity test but not to have any words placed in his mouth. From the standpoint of MGM, the most serious refusal was from Niels Bohr. The early scenes in the screenplay featured him in Europe. As part of the tensions of the race to beat the Nazis to the A-bomb, much was made about smuggling Bohr out of Copenhagen and then bringing him to the United States. Bohr's essentiality for the A-bomb project was more than strongly implied, and he was falsely placed at the Alamogordo bomb test.

In May and July, MGM used John Wheeler of Princeton as their go-between, assuming that Bohr would eventually agree to his depiction, perhaps with a few changes. In October Bohr turned down MGM, citing the many inaccuracies of the plot. Because Lise Meitner had also refused permission, the company assumed she had influenced Bohr. They offered Bohr a private showing of the first version. Not only did Bohr decline, but his lawyers sent warning letters to MGM. By December, at least, MGM was cutting the original film and reshooting scenes, a process continuing into January 1947.[27]

The A-Bomb Decision

Dropping the bomb on Japan provided a problem for dramatic truth, especially after the buildup of tension over the menace of a Nazi nuclear weapon. In the original script Niels Bohr shocks Oppenheimer and others when he brings the news that the Germans are sending atomic experts and materials to Japan by submarines. Another, later scene has General Groves telling President Roosevelt and Secretary of War Stimson that the Japanese will meet an invasion of their homeland with atomic weapons as an argument for dropping the bomb. Bohr's intransigence and Groves' firm denial of his use of that argument eliminated those passages.

The original screen play has a fictional German physicist named Schmidt in Lise Meitner's laboratory. Later, the scenario has a submarine leaving Hitler's doomed Reich carrying Schmidt and sundry atomic secrets, eventually to surface in Tokyo Bay. There Schmidt disembarks amid rejoicing that the two countries could continue developing the decisive weapon for victory. The Japanese rush Schmidt off to a modern laboratory they have established for him in the city of Hiroshima. Schmidt did not survive to the final film version.[28]

An unexpected, vociferous source of objection to the film, in particular the depiction of the decision to drop the A-bomb, was the prominent political commentator Walter Lippmann. A viewing of the first version outraged him. He could not understand how so many eminent scientists agreed to their incorporation into such a banal "American success story." Particularly appalling to Lippmann was what he viewed as a libelous depiction of Roosevelt and Stimson in the scene with Groves. Not only were the men denigrated in his opinion, but so too was the seriousness of the elaborate, careful process preceding the decision. Even worse, to Lippmann, was the one scene with Harry Truman, whose visage did not appear. (Because of that, Truman's permission was not necessary.) As an actor spoke, the camera shot over his shoulder to a listener as the dramatic Truman's voice expressed concern over U.S. losses and unconcern over Japanese deaths: "But I think more of our American boys than I do of all our enemies. . . ." Lippmann' letters of protest forced MGM to reshoot the scene with President Truman simply expressing the belief that dropping the bomb could save Allied deaths in an invasion of Japan.[29] The veritable Truman refused to intervene to avoid the charge of engaging in censorship but in a private letter complained that the brief scene was objectionable in implying he had simply made a snap decision.[30]

Just why Truman's action was important was not specifically elaborated, although the scenes of devastation at Hiroshima at least hinted that this was not simply a bomb bigger than a conventional weapon. Wead's screenplay had the newsreel announcer say at the beginning:

"A message to future generations! Come what may, our civilization will have left an enduring record [i.e., the MGM film] behind it. Ours will be no lost race." That, and the very title, certainly suggested catastrophic possibilities. The implications were not absent from the minds of Sam Marx and his colleagues. For example, when James Franck, T. R. Hogness, and Harold C. Urey on September 9, 1946, protested the use of Lise Meitner in the film even under a fictitious name, Marx agreed but added a reference to John Hersey's piece on Hiroshima in the *New Yorker* "[which] is making many readers feel that the creators of the atomic bomb are the world's greatest war criminals[;] it should be a relief to many scientists that a motion picture of this magnitude is on its way, hailing their achievement as the most magnificent triumph of modern times."[31]

The filmed scenes at the Metallurgical Laboratory made much of monitoring for radiation hazards, preparing the audience for Matt Cochran's fate. In the script and in the released film are dialogue at Alamogordo between Oppenheimer and Brigadier General Thomas F. Farrell, Groves's deputy, about the possibility of the fission "going around the world," that is, converting the planet to a fireball. Oppenheimer rates the possibility as less than one in a million. Farrell tells Groves that those odds are too small, implying at least imprudence on the part of the scientists. After the bomb goes off, when questioned again about whether he had not really worried about the catastrophic possibility, Oppenheimer replies: "In my head, no; in my heart, yes." That concern was in the air after the A-bombs dropped on Hiroshima and Nagasaki, but not among those in the project who had earlier determined it was not possible with a nuclear weapon of that size and nature.[32]

Focusing on a nonexistent peril diverted attention from real perils. The original screenplay treated radiation and burn hazards at the Alamogordo test by announcing that injury to the skin "is best overcome by covering or use of lotion freely." Groves firmly criticized that and the wearing of dark glasses as untrue. In the film only the dramatic Oppenheimer wears goggles with dark lenses to peer at the blast.

Lost somehow to the final film version is a bit of dialogue in the original screenplay between Oppenheimer and Bush. As they watch the mushroom cloud at Alamogordo, Oppenheimer says: "The cloud is dissipating nicely." Bush replies: "That's one fear we can forget."[33] The exchange rings true for that particular point in history when many responsible individuals in the Manhattan Project and OSRD regarded long-term radiation hazards with what we can now describe as a mixture of ignorance and wishful thinking. But well before the end of 1946, the tests at Bikini ("Operation Crossroads") had disclosed disturbing possibilities. In *The Beginning or the End* the film audiences saw a turtle walking across the ground at Alamogordo right after the test, no doubt a dramatic truth symbolizing that life survives nuclear explosions.

Did It Matter?

In a notably jaundiced review of *The Beginning or the End, Time* magazine wrote: "The picture seldom rises above cheery imbecility." The film was a commercial, historical, and artistic flop. At least seventy-five films of that year grossed more at the box office.[34] Although the *Time* reviewer went on to lecture Hollywood—"regard [the] audience as capable of facing facts, even problems which may prove unsolvable, [and] stop treating cinemagoers as if they were spoiled or not-quite bright children"—his conclusion was relatively cheery. There was no harm done unless *The Beginning or the End* stopped the making of the better picture.[35] Of course, no one knows the reactions of the millions who did see the film.

The lack of success of the film probably reinforced the attitudes of the historic actors, shared by the *Time* reviewer. Despite words about bringing truth to the audience, in the last analysis it did not matter or did not matter very much. The Army Corps of Engineers had reached the same position before the release of the movie. In knowingly signing up with MGM, Arthur Compton, Enrico Fermi, J. Robert Oppenheimer, Vannevar Bush, Albert Einstein, and others all agreed to not unreasonable distortions of history (as Bush put it), to deviations from veritable to dramatic truth. What were unreasonable were a small number of gross deviations from fact and violations of perceptions of self and of standards of propriety. Beyond that was little apparent concern for the viewing public.

Perhaps the veritable participants agreed out of a sense of play, of vicariously becoming part of Hollywood's great make-believe world, figuratively rubbing shoulders with individuals accorded almost mythic status not only in the United States but many other countries. Many individuals in the Manhattan Project played themselves, re-creating past events in the March of Time documentary *Atomic Power* released in August 1946 (that is, during MGM's filming). It showed, for example, Conant and Bush stretched out on the ground awaiting the bomb blast at Alamogordo. The two men were filmed on a sand-strewn garage floor in Boston.[36]

Joining the sense of play, perhaps, was a sense that the real "history" was elsewhere. If "history" was elsewhere—not at MGM in Culver City, California, nor on a Boston garage floor—where was it? Perhaps in their memories, perhaps in the scientific and technical literature? Perhaps "history" was over, gone, with no tangible existence?

Perhaps they sensed that "history" was dispersed in their unpublished personal papers and in the archives of institutions. I doubt it. Certainly, many did not like what resulted when historians and others began poking around memories and archives.[37] Although they would fight for freedom of information for the advancement of science, some were quite willing to disregard the strictures of *Time* magazine and consider limiting what the public should be told of the possible horrors of modern warfare. James B. Conant, for example, later wrote a report for the Defense Department along these lines.[38] In all sorts of contexts, it did not matter that veritable truth was compromised. It is hard to avoid the conclusion that compromises with veritable truth are more likely the further the medium from familiar forms of professional communication. And a Hollywood saga was far from the style of the *Physical Review*.

When the first film version was finished, it represented Wead's screenplay as amended slightly because of the complaints of various participants. At a sneak preview, MGM was delighted with the overwhelmingly favorable response.[39] They may have had a hit on their hands before Bohr and the others forced cuts and reshooting of scenes. Given MGM's recorded of popular successes, perhaps the original, unrevised script would have yielded a successful film, artistically and commercially. Who can estimate the effects on tens of millions of viewers if Schmidt had disembarked in Tokyo Bay and left for his laboratory in Hiroshima?

Popularizations have consequences even if not always precisely definable. Because it does matter, we can regret MGM won the cooperation and assent of so many key individuals. The subject was unsuitable for the kind of treatment MGM had in mind. Because it does matter, we can express relief that MGM had to get their assent and consequently had to compromise its original dramatic vision. Think of how Hollywood has warped the perceptions of so many inside the United States and in other countries about the settlement of the American West. Hollywood did not need the written assent of the long dead original settlers and of the American Indians.

Although I have located no direct consequence of this film on the public presentation of science and scientists—at last not in a wide range of later correspondence of nuclear physicists and other scientists and records of scientific organizations—one can speculate about two indirect effects. First, there was a noticeable increase in interest after 1947 in reaching the lay public but in forms where professional control was paramount. The reformed *Scientific American,* television documentaries, and countless books by scientists resulted. But now scientists, engineers, and physicians have attracted the attention of the world of belles lettres with results not always pleasing to fastidious scientists and historians of science. Bertolt Brecht and Arthur Koestler are now joined by far lesser talents implicitly following MGM's precedent.

A second possible consequence hinges on the very difficult matter of the general public's perception of scientific integrity. The aura of secrecy around the bomb soon engendered uneasiness about whether the whole truth was being disclosed. As more and more came out about such matters as the problem of fallout, there was an erosion of confidence not only in governmental agencies but also in scientists as sources of objective truth. The film, if remembered at all, became another instance of evasion by those popularly charged with being bearers of truth. If that is a valid speculation, it might have furthered the trend to a greater professional role in reaching the general public to avoid repetition of such instances.

This incident suggests a troubling moral to a historian. His guild has a decided preference for anchoring generalizations firmly to detailed particulars of contexts. Is it a professional virtue or an act of self-righteousness? Does our abstracting from veritable truth to historical truth inflict any injury on the past, different from but just as real as that resulting from an application of dramatic truth? If so, how can we minimize this injury?

Notes

1.   The Oppenheimer and the Bush Papers are in the Manuscript Division of the Library of Congress. The files on the movie are both in Box 171 of the respective collections. Box 11 of the General Correspondence of Groves in Record Group 200 of the U.S. National Archives contains the record of his dealings with MGM. For Albert Einstein, see 57–147 to 57–173 of the Einstein Papers in the Mudd Library, Princeton University. I have viewed the film in the holdings of the Library of Congress. Through the courtesy of MGM/UA Entertainment Co. I have examined in their records file 1377 on the production of the film and file 7005 on obtaining permissions to be depicted from the participants.

After completion of this paper, I learned of Michael J. Yavenditti's "Atomic Scientists and Hollywood: The Beginning or the End?" *Film and History* 8 (1978):73–88, with a quite different but not incompatible interpretation because of its focus on the atomic scientists' movement and the issues it raised of the public roles of researchers.

2.   E. R. Tompkins to E. Fermi, Jan. 16, 1946, Oppenheimer Papers. See also Alice K. Smith, *A Peril and a Hope* . . . (Cambridge, Mass., 1971 ed.), pp. 293–294.

3.   Carter J. Barron to H. S. Truman, Nov. 21, 1945. President's Secretary's file 112, Truman Library, Independence, Missouri.

4.   As late as March 1946 when he withdrew, Hal Wallis of Paramount Pictures hoped to make the movie and was dangling the possibility of scholarship money before the scientists. J. N. Teeter memorandum, Mar. 7, 1946, and Teeter to Oppenheimer, Mar. 26, 1946, both in Oppenheimer Papers.

5. W. R. Shank memorandum to W. Higginbotham, February 20, 1946; R. Noyes to Oppenheimer, February 26, 1946—both in Oppenheimer Papers. According to a telegram in 7005 (W. Consodine to S. Marx, May 16, 1946) on a conference with Einstein in which H. D. Smyth spoke up for the Federation, MGM sent the organization a check which was returned with a request for a larger sum.

6. Groves Agreement with MGM, Dec. 31, 1945, Groves Papers. This does not appear in Groves's autobiography. Other documents in Box 11 disclose that the political columnist Drew Pearson had scented the deal by late 1946 but did not publish a word of it. This is based on a review of file 7005. For his efforts on behalf of MGM, John Wheeler of Princeton had the company contribute $500 to his university. E. R. Tompkins of the Clinton Laboratory (now Oak Ridge), who actually first suggested the film, received $100.

7. The script used is in the Bush papers in two identical copies. Besides the original versions dated Mar. 22, 1946, there are a number of revisions of specific scenes dated May 10, 1946, inserted at the appropriate points in the original text. All subsequent comments refer to this document.

8. In none of the documents examined are there discussion of these two fictional female characters during the development of the script.

9. Pages 28 and 30 of the screenplay. The *Life* review appeared in the issue of Mar. 17, 1947, pp. 75–81.

10. V. Bush to B. M. Baruch, Dec. 20, 1946, Bush Papers.

11. Groves to W. Consodine, Mar. 12, 1946, and attached documents. Groves Papers.

12. James K. McGuinness memorandum to L. B. Mayer, July 16, 1946, enclosure to 57–154 in Einstein Papers.

13. Based on a comparison of the screenplay with comments in various letters in the Groves Papers, particularly the materials cited in 11.

14. W. Consodine to Groves, Mar. 7, 1946, Groves Papers.

15. Telegram to David Hawkins, April 29, 1946, Oppenheimer papers. Hawkins authored the official history of Los Alamos. He and H. T. Wensel of the National Bureau of Standards, who worked on the S-1 for the OSRD central office, were in Hollywood during the filming as technical advisers, in effect replacing the individuals from the atomic scientists' movement.

16. Marx to Oppenheimer, Apr. 2, 1946, and Cronyn to Oppenheimer, Apr. 30, 1946, Oppenheimer Papers. Oppenheimer signed his waiver on May 8, 1946, subject to receiving assurance in writing about desired changes. The waiver form, in the same location, reads, in part: "I understand that although you will attempt to show the historical facts with accuracy, you will however, have to dramatize your motion picture story, and I have no objection thereto, and you may rely on my personal irrevocable consent to proceed."

17. Oppenheimer to J. J. Nickson, May 29, [1946] replying to Nickson's letter of May 17, 1946, Oppenheimer Papers.

18. Groves to Bush, June 3, 1946, Bush Papers.

19. Notes of conference of November 15, 1948, with Carter Barron and Groves to Barron, Nov. 19, 1946, Groves Papers.

20. Ibid. Groves was sensitive to being portrayed as treating industrialists rudely.

21. Albin E. Johnson to Groves, [Jan.–Feb. 1947], Groves Papers.

22. Bush to F. G. Fassett, Jr., July 15, 1946, and Marx to Fassett, July 19, 1946, Bush Papers. Fassett worked for Bush at the Carnegie Institution of Washington and helped him with his writings and with public relations.

23. Fassett to Marx, October 26, 1946; Marx to Fassett, Nov. 13, 1946; Fassett memorandum to Bush, Nov. 13, 1946, Bush Papers. The quotation is from the Oct. 26 letter.

24. Bush to Baruch, Dec. 20, 1946, Bush Papers.

25. The declinations occurred on May 26 and June 24, 1946. The July events are in exchanges between Marx and Einstein, 57–156 to 57–161. The Szilard telegram and Einstein's permission are in 57–163 and 57–164.

26. Barron to Groves, Dec. 4, 1946, Groves Papers.

27. The Bohr account is based on copies of correspondence in the Einstein Papers (Bohr to Wheeler, May 16 and July 25, 1946; Bohr to Consodine, Oct. 9, 1946) and in the Oppenheimer Papers (Marx to Oppenheimer, Oct. 21 and Dec. 4, 1946); file 7005 has much on the unsuccessful MGM efforts.

28. From a memo of Apr. 24, 1946, of an MGM conference in file 1377 there was agreement that the account of Schmidt's trip was fictional, although Consodine insisted such a submarine was sent but turned back. Meitner insisted there were no Nazis in her laboratory. Schmidt, it appears, was inserted in the script so that the audience would not confuse Niels Bohr and Otto Hahn and that Hahn would not be labeled as a Nazi. For Groves's objections to other aspects of this part of the screenplay, see Groves to Barron, Apr. 15, 1946, Groves Papers.

29. Lippmann to F. Aydellotte, October 28, 1946, Oppenheimer Papers. Traces of Lippmann's vehement reaction occur elsewhere. See Bush to J. B. Conant, Nov. 27 and Dec. 6, 1946; and R. P. Patterson to Bush, December 2, 1946, Bush Papers. The original dialogue with Roosevelt, Stimson, and Groves is on 163 of the Wead screenplay; changes dated April 30, 1946, have the Truman words Lippmann objected to on 83–83A.

30. In Bush to Baruch, Dec. 20, 1946, re censorship. Truman to Roman Bohnan, Dec. 12 1946, President's Secretary's file 112, Truman Library, Independence, Missouri. From file 1377, we know that Truman did see an early version of the script, as did his secretary Charles G. Ross, the source of Bush's information. (Ross to Barron, Apr. 19, 1946.) Ross saw the revised script, but neither Truman nor his family saw the first film version. (Strickling to Barron, December 6, 1946 and Ross to J. K. McGuinness November 4, 1946.) The sources are unclear as to whether Truman saw the revised script.

31. The opening words indicated the date 2446 for the expected opening of the buried cache of the reels containing the film. The "twenty-fifth century," was selected, I suspect, for that future date was evocatively familiar, as in the popular pre–World War II comic strip, "Buck Rogers in the Twenty-fifth Century." File 7005 has the letter of the three scientists and Marx's draft of his reply [Sept. 17, 1946].

32. The point here is that uncertainty existed about the nature and extent of the explosion at the test site. The men involved had calculations of what might occur but could not vouch that the explosion would exactly match the calculated effects. Later work indicated that the estimated power of the explosion at Hiroshima, 20,000 tons of TNT, was too high.

33. Page 99 of the screenplay.

34. *Variety,* Jan. 7, 1948. Of 369 releases, only 75 grossed $2 million or more.

35. The *Time* review is in the issue of Feb. 24, 1947.

26. Raymond Fielding, *The March of Time, 1935–1951* (New York, 1978), pp. 291–293. Bush figured prominently in the March of Time film. His papers contain part of the script with a number of jaundiced comments. Bush refused cooperation until the script was satisfactory. *Atomic Power* was far shorter than the MGM film and brought the story up to the current debates of early 1946 on atomic policy.

37. For example, see Oppenheimer to G. Seaborg, May 3, 1962, Oppenheimer Papers, for comments on Hewlett and Anderson's *The New World— 1939/1946* (1962), the first volume of the Atomic Energy Commission's official history.

38. Secretary of Defense Louis Johnson to Truman, Aug. 22, 1950, White House Official file 192, Truman Library, Independence, Missouri.

39. Marx to Oppenheimer, Oct. 16, 1946, Oppenheimer Papers. Up to the world premiere before a gala audience in Washington, at least some MGM officials were certain they had a big hit on their hands (Barron telegram to E. J. Mannix, Jan. 7, 1947, in file 1377).

HISTORY AS CRITICAL

BATTLEFIELD

15. Uniformity as Hidden Diversity: History of Science in the United States, 1920–1940

I wrote this paper for an occasion, a symbolic event, a joint session of the American Historical Association and the History of Science Society at Chicago on December 29, 1984, commemorating the centennials of the founding of the Association and the birth of George Sarton. In all his years at Harvard, Sarton had never had an official relation with its History Department nor did he have much, if any, interest in the concerns of the Association and its membership during that period. As for myself, I had only briefly glimpsed the great man at a luncheon a few years before his death. Nor was I enamored of much he had written and some of its consequences.

But in Chicago, we were celebrating ourselves and what we had become in time, not necessarily the prickly particulars of the past. A coming together was proclaimed, not any divergencies. The organizer of the session believed that contemporary history of science represented the fusion of two traditions: the science-derived one represented by Sarton, and the Americanist tradition with its concern for integrating scientists and their knowledge into the history of a national culture. I had already expressed my qualms about that interpretation in my introduction to a collection of *Isis* articles, *Science in America Since 1820* (1976). Basically, I simply doubted our writings had such impact. Nor did I regard the Society's membership as uniformly converging. A number of my acquaintances in the field were in a time warp, perhaps somewhere around 1968.

Years before, the intellectual history of the period 1900–1940 in the United States had a great fascination for me. Although I had gone elsewhere in my research, the interest persisted, leading to occasional forays into the primary and secondary literature. When asked to participate in the Chicago session. I could not resist the temptation. Initially, I compiled a list of every possible relevant author and work I could think of. When I dived into that pile of print, two things occurred. First, sometimes my memory of items proved wrong or incomplete. It was disconcerting, to say the least, to have to revise glib certainties. Was my memory at fault, or did my eyes see differently after all these years?

More fundamental perhaps, was the second, the snowball effect of my research. More authors and more titles turned up. An elephantine paper became a possibility, but an absurd one given time limitations. I had to stop the cancerous growth of my note slips. But I had learned something by then. Most of these writings did not deal with science strictly speaking. I mean both in the sense of good old internal history or the newer social mode. Most of the authors, even Sarton, ruminated over Progress, sang paeans to Western civilization, or celebrated the current enlightenment of particular disciplines. The present cover of *Isis* proclaims itself "An International Review devoted to the History of Science and its Cultural Influences." That was the phrasing of its second editor, I. Bernard Cohen. His teacher, George Sarton, the first editor, wrote: "An International Review devoted to the History of Science and Civilization." "Science" came first, civilization followed.

This paper is an attempt to show the main trends in that period and is a contribution to the story of the growth of a scholarly field. My address does not, however, make an important point: what I think converged in the generation of historians of science prior to the centennial celebration in 1984 is their identification with the historians. The professionalizing historians of science, whether Americanists or not, necessarily found useful attitudes and techniques in the general community of historians. Less and less did science and philosophy provide frameworks and tools for the field. When "external" topics became more fashionable, younger historians had little difficulty in leaving the cliches of their mentors. Some mentors, in fact, had already taken tentative steps away from the faith of their own mentors. Even many historians committed to the histories of particular disciplines now openly accepted the necessity for research on the personal and institutional environments of the scientists being studied. Many no longer worried about maintaining the autonomy of the sciences and became actively interested in exploring relations with applications.

As the Editor of *Isis* refused to consider this paper for publication, it was sent to England and appeared in the November 1986 issue of the *British Journal for the History of Science*.

Between World Wars I and II an extensive body of writings appeared in the United States explicitly or implicitly on the historical development of the sciences. I am not referring to the vast literature of popularization in magazines and newspapers but to substantial works, often in book form, coming from various intellectual and scholarly traditions. Only a few examples are classifiable by later standards as professional history of science. Following Arnold Thackray, one can designate some authors as "proto-historians" of science.[1] Most of the writings, including those of the "proto-historians," have distinctive attributes—methods, attitudes, and goals—reflecting traditions other than professional history of science or even the general history exemplified by the American Historical Association's membership of that era. What follows is a bird's eye view of a past of interest for its own sake and for clues about the professionalization of history of science after 1950.

To start, I will sketch briefly an underlying cosmic pattern and a moving principle (a phrase taken from Merz) present in varying degrees of explicitness in much of this literature. Then a survey of various intellectual sectors starting with philosophy, moving next to a consideration of how scientists themselves entered into the creation of the pasts of their guilds. This is where George Sarton enters the account. I shall frankly lay my cards on the table at this point and announce that I am mildly puzzled at the celebration of his centennial— beyond credit for the founding of *Isis*. To my mind, to attain a real history of science required getting far away from the viewpoints and methods of our honored founder. Next, will come a brief excursion into the sociology of science, primarily Robert K. Merton, but also a glance at Edgar Zilsel. A fair number of historians touched on science's past, in addition to the histories coming from scientists, some rather surprising in retrospect. The paper concludes with a brief review of the treatments of science within the United States by past authors. While I do pay particular but not exclusive attention to writings on the United States, my intention is to point up larger issues still present in the discourse of contemporary historians of science. Medicine and technology are not treated in the depth they deserve but appear significantly at a few points.

Cosmic Pattern and Moving Principle

Henry Fairfield Osborn's *The Origin and Evolution of Life* (1917) was probably the last purportedly serious scientific work (at least in English) in a great tradition going back to antiquity. Its obvious immediate predecessor was Alexander von Humboldt's *Kosmos* in the previous century.[2] Writings in this tradition start from the most general and abstract, successively treating the more specific and detailed. From the

Diety or from energy or force, chapter after chapter carry the elaboration of the cosmic pattern to the origins of life and to the manifestations in the course or evolution. Osborn, a paleontologist-naturalist, stopped before reaching the appearance of man, let alone the specific, detailed accounts of that species' course on this planet. In a collection of semipopular essays edited by H. A. Newman coming from the University of Chicago, *The Nature of the World and of Man,* 1923, the treatment extended to man but in a highly abstract, timeless, analysis of mind in evolution by C. H. Judd, a social scientist.[3] All the writings presenting variants of the cosmic pattern were reductionist. Logically and chronologically, everything on the Earth, living and nonliving, arose from the most abstract general forces and principles as these developed over time. Evolution, whether Darwinian or otherwise, carried forward the process of life over time. Man represented a tough problem for this intellectual mode. Social science laws modeled on the physical and biological sciences simply could not handle the complexity of human history.

Doing a reductionist history was a problem, as exemplified by a highly influential work of high popularization from Great Britain, H. G. Wells's *The Outline of History.*[4] Wells starts out with the earth in space and time, followed successively by rocks, natural selection leading to land species (reptiles and mammals), and finally prehistoric man. What follows afterward, however, is largely consonant with the conventional writings of that day with two general exceptions. Wells is outstanding in his effort to achieve a worldview, not one limited to Western civilization. Most of his treatment of the Western world is heavily weighted toward the topics commonly treated by historians in the United States—wars, politics, constitutional development, and the like. What is exceptional are the strong, idiosyncratic interpretations of Wells. Physical and biological factors are barely noted after the prehistory chapter. *The Outline of History* is not in the tradition of geographic determinism present today in the *Annales* school. Nor is it, despite Wells's scientific background, notably more detailed or incisive in the treatment of the development of science than what appears in sundry works by contemporary historians and scientists. Like them, Wells could do little more than to present the facts as known then and to assert first the importance of the physical and the biological, and then accept the essentiality of the growth of man's knowledge of nature.

Wells and his contemporaries did not have to do more (although some successfully went further) because they largely subscribed to a moving principle, the Idea of Progress. Despite doubts about Progress around the turn of the century because of disenchantment with the results of industrialization and despite doubts arising from the horrors of World War I, a fair number of authors in the United States still believed in Progress, a process resulting in modern Western civilization. These writers spun out many works on the making of modern minds, the origins of modern culture, and the growth of civilization, meaning the Western variety. Unlike the Osborn and Newman volumes, the physical and biological were evaded. Like Wells, they tacitly assumed that Progress was the moving principle in which the sciences were important. Above all, the triumph of science promised control of physical and biological factors, a method applicable to the affairs of man, and a guarantee of further movement away from an unprogressive past of brutish existence and prescientific thought. Even after the Depression, it was assumed that Progress would restart if only the right minds took control. That is, Progress was no longer an inevitable forward thrust.

Certainty about the value of the procession to the modern mind produced a historiographic complacency. Progress was the Law. A necessary characteristic of Law was uniformity. History supposedly showed efforts to impede progress; proper history showed how impediments were attenuated and removed. Current impediments required reform actions. The uniformity of Law, whether scientific or historical, masked an incompleteness both as a description of process and as a tool for organizing the diversities of past historical epochs. Progress did not even cover all of the relevant historical experiences of Western man but represented the beliefs of an articulate section of the intellectual community. Its incompleteness masked the diversity of the Western tradition. The very certainty with which Progress was held guaranteed a limited vision of the past of science, even among "proto-historians" of science coming from the scientific community.

Philosophy

In Europe the history of science often appeared linked with philosophy. Many professional scientists and philosophers assumed that science's history was the unfolding of an objective truth coincident with the deepened understanding of proper method. History of philosophy in this dispensation had an overlap with the history of science. Among scientists, particularly in the German intellectual tradition, science had an essentially philosophic aspect, an exemplification of both sound epistemology and ontology. In either sense, history of philosophy and of science could serve as evidence for Progress for those so inclined. And the linkage of history and philosophy effectively eliminated or downplayed aspects of science's past outside the scope of the history of ideas seen in isolation.

Two quite important books for the history of science appeared in the

United States between the two World Wars from the philosophic fraternity: E. A. Burtt's *The Metaphysical Foundation of Modern Science* (1924) and A. O. Lovejoy's *The Great Chain of Being* (1936).[5] The former is a splendid example of an older style of intellectual history. Based on a thoughtful use of contemporary printed sources of the seventeenth century, Burtt's book considered basic issues of the status of the knowledge provided by science, raising questions largely evaded to that date by scientists, historians, and others, particularly individuals under the spell of grand terms like Progress, modern culture, or Western civilization. In comparison, Koyré's later *From the Closed World to the Infinite Universe* (1957)[6] is still in the earlier tradition with its complacent acceptance of the necessary superiority and truth of what science wrought. Burtt's writings fitted in quite well with the post–World War II mood in the U.S. history of science and philosophy of science.

Lovejoy was a liberator from the tacit linkage of Progress with evolution. *The Great Chain of Being*, like Burtt's work, was much admired in the 1950s. Yet neither work became models for the practice of historians of science largely because they remained rooted in the traditions of another learned guild, philosophy. But the two books reinforced the internalist-intellectualist trends in the historical practice of the 1940s and 1950s. Particular examples of influences abound. For example, in his recent book Ernst Mayr notes that he turned to the study of the history of biology after reading Lovejoy.[7]

Science

Despite the quality of Burtt's and Lovejoy's work, philosophers had a minor role in the subsequent development of history of science. By far the most influential stimulus came from the scientific community itself. Basically, neither the historically minded scientists nor the young historians of science after 1950 wanted to emulate the products of the guild of philosophers. What some of these individuals wanted is exemplified by a man, Alfred North Whitehead, who was both from the scientific community and a notable philosophic thinker in his own right. His fellow Britons Arthur Eddington and James Jeans were also role models as superb purveyors of high popularization—sophisticated, nontechnical discriptions of the abstruse. There is a very old tradition in science of writing nontechnical explanations of the content of science and of its methods. The scientists did not feel any necessary need for the enlightenment of philosophy.

Whitehead qualifies for this essay because *Science and the Modern World* largely consists of the Lowell Lectures delivered the previous year.[8] To the Lectures, Whitehead added chapters on mathematics, on religion and science, plus two more on abstractions and on God before the ultimate one on "Requisites for Social Progress." The book is largely chronological and avowedly historical, with the more general themes appearing at the end. (Mathematics is the sole exception.) Whitehead reverses the order of Wells, so typical of much writing of this period. What impresses at this date is not only the sophistication of the discourse but the skillful way Whitehead brushes aside the kind of issues raised explicitly by Burtt and implicitly by Lovejoy to affirm a kind of idealistic commonsense theory of science. This is history affirming the validity of the practice of research. The general chapters at the end are not presented as following from the triumphs of science in a kind of mirror image of the reductionist approach. *Science and the Modern World* is much more sophisticated, perhaps representing a differing national mode.

For example, in 1923 Oxford University Press and the Yale University Press published two related yet divergent works. The latter publisher's *The Development of the Sciences* consisted of six lectures by as many Yale professors on their respective fields.[9] Thoroughly orthodox in content and interpretation, *Science and Civilization*, edited by F. S. Marvin, has Whitehead as one of its contributors.[10] Most of its chapters were conventional treatments by field and period, perhaps somewhat more sophisticated historically than the Yale volume by today's standards. The work concluded with a series of "Science and . . ." chapters, much like the pattern of Whitehead's later book.

Both the British and the American scientists wrote history largely in the service of science in general or of particular fields. In contrast to the British, the Americans tended to avoid the explicit juxtaposition of history of science with questions of philosophy, social policy, or the nature of modern society. Not that such writings did not come from their pens or that American scientists necessarily differed on particular issues from their British counterparts. In part, the scientists in the United States followed the reductionist pattern, a pattern minimizing the need for explicit extended treatments of topics extra to their historical accounts of the actions of scientists and the generation of the content of fields. In part, a differing literature resulted because of a specialization arising from a tacit division of professional responsibility. External issues in the United States largely were within the domain of humanists and social scientists.

The literature, historical and quasi-historical, produced by scientists mostly fell into certain genres. Like the Yale volume noted above, collections of essays were common, either by field or on notable scientists. Sometimes a volume of essays commemorated an anniversary, as in the 1918 publication from Yale, *A Century of Science in America, with special reference to the American Journal of Science, 1818–1918*, a splendid example of how not to do the history of science.[11] Sometimes a great predecessor was memoralized by a biographical volume. Far less common were monographic descriptions of the current scene with explicit attempts at providing historical background. Robert Thompson Young's *Biology in America* has two opening historical chapters and a later scatter of historical references.[12] The history is utterly trite but the rest of the volume, a "general account of biological progress," is still a valuable description of the American scene in the life sciences after World War I. Historical accounts of fields in this tradition often turned out to be little more than current knowledge given in the form of a thoroughly unhistorical review of the literature.

Similar to Young are the many writings conveying history to students as part of the pedagogy of specific scientific fields. These could range from casual identifications of the individuals for whom laws, phenomena, and instruments were named, to full chapters on the development of a science, or to more ambitious productions. As an example, consider Harvey Bruce Lemon's *From Galileo to Cosmic Rays* (1934),[13] a first–year introductory college physics text. Organized by physical topics with the assumption of a correlation of chronological appearance and of logical priority, Lemon provides a running patter of historical references to the presentation of the physics. Most scientists in this tradition of practitioners' history had some awkwardness in comprehending real history of science.

Full-scale histories were produced by scientists, such as William A. Locy's *The Growth of Biology* (1925). My impression is that none of these works are consulted and cited by today's scholars. Later secondary sources are regarded as more reliable and as more sophisticated in viewpoint. Even though Locy consulted primary printed sources, so much more is now available, including manuscripts, that a Locy simply does not attract.

An easy step from that genre of scientific literature was the preparation of historical bibliographies. More than Thackray's assumption of antiquarianism was involved in the appearance of historical bibliographies and the collecting of old books and autographs. Most of the historical writings by scientists were not based on primary sources but on secondary works embodying conventional accounts. For the purposes of scientists—primarily inculcating a received account to reinforce group values—such secondary works sufficed, perhaps with rare dips into an original text of one of the great giants. Over time there was a trend to increasing reliance on primary printed sources. In a few rare instances a scientist-historian actually used manuscript sources. Slowly, almost imperceptibly, a few members of the scientific community became fascinated by the sources and perhaps went beyond that to a concern for the past for its own sake apart from any didactic value in the upbringing of scientists. If some did so significantly—a debatable point—there is hardly any evidence suggesting real insight into the problems of historical research.

While no field of the pure and applied sciences was devoid of its practitioners of history, and historical courses appeared in a number of colleges and universities within various scientific departments, three fields were notable in producing quasi-historical writings. In taxonomic fields (that is, in the area of natural history), the naming function generated a concern with precursors. Botany was possibly the most obsessive, producing an extraordinary scholastic or antiquarian fixation among many in the field. The concern with priority extended to the point, legend has it, of adding a name of a scientist to the formal Latin species designation, even where the identification was never published in the past and only existed in a cryptic note in an obscure manuscript source.

Added to the concern for proper priority credit was the fascination—no other word does justice—with the scientific exploration of the natural history of the United States. The "romance of the West" was very much alive in natural history circles. To this day scholars are indebted to Max Meisel's three-volume bibliography of "the pioneer century, 1769–1865."[14] Biographical works on the paleontologists O. C. Marsh and E. D. Cope are still very respectable in their use of primary sources.[15] They even attempt a fair appraisal of the personalities and actions of their two subjects. Added to the considerable contributions of the geologist G. P. Merrill, the writings on natural history fitted easily into the conventional historical theme of the settlement of the frontier.[16] Unintentionally, the output from natural history tended to obscure the work in other areas in the national past—the 40 percent of the total in the last century contributed by physical scientists. Chemists, astronomers, physicists, and meteorologists were simply less obsessed by their predecessors.

Not so the physicians. The ritualistic act of naming was not involved. Instead, a curiosity about older clinical practices and about the history of the guild as such produced a fair amount of historical prose by physicians, not all of which can be disregarded even today. No other scientific or technical field could boast the state achieved by history of medicine in the United States before World War II. Two full-scale histories of medicine in the United States had appeared, by Packard and by Sigerist.[17] A special institute for history of medicine existed at Johns Hopkins University. George Sarton, struggling to establish institutions upholding his vision, could only protest vehemently in *Isis*.[18]

Sarton raised the banner of mathematics in his complaint about the history of medicine. Mathematicians loomed very large in what passed for history of science in the interwar years. Two, David Eugene Smith of Columbia and Louis C. Karpinski of Michigan, were early presidents of the History of Science Society. Mathematicians and botanists were alike in their concern for predecessors. The historically minded in both fields behaved as though past mathematicians and botanists were living contemporaries. Naming a dead colleague as the first discoverer of a species was functionally like trying to prove a mathematical conjecture from the past. In both fields, historical distinctiveness was obliterated by a sense of intellectual timelessness. Results existed independent of place and date. Despite the naming of botanical species and of mathematical precursors, the named individuals largely vanished. Their achievements replaced the historical actors. Where a historical writing consisted of a literature review giving a précis of published texts, ahistoricity is markedly evident. The best example of this from an American mathematician is Leonard Eugene Dickson's three-volume *History of the Theory of Numbers*.[19] Despite its title, Dickson's book is a bibliographic compilation completely devoid of historical analysis and explanation. Nevertheless, it is a useful introduction to that literature.

Historically minded mathematicians shared Dickson's Whiggish preference for showing the path to modern mathematics. Among some, the fascination with old books and manuscripts produced an antiquarianism we can describe as modest steps to a professional history of science. In 1925 Karpinski published a *History of Arithmetic* "as a vital and integral part of the history of civilization."[20] As the book was aimed at teachers (presumably in the elementary grades), Karpinski gave particular attention to what was still being taught, leading this modern reader to wonder what was being left out. The spirit of the work markedly differed from Patricia Cline Cohen's recent *A Calculating People: The Spread of Numeracy in Early America*, which cites Karpinski and is also much concerned with arithmetic.[21] Karpinski later produced a fine bibliography of early printed mathematical works in the United States.[22] Karpinski's bibliography was one of the sources of my later dissatisfaction with a historical orthodoxy overwhelmingly stressing the role of natural history in the course of the sciences within the United States.

An avid promoter and a prolific contributor to the history of mathematics, David Eugene Smith was also a collector. His papers at Columbia contain numerous and very interesting historical manuscripts. Smith was a true pioneer for that date, considering he produced histories of mathematics in both Japan and America (the latter to 1900 with Jekuthiel Ginsburg). Smith and Ginsburg concentrated on pure mathematics, with a few excursions to physics and astronomy as examples of applications. Smith and Ginsburg used primary printed sources and even some manuscripts. They were concerned to show the rise of modern mathematics in this country, concluding that a great increase in worthy work occurred before 1875 and the next quarter's notable expansion in the early age of the graduate school. Satisfactory explanations of the reported upsurge are absent. Smith and Ginsburg simply lacked any sense of strategic issues and mechanisms in history. Faced, for example, with a late-eighteenth-century dispute between DeWitt Clinton and sundry Britons, the very bread and butter of current history of science and American cultural history, all they could say was "From these statements we may fairly conclude that modern mathematics was substantially unknown in America in the eighteenth century, but that questionable manners then and later existed both in Mother Country and in her offspring."[23]

By far the most eminent of the "proto-historians" from mathematics was Florian Cajori of the University of California, Berkeley. In retrospect his work in the interwar years is dated, but clearly within or well on the way to the professional practice of history of science. Given the recent upsurge of interest in the history of science and technology in Latin America, his book *The Early Mathematical Sciences in North and South America* (1928) is interesting even though it does not successfully compare the distinct cultural worlds of Ibero-America and British America.[24] For me, it came as a pleasant shock in preparing this essay that Cajori in that work made a point I argued for roughly forty years later. After satisfying practical needs, Cajori states: "But early in American history the cultivation of the mathematical sciences passed beyond immediate necessities. There was evidence of a desire to cultivate science for its own sake. There was a craving for higher things."[25] Cajori's brief biography of Hassler, although flawed in understanding of historical context, was one of the rare instances where a scientist-historian ventured significantly into manuscript sources.[26]

To a great extent, Sarton's exoticism masked how close he was to individuals like Dickson, Cajori, David Eugene Smith, and Karpinski. Belgian origins, pleas for a "new humanism," assertions of a vaguely defined "unity" of science—all mingled with an evangelical fervor for history of science—made Sarton a kind of icon, not at all like the other mathematical practitioners of history. Like Sarton, they believed in the centrality of mathematics, not only in the history of science but even in the total history of mankind.

In his 1935 rejoinder to the pretensions of the history of medicine, Sarton criticized conventional academic history for neglecting *the* main feature of history: "The history of science should form the core of every history of humanity because the elements which it describes and discusses are the ones which reveal most clearly the rational, progressive and cumulative tendencies and explain in the best manner the function and purpose of man in the general scheme of things." The core of history of science is history of mathematics and the mathematical sciences, "the other branches of science being dealt with in the order of their mathematical contents." Of course, Sarton is giving Comte's hierarchy, making quite plain that this order of fields of knowledge corresponds to human progress. To Sarton, "Mathematics forms the very core of human thought, and hence of human life, but it is a *secret* core . . . "

whose purpose "is purely intellectual, aesthetical, disinterested. It is a great joy to the mathematician, comparable only in its purity and fervor to religious ecstacy." Administering the coup de grâce to medical history, Sarton concludes that physicians do not "understand the world and man's place in it" because they study the macrocosmos while mathematicians and mathematical physicists study the universe and man's relationship to it.[27]

In the next year, 1936, George Sarton summed up his views of science, a view in which mathematics had already been given the central place. After defining science as systematized positive knowledge, he wrote the following:

Theorem   The acquisition and systematization of positive knowledge are the only human activities which are truly cumulative and progressive.
Corollary   The history of science is the only history which can illustrate the progress of mankind. In fact, progress has no definite and unquestionable meaning in other fields than in the field of science.[28]

In Sarton's practice of history of science, the theorem and corollary were not only intertwined with the idea of progress but typically came out as a recital of great heroic personages.

Although attractive to scientists and others, Sarton's antiquated viewpoint had both limited interest and utility as a basis for historical practice. The "new humanism" of Sarton involved ruling out much of what most contemporary intellectuals in the United States savored as humanistic; law-like uniformity here resulted in a deliberate impoverishment of the historical record. The unity Sarton proclaimed was a chronological juxtaposition of fields to illustrate the cumulative progression of a mathematized worldview. Chronological interactions between fields or with the external environment were simply not pursued in any analytical fashion. Above all, a man who proclaimed that mathematics and mathematical physics studied man's relationship to the universe had substituted ideological rhetoric for realistic description. The fervor with which Sarton espoused his new humanism, his unity of science, and his insistence on the scientific competence of historians of science arouse suspicion. Was he reacting to modern developments not amenable to his vision? Was Sarton trying to turn the clock back, perhaps to a past that never was?

Sociology

If medicine as practiced historically was a threat to Sarton's worldview, history of technology was so modest an endeavor that Sarton could publish Robert K. Merton's "Science, Technology and Society in Seventeenth Century England" in *Osiris* in 1938.[29] Merton completely omitted medicine. Today many are inclined to see Merton's work as a precursor of the externalist strand in the history of science. A recent article correctly stresses the motivating role of the drive for sociological law.[30] While Merton does cite Hessen's Marxist interpretation, the work is more indebted to Weber and Tawney. The work is clearly not an externalist history as understood by the guild of historians. Although severely criticized by historians, particularly those reacting to Hessen and responding to Koyré's views, Merton's book was influential and is a classic example of historical sociology.

Merton cited Sarton's words as support of his thesis: "There is no doubt that mathematical discoveries are conditioned by outside events of every kind, political, economic, scientific, military, and by the incessant demands of the arts of peace and war. Mathematics did never develop in a political or economic vacuum."[31] But Merton neglected to note that these phrases occurred within the context of an expression of doubt about Hessen, an assertion of the existence of an internal necessity in science, and specific repudiation of externalism: "The main sources of mathematical invention seem to be within man rather than outside of him, his own inveterate and insatiable itching for intellectual adventure."[32]

I suspect Merton's work was acceptable to Sarton because of its emphasis on external law-like factors—exemplified by the use of quantitative evidence—to explain changes in fashion in the support of research fields. Merton is rather ambiguous as to whether or not external economic or other factors actually determine the specific content of seventeenth-century sciences. That is what Sarton cared about. As long as Merton was perceived as concerned with "cultural roots" and "cultural values," as long as he distinguished between "subjective attitudes of individual scientists [and] the social role played by their research," Sarton was not likely to take offense. To put it in contemporary terms, Merton in 1938 (and today) was not a sociologist of scientific knowledge.

Merton had great impact but of a negative kind. Following Koyré, the emerging professional history of science tended to shun his viewpoint. The English historian A. R. Hall used Merton's citation of ordnance as the jumping-off point for a brilliant monograph showing that history did not follow Merton's sociological formulation, specifically that Merton simply did not know the facts. He followed that up with an influential critical essay, "Merton Revisited."[33]

Because of the importance of the seventeenth century for both history of science and for British history, a considerable body of research appeared after 1938, starting with the British scholar G. N. Clarke and continuing to this day.[34] Clarke pointedly called attention to the absence of medicine, a deficiency Merton acknowledged. Merton necessarily had a significant historiographic position in the writings on the intellectual consequences of the Puritan revolution and on the origins of the Royal Society. From Clarke onwards, historians of the period generally ended up criticizing his loose use of terms, fairly consistently concluding that his was an oversimplification of the true complexity of the past. Merton's quantitative approach was not convincing to later historians of science, perhaps even to those who later lapsed into prosopography. Merton's use of a biographical compilation, the *Dictionary of National Biography*, resulted in the *Dictionary of Scientific Biography* warning against such uses of its entries.[35] Merton's main source, the *DNB*, is secondary, as is the *Dictionary of Scientific Biography*. Despite all of the adverse comments, Merton's book had more influence on history of science than any other product of the interwar years. Perhaps there is a moral here.

A refugee from Germany, Edgar Zilsel, had a very brief moment of influence with a thesis quite different from Merton's but often seen as closely linked. Zilsel postulated that modern science did not come from intellectuals at the universities of Renaissance and early modern Europe. They were too immersed in orthodoxies hostile to modern science. Zilsel believed that the experimental spirit derived from the practice of artisans and was picked up by individuals outside of the then dominant intellectual circles. Very little original research buttressed Zilsel's papers. The growing concern among historians of science with the role of theory and of intellectual climate, reinforced by historical work on medieval thought, limited Zilsel's influence among historians of science in post–World War II America.[36]

History

Merton's book represented, at least to some, a forward thrust of the positivistic course of human progress so ardently called for by Sarton and others. If Merton and his colleagues succeeded, more and more of general culture would fall within the desired pattern of rational law and, therefore, the hegemony of science. Unlike Sarton, historians and social scientists did not define civilization or modern culture in quite the same spirit. They were less interested in the details of scientific knowledge than the scientists. Historians were usually willing to accept, as given, the triumphs of science. What most concerned these individuals was the social, the humanistic, and the moral.

Even before World War I, Americans had produced historical works, clearly precursors of professional history of science. Dorothy Stimson in 1917 published her study of *The Gradual Acceptance of the Coperni-*

can Theory of the Universe, notable for its translations of rare texts.[37] A future president of the History of Science Society, Dr. Stimson spent many decades researching the origins of the Royal Society, a scholarly cottage industry after Merton. Martha Ornstein's thesis on *The Role of Scientific Societies in the Seventeenth Century* appeared in 1913 and was reprinted in 1928.[38] Her early death in 1915 cut short a promising career. Most important of these early works was the 1905 dissertation of Lynn Thorndike, *The Place of Magic in the Intellectual History of Europe,*[39] the precursor of his eight-volume study, *A History of Magic and Experimental Science* . . . (1923–1941).[40] From Thorndike came the strong tradition of studies of medieval science in the United States.

Stimson, Thorndike, and Ornstein did their dissertations at the suggestion of James Harvey Robinson of Columbia University. (Parenthetically, Columbia University had a great role in the interwar years in the study of science's past, in marked contrast to its quite modest contributions after professional history of science appeared.) In 1921, Robinson published *The Mind in the Making: The Relation of Intelligence to Social Reform.* Quite like the pattern in H. G. Wells's *Outline of History, The Mind in the Making* starts with the animal heritage and the savage mind, traces the growth of critical thinking to and after the Scientific Revolution, but concludes with social problems awaiting solution. Perhaps more reform-minded than many of his fellow historians, Robinson was less interested in the details of the scientific venture than in how the spirit of that venture could solve ills in the society. Progress still had further to go.[41]

Robinson's student Lynn Thorndike only carried his work forward to Newton's triumph. During World War I he started writing a popular book "determined to do what little I could to keep civilization alive." *A Short History of Civilization* (1930) is in the spirit of Robinson's *Mind in the Making* but also has a very interesting attempt for its day at coverage of non-Western civilizations.[42] Thorndike's great eight-volume work is quite explicit on his point of view. The 1905 thesis limited its scope to "magic only as connected with science and with learning— only as accepted by educated men."[43] In the first volume of *History of Magic and Experimental Science* . . . he further explained his concern was with magic as a system of thought. When that system was abandoned by the learned and the educated, magic was "degraded to the low practices and beliefs of the ignorant and the vulgar"—and only fit to be taken up for study by anthropologists.[44] Thorndike's frame of reference was quite unlike Keith Thomas's recent work.[45] Practically all of the interwar literature was great man and great idea centered.

Not that Thorndike was unanalytical or incapable of critical comments at variance with a simplistic framework of Progress fueled by science. For example, in his chapter on the mid-seventeenth century, while citing the growth of experimentation, skepticism, and rationalism,

Thorndike concluded of those he was discussing, "their spirit was curious and credulous rather than skeptical and critical."[46] Given his basic orientation, Thorndike simply was severely constrained in the range of possible historical questions. Two quotations, one from his first volume, the second from his last, illustrate his frame of reference:

> But fortunately it [sources for history of thought and art] is more reliable, since the pursuit of truth or beauty does not encourage deception and prejudice as does the pursuit of wealth or power. Also the history of thought is more unified and consistent, steadier and more regular, than the fluctuations and diversities of political history. . . .
>
> The physical world was reduced to uniformity, and all wonders and monsters had to cease, except for the solitary marvel of Newtonian physics. . . . The whole pack of nonsense fell off the shoulders of experimental science as it passed through the narrow wicket gate of the Newtonian physics.[47]

The full-scale work in the history of science best exemplifying attitudes of men like Sarton, Robinson, and Thorndike is Joseph Mayer's *The Seven Seals of Science,* whose subtitle tells it all: "An account of the *unfoldment* of orderly *knowledge* & its influence on human *affairs.*"[48] (The italicizing is Mayer's.) A social scientist at Tufts, Mayer got his title from Robinson's *Mind in the Making.* Robinson, of course, found the phrase in chaps. 5–8 of the Book of Revelation in the New Testament. Robinson and Mayer used the seven seals image to underscore the concept of cultural lag. The seals on the books of nature were opened earlier than the seal on the last book, the book of man. Mayer's text was in the logical order of the opening of the books, corresponding (with overlaps) to the chronological order of the breaking of the seals: mathematics, astronomy, physics, chemistry, geology, biology, and psychology. Aiming at a popular and accurate account, *The Seven Seals of Science* is a very creditable work by the standards of its day and compares favorably in some respects to later one-volume histories by the British authors Dampier-Whetham, Pledge, and Mason.[49] Comte's spirit may have been alive at Tufts in 1927, but one wonders how he would have reacted to a Protestant imagery implying the end of history when the sciences of man came into existence.

Neither Robinson nor Mayer literally expected the Day of Judgment to follow hard upon the sciences of man, merely reform. What that meant is not too clear programmatically. What is clear is that reform meant science and men of science directing human affairs. Men of science included both natural and social scientists, as Thorndike made explicit in the *Short History of Civilization:*

> The scientist has all the serene courage, inner peace, and trained mind of the ancient philosopher. . . . [T]he student of historical and social sciences is consequently more open-minded and better balanced than his fellows. If society can

be imbued with this attitude, its restless waves of passions, superstitions, harmful habits, idle thoughts, incompetant, unscientific leadership, and woeful ambitions may soon be stilled.[50]

This theme appeared earlier and persisted. Joseph Mayer cited H. G. Wells to that effect in his book.[51] And in the edition of his popular history, *The Story of Mankind,* appearing at the outbreak of World War II (1939), Hendrick Willem van Loon called for letting science and scientists rule.[52] World War II shattered the last remnants of the belief of Progress in human affairs even as it raised acute questions of the necessary beneficence of scientific advance and the omnicompetence of scientists.

Doubts of Progress and of science's inherent goodness existed in the interwar years in the United States, even on Morningside Heights. In 1926, the philosopher John Herman Randall, Jr., published *The Making of the Modern Mind: A Survey of the Intellectual Background of the Present Age,* a very influential work in its day.[53] Randall gave a straightforward old-fashioned history of ideas account quite in the spirit of his colleagues Robinson and John Dewey. That is, Randall believed in the progression to a modern mind and assumed the possibility and desirability of the modern mind effecting improvements in society.

Three years later Randall published a quite different and far more interesting book dedicated to John Dewey. *Our Changing Civilization: How Science and the Machine are Reconstructing Modern Life* contrasts the world of the Umbrian peasant with what Randall observed in the Deutsches Museum in Munich. Randall saw the pressure of science and the machine on the traditional world and its desired values, including religion. Noting the absence of the life sciences, Randall described natural science as in the spirit of Bacon and Descartes, "natural to a society bent on mastering nature by mechanical means." Unlike his colleague Thorndike's view of Sir Isaac Newton written at the end of his multivolume work, Randall believed the Newtonian system "built up a system of beliefs useful for trade and commerce, a technique of prediction and control, of manipulation."[54] The connotations are decidedly non-Sartonian.

Such qualms were absent from the most distinguished work dealing with science by an American written during the interwar years, Cornell University's Preserved Smith's two-volume work, *A History of Modern Culture* (1930–1934).[55] A modern European historian and another of the students of James Harvey Robinson, Smith's intention was very much like his contemporaries dealing with modern minds and the like—to explain the triumph of Western civilization. More than science is covered, but Smith actually studied various classics of early modern science in his first volume on "The Great Renewal" and cited both Jeans and Heisenberg in his second, "The Enlightenment." Despite the passage of a half century, Smith's book retains a certain freshness, even a degree of sophistication, largely absent from the chroniclers of the onward thrust of a Progress propelled by triumphs of positive knowledge.

For one thing, Smith defined culture as intellectual life, noting that now the differences between the cultures of various classes is more important than the differences between the various nations of the white race: "In a sense, a history of culture is really a history of the intellectual class. Civilization is imposed by the leading classes on the masses, often against their stubborn opposition, generally without their full knowledge of what is taking place, and always without their active cooperation."[56] Smith is postulating a class-based social process for the historical development of the modern mind. Consequently, he pays explicit attention to social control, specifically the actions of school, church, and state.

Although some of the language resembles the Sartonian pieties, Smith somehow manages to give a different twist to his formulations:

If we may speak of their [intellectuals] mentality as the modern spirit, it is, as distinguished from the spirit of earlier ages, rational, free, forward looking, and self-conscious. To trust reason rather than tradition or authority, to assert the liberty of the individual, to look to the future rather than to the past, to regard the truth as relative and subjective rather than as absolute and objective, are the notes of the modern spirit.[57]

(Compare "relative and subjective" with the previously quoted words of Sarton.)

Despite his concentration on the intellectual class, Smith is very American in his concern for how the modern culture is spread to the masses. A specific chapter is on the "Propaganda of the Enlightenment." He postulates a process (probably not original to him) somewhat different from the intellectual-imposed cultural order of his opening remarks quoted above. It is more conventional, yet comes up with a conclusion reflecting a different set of concerns from his fellow authors. Smith believes the progress of science has two secondary consequences.[58] First, each increment of knowledge sooner or later leads to a technological advance. In the same year that Smith's second volume came out, Lewis Mumford noted of his neotechnic phase that "the main initiative comes not from the ingenious inventor, but from the scientist who establishes the general law: the invention is a derivative product."[59] Neither Mumford or Smith are original in this belief (a great vexation to many of our contemporary historians of technology). Unlike Smith's and Mumford's contemporaries who simply assumed that science yields technology or that the textile machinery of the first industrial revolution appeared as responses by ingenious artisans to economic pushes,

Smith and Mumford apparently need to assert the primacy of the cultural, the general idea, over the particular practical product. Was this a tactic to evade the kind of conclusions Randall drew from the displays at the Deutsches Museum?

Smith's second consequence of scientific advance returns to the problem of popular diffusion. Scientific advances produce altered worldviews. Somehow, these spread: "I mean [science's] diffusion into ever wider social circles, until what once had been the arcanum of a tiny group or even of a single thinker, becomes the commonplace of the man in the street and of the child in the school." Propaganda and the institutions of social control, one notes with interest a half century after Smith wrote, can work both ways in this author's worldview—for the attainment of the liberal vision of Progress or to stifle dissent. Yet, Smith inclined to the optimistic. On balance, Progress had occurred in the Western world, particularly among his fellow intellectuals.

The Sciences in the United States

Perhaps Smith's relative optimism had nationalistic roots. Not by chance did he end his second volume on the Enlightenment in 1776. The two volumes have characteristics of tone and emphasis different from much of the writings giving the origins of the preferred modern culture. Smith departs in his views in many essentials from those in the conventional and widely used modern European textbook of Columbia's Carlton J. H. Hayes, *A Political and Cultural History of Modern Europe*.[60] Quite early, Preserved Smith praises the Puritan Commonwealth as "most glorious" and characterizes Charles I of England as "justly punished."[61] As to the Enlightenment he gives primacy to the British contributions, conceding to the French only preeminence in the role of propagandizers.[62] The framework of the two great volumes is curiously Puritan (in pride of ancestry), Anglophone (in intellectual heritage), and American (in seeing diffusion among social classes as a problem).

Smith has warm praise for the colonial Americans:

Though in their infancy, the colonies could contribute little directly to the culture of the world, it is remarkable that even in the seventeenth century interest in science and in literature flourished mightily beyond the Atlantic. . . . [S]he [America] established schools and universities as good as those of the old country, and frequented by a much larger proportion of the population, and she eagerly bought and read the best books produced by Europe. Nor was she unable to add to the common stock of ideas. In Jonathan Edwards she produced one of the profoundest religious philosophers, and in Benjamin Franklin one of the greatest of the illuminati of the century.[63]

Smith's praise of colonial culture was part of a tradition leading to the false assumption of a golden age of colonial science followed by a precipitous decline after independence. During the 1930s, Theodore Hornberger[64] and others were beginning to work on colonial science and culture, eventually uncovering a more complex past than Smith perceived. Most important in retrospect was Smith's upbeat tone derived from the general progress of the national society, not from a summation of great advances presumably successively approximating the Truth. Smith was placing the sciences in the United States firmly within the context of the national history. The contrast with Sarton could not be greater.

Consider the treatment of the sciences within the United States by Joseph Mayer, the chronicler of the seven seals. To him, in two separate but repetitive articles, the problem of the history of science in the United States came down to a determination of the rates of production of great men in science.[65] Mayer had to counter the thesis of American indifference to basic research and also offer arguments for increased support of an obviously expanding research establishment. By statistics spurious even for a social scientist, Mayer "proved" that the United States had attained parity with the leading European powers. Mayer's work is interesting in three respects. First, he increased his American numbers by noting contributions in astronomy, a point not widely recognized among many conventional historians of science as late as the 1960s in my experience. Second, in a burst of nationalistic enthusiasm, Mayer proposed unsuccessfully that the History of Science Society and the American Association for the Advancement of Science sponsor an exhaustive history of the contributions of his great ones. How fortunate we, the students of American science, were in not being hobbled by such an ill-conceived, premature venture. Our colleagues studying the sciences in western Europe were not so fortunate in the literature they had to overcome. Third, Mayer's apparent obliviousness to historical context eventually became a real and growing gap between the older protoprofessionals and the younger, emerging professional historians of science, not to mention their colleagues from American history who brought a different agenda to what should have been a common venture.

There was a growing literature, historical and quasi-historical, on the sciences in the United States during the 1920s and 1930s. The natural historians and the mathematicians were not the only scientists looking at the national past. Impressed by what he saw and seeking a model for reform in France, Maurice Caullery had already written an account of universities and research institutions.[66] Not a history, yet, if read acutely, Caullery's work posed historical problems. Another foreigner, J. G. Crowther, produced an interesting collection of biographical essays.[67] From the guild of historians, besides Hornberger, came

Nathan Goodman with work on Rush and Franklin.[68] Courtney Hall contributed a workman-like biography of the early scientist, Samuel Latham Mitchell.[69] And from the framework of southern regionalism, T. C. Johnson contributed a pioneering but irritating volume on the scientific life in the Old South.[70] And there were others.

Perhaps the most unexpected work was from the tradition of current description of science, Bernard Jaffee's *Outposts of Science* (1935).[71] A skillful, conscientious popularizer, his later work, *Men of Science in America* (1944), briefly served as the standard introduction to the subject at the start of the period of professional history of science. At first glance *Outposts of Science* looks like just another collection of biographical essays. Nor was it encouraging, at least to this reader, to have this account of visits to eminent living scientists characterized as "a scientific pilgrimage to the firesteps and outposts of the present battlefronts of science in America."[72] Behind this image of World War I trench warfare was considerable background research, interviews with the eminences and their associates, and an aim, not always achieved, to place research in larger contexts. The text for the chapters were checked by each scientist, resulting in an approved version, interesting as a source for today's scholars. Jaffee had a realistic view of the personal characteristics of his subjects, something absent from the pages of *Isis* of that period. We learn that some of Jaffee's subjects changed his original text:

Several of the chapters came back to me with personal notes tragically blue-penciled. In general there was an objection to any attempt at glorification of their own personalities or overestimation of their achievements. Personal vanity, however, was by no means lacking in some. One man erased every reference to rivals in his field on the ground that they were "unimportant." Another made such voluminous additions to his own life and work that only an obituary would be obliged to print them.[73]

Jaffee produced what we now call contemporary history. It is a work with valuable information and commentary, some surprising. Not many authors in 1935 would have published a glowing estimation of Barbara McClintock's achievements.[74]

Despite the increasing evidence of interest, the history of science did not become a recognized subfield within U.S. history in the interwar years. As in the case of European history, the sciences had a recognized but minor place and remained underdeveloped historiographically. For example, Arthur M. Schlesinger, Sr., and Dixon Ryan Fox edited a thirteen-volume series, *A History of American Life* (1927–1948). Very influential in its day, the series is notable in retrospect for the attempt to go beyond the political, the constitutional, and the military to a broader view including the social, the intellectual, and the economic.

The sciences appear in all but one of the volumes.[75] Treatments range from entire chapters in several volumes to five pages or less in many other volumes. As a whole, the passages are more interesting as an index of current knowledge and of good intentions than as historical accounts of their ostensible subjects. The elder Schlesinger had some reservations about the growth of specialization in general: "Withdrawn from the market place and the forum, closeted with the idea, brooding and moiling in monkish seclusion, the specialist wedded a skeleton bride whose osseous kiss and rattling embrace rewarded him with an ecstacy beyond Helen's."[76] Yet, he and few like-minded contemporaries encouraged a small number of their students to study the sciences in the American context. Because of that, Americanists (historians of the sciences in the United States when the professional specialty arose) had a peculiar advantage over colleagues studying European science who mostly fell under the sway of the practitioners' history of the scientific community or of derivatives of the Sartonian order. The peculiar advantage—hardly an unmixed blessing—was that the Americanists were prematurely working in modes that later became quite conventional within the history of science.

Perhaps that is best illustrated by reference to a work published early in the start of the professionalizing period, one not within the canon of the history of science. It was 1950 when Crane Brinton of Harvard noted the cumulative nature of scientific knowledge in his *Ideas and Men: The Story of Western Thought*.[77] Elsewhere in Cambridge, T. S. Kuhn was thinking contrary thoughts at that time. Brinton had been the thesis adviser of Henry Guerlac of Cornell. The book was much like the products of the interwar years concerned with the modern mind, modern culture, Western civilization, and the like.

But Brinton introduced a new note, implicitly carrying forward the hint in Preserved Smith's great earlier work. Brinton tried to define a new kind of intellectual history, one not exemplified in this book nor in his other writings known to me: "To try to find the relations between the ideas of the philosophers, the intellectuals, the thinkers, and the actual way of living of the millions who carry the tasks of civilization . . . [or the connection between] what a few men write or say and what many men actually do."[78] Brinton saw an intellectual history with a spectrum close to philosophy at one end and at the other close to the social historian "or just a plain historian concerned with the daily lives of human beings."[79] The spirit, if not the letter, of Crane Brinton's formulation is now widely spread in the American historical community. Historians of science who came from American history had no trouble with this formulation.

Essentially, Brinton's formulation influenced them, not any exclusive adherence to externalist history of science. I do not mean that they literally gathered around a banner bearing Brinton's words. Rather, many trends within academic history in the United States (not only the specialty of American history) resonated with the spirit of Brinton's position. It was a comfortable position to young historians entering upon the study of the sciences within the United States in the two decades following World War II. Brinton's formulation was utterly alien to the early professional historians of science within Europe in their self-consciousness about their relationship to the scientific community, some with lingering traces of a Sartonian pride in the greatness of science, and often with a need to prove they really knew science. As professional history of science in the United States matured, inevitably its practitioners became closer to the outlooks of colleagues in history, even to some of the social sciences. Brinton's spirit appears today among historians of science but is hardly omnipresent. The formulation of *Ideas and Men* is the unfinished agenda of the discipline. Diversity, not uniformity, is the emerging characteristic of contemporary history of science.

Notes

1. Thackray has three important articles on the origins of professional history of science in the United States: (with R. K. Merton) "On Discipline-Building: The Paradoxes of George Sarton," *Isis* 63 (1972):473–495; "The History of Science Society: Five Phases of Pre-History," *Isis* 66 (1975):445–453; "The Pre-History of an Academic Discipline: The Study of the History of Science in the United States, 1891–1941," *Minerva* 18 (1980):448–473.

2. Osborn's book originally appeared as the Hale Lectures of the National Academy of Sciences. *Kosmos* appeared in five volumes from 1845 to 1855. Both *Kosmos* and the Hale Lectures were concerned with a concept of unity quite at variance with the growth not simply of specialization but of knowledge of the complexity of physical and biological phenomena.

3. *The Nature of the World and of Man* was published by the University of Chicago Press, apparently with great success. The essays are fine examples of high popularization, to make a distinction with the articles in the popular press and magazines.

4. I have used the one-volume edition of 1931. Wells's book sold well in many editions. It is a considerable achievement. Its author merits a careful reconsideration.

5. Burtt was primarily concerned with the philosophy of religion and related topics. Lovejoy stimulated a considerable body of work in the history of ideas notably deficient in its concern for the context in which the ideas originated and were accepted.

6. Like Lovejoy, Koyré approached his subjects as ideas carefully abstracted from contexts of personality, place, and time. I had the privilege of

hearing him give a series of seminar talks in 1951/52 from which the book derived. Although greatly impressed, I then sensed (and still do) an incompleteness in such works.

7. E. Mayr, *The Growth of Biological Thought* . . . (Cambridge and New York, 1982), p. 17.

8. Cambridge, Mass., 1926.

9. E. W. Brown et al. (New Haven, 1923).

10. Oxford and New York, 1923.

11. New Haven, 1918. Perhaps I am overreacting since a great deal of effort by myself and fellow historians was expended to get beyond and around the essays in that work.

12. Boston, 1922.

13. Another product of the University of Chicago Press whose scope then apparently differed from today's.

14. The full title merits repetition: *A bibliography of American natural history; the pioneer century 1769–1865; the role played by the scientific societies; scientific journals; natural history museums and botanic gardens; state geological and natural history surveys; federal exploring expeditions in the rise and progress of American botany, geology, mineralogy, paleontology and zoology* (Brooklyn, 1924–1929), 3 vols.

15. H. F. Osborn, *Cope, Master Naturalist* (Princeton, 1931); C. Schuchert and C. M. LeVene, *O. C. Marsh, Pioneer in Paleontology* (New Haven, 1940).

16. G. P. Merrill, *Contributions to a History of American State Geological and Natural History Surveys* (Washington, D.C., 1920), and *The First Hundred Years of American Geology* (New Haven, 1924).

17. F. R. Packard, *History of Medicine in the United States* (New York, 1931). H. E. Sigerist, *American Medicine* (New York, 1934).

18. George Sarton, "The History of Science versus the History of Medicine," *Isis* 23 (1935):313–320.

19. Originally published in 1919 by the University of Chicago Press. I have used the New York reprint of 1952.

20. Chicago, 1925. I find it hard to believe that this nice little antiquarian volume could entrance teachers of artithmetic, let alone improve their teaching.

21. Another University of Chicago Press book, 1983. Cohen is interested in the development of a statistical spirit but covers some of the same ground as Karpinski.

22. L. C. Karpinski, *Bibliography of Mathematical Works Printed in America through 1850* (Ann Arbor, 1940).

23. David Eugene Smith and Jekuthiel Ginsburg, *A History of Mathematics in America Before 1900* (Chicago, 1934). The quotation is from p. 64.

24. Boston, 1928.

25. Ibid., p. 3.

26. See my introduction to the reprint edition of *The Chequered Career of Ferdinand Rudolph Hassler* (New York, 1980). The biography appeared originally in 1929.

27. See Sarton, "History of Science vs. History of Medicine," pp. 317–319.

28. In 1936 Sarton published *The Study of the History of Mathematics* and *The Study of the History of Science*. I have used the Dover combined reprint of the two in 1954. The quotation is from p. 5 of *Science*.

29. 4 (1938):360–632. For a comment on Sarton and technology, see Lynn White, "Science in China," *Isis* 75 (1984):171–179.

30. Gary A. Abraham, "Misunderstanding the Merton Thesis: A Boundary Dispute Between History and Sociology," *Isis* 74 (1984):368–387.

31. This should be read in the context of Merton's chap. 10.

32. The quotation above and the material of the last few sentences is from Sarton, *Mathematics*, pp. 14–17.

33. *Ballistics in the Seventeenth Century* (Cambridge and New York, 1952); "Merton Revisted, or Science and Society in the Seventeenth Century," *History of Science* 2 (1963):1–16. Hall takes an internalist position like Sarton and Koyré. In the 1978 reprint of his book, Merton has a new preface dated 1970 whose last paragraph, without citations, concedes that "both science and religion also developed under their own steam." How this squares with his other points in the preface and the monograph is unclear. Merton reluctantly concedes a problem with his treatment of Puritanism but affirms his adherence to the point about technology.

34. Two recent examples are John Morgan, "Puritanism and Science: A Reinterpretation," *The Historical Journal* 22 (1979):525–560, and Michael Hunter, *The Royal Society and Its Fellows, 1660–1700* . . . (Chalfont St. Giles, England, 1982).

35. *Dictionary of Scientific Biography*, vols I, p. ix.

36. For example, see his "The Sociological Roots of Science," *American Journal of Sociology* 47 (1941–42):544–562 and "The Origin of William Gilbert's Scientific Method," *Journal of the History of Ideas* 2 (1941):1–32.

37. New York, 1917.

38. And reprinted subsequently. Time has not been too kind to this work, very useful in its day.

39. New York, 1905. The thesis is limited to classical antiquity.

40. Thorndike initially planned to cover "The First Thirteen Centuries of Our Era" but managed to reach Newton.

41. New York, 1921. This is a very limited look at a very interesting, influential figure in the American historical profession.

42. New York, 1930.

43. Ornstein, *Scientific Societies*, p. 36.

44. Thorndike, *Place of Magic*, vol. I, p. 4.

45. *Religion and the Decline of Magic* (New York, 1971). Thomas's book is on the period in England of the foundation of the Royal Society and of Newton's career. It presents a picture of the persistance of older ways just at the point when rationalism and science became accepted by the learned and upper classes. Thomas's work makes Newton's role as the "last magician" (in Keynes's words noted by Thorndike) more interesting and credible.

46. Thorndike, *Place of Magic*, vol. VIII, p. 261.

47. *Ibid.*, Vol. I, p. 3; Vol. VIII, p. 603.

48. New York, 1927.

48. New York, 1927.

49. By this I simply mean that the passage of a generation had not resulted in any marked improvement over Mayer except in minor details.

50. Robinson, *Mind in the Making*, p. 565f.

51. *Seven Seals*, p. 213.

52. Pages 404–405.

53. Boston, 1926. Randall's book remained a staple of undergraduate reading lists into the 1950s.

54. New York, 1929, pp. 119, 183.

55. New York, vol. II, pp. 930–934.

56. Ibid., vol. I, p. 4.

57. Ibid., vol. I, p. 7.

58. Ibid., vol. II, p. 121.

59. *Technics and Civiliaztion* (New York, 1934), p. 217f.

60. I have used the two-volume edition of 1932–1936. I do not mean to denigrate Hayes, whose work is quite solid and interesting in its attempt at depth of coverage, including treatments of modern science and on cubism at the end of the second volume.

61. Preserved Smith, *History of Modern Culture*, vol. I, pp. 13–14.

62. Ibid., vol. II, p. 14.

63. Ibid., vol. I, p. 14; vol. II, p. 9.

64. Theodore Hornberger, *Scientific Thought in the American Colleges, 1628–1800* (Austin, Tex., 1945).

65. "America's Influence on the Development of the Sciences," *Scientific Monthly* 26 (1928):60–69; "Can Americans Be Scientists: Our Contributions to Physical Science," *Technology Review* 32 (1930):233–236, 254, 256.

66. M. Caullery, *Universities and Scientific Life in the United States* (Cambridge and New York, 1922).

67. *Famous American Men of Science* (New York, 1937).

68. Nathan G. Goodman, *Benjamin Rush: Physician and Citizen* (Philadelphia, 1934).

69. *A Scientific in the Early Republic* (New York, 1934).

70. *Scientific Interests in the Old South* (New York, 1936).

71. New York, 1935.

72. Jaffee, *Outposts of Science*, p. xv.

73. Ibid., p. xxi. See also his comments on p. xx.

74. Ibid., p. 31.

75. The exception is the volume by Ida Tarbell, *The Nationalizing of Business* (New York, 1938).

76. *The Rise of the City, 1878–1989* (New York, 1933), p. 220 (vol. X of *A History of American Life*).

77. New York, 1950.

78. Brinton, *Ideas and Men*, p. 8.

79. Ibid., p. 9.

16. Science, Scientists and Historians of Science

"Science, Scientists, and Historians of Science" appeared in 1981 in *History of Science* and is reprinted with the original preface.

Author's Preface: The 25 January 1980 issue of *Science* carried a news item entitled, "History of Science Losing Its Science," which opened as follows: "Once a highly respected field that focused on the conceptual evolution of scientific ideas, the history of science is losing its grip on science, leaning heavily on social history, and dabbling with shoddy scholarship."[1] Rather a strong indictment, particularly as its source was an address by Charles C. Gillispie of Princeton University recently given at the 1980 annual meeting of the American Association for the Advancement of Science at San Francisco, California.

The story concluded with an injunction by Gillispie to scientists to watch history of science closely "lest the field fall prey to those who would use history against science." The principal example supporting that injunction was a round table at Princeton on the history of atomic weapons whose participants Gillispie described as unknowledgeable on the technical issues and, therefore, on the history. Social history of science was described by Gillispie, according to the reporter, as devoid of "hard" science, with a "lust" for the personal and anecdotal, saying that "students of science history have tended to be political scientists," whose lack of technical background produces a "depiction of scientists as hucksters of weapons and research."

The historians of science in the United States are still a rather small tight community, and word of the address spread even before the *Science* article. Although no text was released by its author, those in attendance (and rumor has it a few tape recordings) confirm the *Science* account, which dealt only with a small part of a larger and less controversial body of reflections. What surprised me was the firm support of a few historians in private conversations—even those who knew that new Ph.D.'s in the field were not "political scientists." One colleague thought it "long overdue" and launched into a strong attack against unnamed historians of science who thought of themselves primarily as historians, not as scientists. As I fall into that category, the conversation became rather stilted. On the other hand, opponents of Gillispie's views who spoke to me were more astonished than outraged. The bare news story in *Science* did not make sense to them.

Further reverberations in print occurred. A nationally syndicated columnist, Edwin M. Yoder, Jr. did a piece entitled "Narcissism and the History of Science,"[2] apparently based on more of the text than given by *Science*. Yoder's expressed feelings against social history and the social sciences told less about those subjects than of the intellectual limitations of American journalism. Coincidentally, the editors of the two leading U.S. journals in the history of medicine, Leonard Wilson and Lloyd Stevenson, came out with strong statements first believed by many to represent reactions to the AAAS talk, a possibility ruled out by simple chronology in Wilson's case.[3] Yet, the generic relation was striking. There was concern with the downplaying of hard content, a strong aversion to the historical conclusions of much recent social history of medicine, particularly as displaying "condescension" and "hostility" to physicians. As in the case of Gillispie, there was an implication of differences over current issues. Similarly, there was an implication of the need for medical awareness of the activities of the historians in question. Combined with the *Science* piece, this was rather heady stuff for a field hitherto more noted for academic teapot tempests (at least in the United States).

A strange public dénouement followed. The 29 February 1980 issue of *Science*[4] had two letters on the original news story. Not surprisingly, one letter (from Robert E. Kohler of the University of Pennsylvania) strongly defended the current status of history of science. The other letter was from Charles Gillispie, who now withdrew his previous aspersions on the competence of the speakers and the audience at the Princeton colloquium, corrected a factual error in his text, and noted that scientists doing history (as well as social historians of science) paid attention to "soft" topics.

Even a tempest in a teapot can make waves. In this case, the waves were disquieting. At a meeting of the board of a leading university press after the first *Science* story, a scientist-member unsuccessfully moved to halt the publication of a work in the history of science on the grounds that the field was not really "science." Shortly afterward a distinguished scientist decided against giving his personal papers to a major university library because of the "state" of the history of science disclosed by the address to AAAS. Fortunately, waves from teapot tempests do not carry much energy. Most historians of science continued to function as before, for better or worse; most scientists apparently never heard of the complaints or, if they had, were not moved to extreme measures. Yet, for a moment, the possibility of polarization of the field existed, as well as the estrangement of historians of science from scientists. Happily, that moment was short-lived.

Even before 29 February, the planners of the 1981 AAAS session at Toronto had decided to pick up the question of the status of scientific content in the current practices of the historians of science. What follows is one of the three papers at that session. Professor Kohler's views are well expressed in his letter. F. L. Holmes's position is essentially given in a recent paper in this journal.[5]

A personal note. I was surprised when the session organizer invited me as someone representing a moderate position. My interests are heavily in institutional and social history; but I qualified as a moderate because the *Henry Papers* also carefully document Joseph Henry's experiments. Classification as a moderate after all these years was a bitter pill. What follows is a very personal statement slightly revised from the original oral version. It is moderate only in the sense that it was in no way aimed at Professor Gillispie because of my concern with the larger issues and because of my respect for him personally and professionally.

How extraordinary it is to be discussing the supposedly vanishing role of science in the history of science. The topic was unthinkable not too many years ago. I consider it absurd now to argue whether or not the ideas and data of the sciences in the past are in danger of vanishing from the practice of historians of science. For one thing, a significant portion, easily a majority, of the teachers and graduate students in the field have scientific backgrounds which inevitably influence their work. Almost everyone agrees that scientific content is important because that is what makes scientists different from lawyers, clergymen, and other professional groups. Neither I nor any other external or social historian of science known to me can perceive a future history of science devoid of internal history. I use the terms "internal" and "external" not because I believe in them, as I shall make clear, but because they represent traditional polar alternatives.

Yet, there is this feeling of a loss or impending loss of technical content. Judging by crude measurements, at least one-half to two-thirds of all English language work in the history of science is concerned in various senses with the ideas and data of the sciences. Perhaps complaining scientists and historians of science are reacting to a change from an older, preferred state. It would not surprise me if someone reported that a generation ago, let us say from 1955 to 1960, 80–90 percent of the work in history of science was internalistic. Perhaps a preferred balance is upset, resulting in a wrong mix in research, even a mix done in an impermissible manner. Or is loss of a near monopoly being equated with extinction?

Some complaints about history of science are improperly leveled at historians of science. Science is a multifaceted activity of great importance attracting a wide variety of professions other than the sciences or history of science, each asking their own questions for their own purposes. Being outside a totalitarian system rigidly defining intellectual territoriality, we have to live with the incursions into science of economists, lawyers, sociologists, general historians, philosophers, anthropologists, political scientists—even emissaries from belles lettres. We also have to recognize that the incursions are both proper and understandable given the great importance of the sciences in our civilization. A studied disregard would indeed indicate a serious problem in our society. Since we have all benefited from the contributions of other intellectual fields, do not blame historians of science when the results are displeasing.

Considerations of history of science by outsiders to the guild—including scientists—can include use of historical "facts" or even full-scale forays into historical research. While these are often impeccable or at least harmless, what is the historian of science to do when "facts" and interpretations are false, misleading, or imcomplete? Or when major research efforts have serious flaws? I certainly do not want to set myself up as the Lord High Pedant. Some scientists do not like being corrected, because they feel it not only a reflection on their mastery of their field but a violation of proprietary interests.

If the historical errors committed by others are serious enough, historians of science should do more than gripe and carp. That will never convince the offenders or interested readers. Far better is an attempt to give a cogent analysis of the supposed errors, complete with supporting evidence. Better still, historians of science should, by example, demonstrate the right way. If other disciplines are moving into areas neglected by historians of science, they have little right to complain. After all, the history of science in the United States until recently was largely in the hands of scholars stemming from general American history.

The changing mix in the practices of historians of science is best understood as the result of an evolution from older forms of the discipline to what I believe is best characterized as an emerging consensus. Much of the history of science as a field originated in the interests of practising scientists, with lesser, but significant, inputs from philosophy and from history. To this day, there are enough practising scientists with historical concerns to generate an appreciable literature.

That some of the products of practitioners' history are solid or notable for bio-bibliographic contributions should not obscure the limited, if not amateur, nature of most writings in this genre. I am indebted to and very respectful of Partington, but want a different history of chemistry.

The founding generation of professional historians of science in the United States (some still active) took over many of the attitudes of practitioners' history even as they amended its practices to define history of science as a form of intellectual history or history of ideas. At its worst, practitioners' history was (and is) a recital of approved content with a modest or trivial veneer of personal anecdote and institutional asides. Related to heroic positivistic myths, it displayed an unambiguous path to present truths. By centring on the content, there was a tendency to merge the identities of the scientists with the science. This has affected internal history of science to this day. One can find examples in the literature of the last three decades in which substituting letters of the Roman alphabet for names of persons; letters of the Greek alphabet for place names; letters of the Hebrew alphabet for institutions; and $t_1 \ldots t_n$ to indicate temporal order will produce little or no change in the published accounts. Here the fixation on content has produced second-rate science pretending to be history.

Most of the early professional historians of science in the United States were acutely aware of the perils of practitioners' history. Perhaps the best known reaction is the historiographic musings of T. S. Kuhn. In their research, these historians of science began uncovering complexities outside some postulated unambiguous path to the present; more important, they began to present a history of the past of interest for its own sake. More and more, historians of science in the last two decades became more like general historians and less like scientists.

But that meant, of course, that historians of science no longer could be relied upon to provide an unquestioning time dimension for an ideology of science. By one of those wonderful accidents of history (or is it an accident?) which makes the subject worth doing, this development within the history of science roughly coincided with the troubled years of the Vietnam War. To be more precise, the development antedated the years of unrest which undoubtedly accelerated the trend. The atom bomb had dramatically linked basic science with death and destruction; the Vietnam years were disturbing to many academics because they produced a disregard of or hostility to previously little-questioned norms and beliefs. Established historians of science encountered a new generation greatly interested in "external" policy issues. Like some scientists, some historians were anything but amused.

If we look back at the development of professional history of science in the United States from 1955 to the present, we can discern an evolution from a small, talented group to a larger, more diverse body of investigators. The early professional historians of science contributed much to our knowledge. Although they had very strong ideas about content and method, they were not a monolithic group. Many produced fine pioneering examples of "external" history.[6] Their power of enforcing conformity, although real, was strictly limited. As I look back from the vantage of having been an outsider to the in-group, I see their principal concern was not doctrinal conformity but to build an autonomous discipline between science and history, somehow benefiting from the best of both worlds.

Once having committed the field to being really history and not warmed-over science with a sprinkling of human interest anecdotes, historians of science as a group followed the trends associated with history in general. The more rarefied history of ideas—a study of the idea in isolation in the Lovejoy tradition—gave way inevitably to the study of ideas in social contexts in the history of science. The trend affirmed the truism that history is about the behavior of humans—either individuals or groups. Things, whether intellectual or material, are of historical interest only insofar as they relate to the hopes, fears, actions, and fates of humans.

History as an intellectual enterprise has attributes differentiating it from certain conventional views of both the natural sciences and the social sciences. History tends to the holistic—to see the past in the round, to deal with detailed specifics of time, place, and person. Quite commonly, histories are praised for conveying a "sense" of reality; a "feel" for the past; or a "texture" of a society. Being holistic in history means conveying not only knowledge but something ontological and even aesthetic.

I think this arises from the nature of the documentation historians grapple with. For the modern period especially, but even for older eras, the evidence is often detailed and diffuse, sometimes voluminous. It conveys a sense of complexity while frustrating inquiry by clearly being incomplete. Wonderful problems are presented in ways guaranteeing their remaining mysteries to those seeking ultimates. The sense of wonder leads historians to want to present all of their findings, even at the risk of blurring the clarity of their explanations.

History, consequently, departs from the practices of the sciences and the social sciences in its explanatory modes. The sciences strive for statements of law-like regularities of universal scope based on assumptions of uniformity. The social sciences tend to exaggerate both the uniformity and the law-like regularities of their findings. Because of its holistic tendencies, history encounters too many variables, resulting in great caution in generalizing. It is exceedingly difficult to specify the universe from which the large, diffuse sample of sources come, in the absence of any convincing principle of uniformity. Typically, historians react by obscuring their generalizations in narratives and descriptions, where they are later uncovered by critics and by philosophers. But that is changing as the history of science, like general history, is now responding to the influences of the social sciences by being more specific as to methods, concepts, and conclusions.

The emerging consensus in the history of science reflects an array of influences from history, the social sciences, and the philosophy of science. Spencer Weart's *Scientists in Power* (1979) is a spendid example of this emerging consensus.[7] The most notable aspect is the tacit resolve to avoid "hard-line" positions on the division of "internal" and "external" history. It is an obvious consequence of the holistic tendency to want to give both. There is a widespread recognition that the distinction is sterile; that historians need to understand both; that their teaching should convey both; that while economy of research effort may require concentration on one, the ideal is to bring both into any description and explanation and that the distinction is probably fallacious. Attempts to give percentages in these terms ultimately fail because historical writings are now often both "internal" and "external."

Had the history of science started out otherwise, as an exercise in external history, I believe the trend would have been to introduce more technical content and philosophic analysis. I cannot predict a future in which the present trend will continue to the extent of extirpating internal history. Nor will I predict a reverse of the pendulum to the status of 1955–1960. What I will predict is that having tasted the fruits of real history, historians of science will not return to any position in which science is viewed one-dimensionally.

I am against any formulation which gives a particular subject or method a privileged position within the history of science. Science is a complex, multifaceted human activity; its history should not be an oversimplified venture. What I do insist on is that choices by historians reflect the state of the art and the nature of their problems, not some ideologically predetermined hierarchy. As the state of the art and the problems will change in time, the mix of subjects and methods will also change.

What concerns me most is not the hostile reaction of a few historians of science to the emerging consensus. Time has a way of dealing with problems of generational succession. I am sensitive to the danger that complaints from within the community of historians of science may receive prominence via reactions among the minority of scientists seriously concerned with history and those unduly sensitive to the possible effects of scholarly writings on current policies. Complaints about loss of content in history of science are often coded messages calling for censorship. Perhaps I am naive, but I cannot believe that the majority of the scientific community really wants historians to remain quiet or to rewrite their findings to avoid disturbing comfortable assumptions.

What I perceive is a pattern in which historical results tend to an indifference to the passions of the moment, whether of the scientific community or of its critics. Both groups must be infuriated at times by a scholarly community which persists in showing that past events are unique or are cryptic variations on a recurrent theme. In either case there is an implication of a genuine element of contingency not conducive to drawing contemporary lessons. One thinks of the reactions to Darwin and suspects that the apparent indifference to the passions of the moment is evidence of basic soundness. I say "apparent" because the historians are not hermetically sealed from society. Nor were the scientists they study.

The reactions of the scientific community are important. A significant fraction of the community of historians of science is in science departments or organizations. The old instinct of getting the best of both worlds—science and history—is still sound. No one will gain from a rupture of relations. Problems typically arise when historians of science, like the general historians, move into contemporary history. There is no difference, in principle, between studying the politics of science in Revolutionary France to explain the founding of the Musée d'Histoire Naturelle and studying the scientific politics of nuclear warfare. While few living scientists will man the barricades over interpretations of the phlogiston theory, many individuals can be cut close to the bone by contemporary history. Knowing this, I prudently avoided contemporary history until recently so that I could have the illusion of objectively studying my scientists as one would study animals in a laboratory. It is not good experimental form for lab animals to talk back to the principal investigator.

This is a real problem not avoided by such techniques as oral history. The historian of science and his subjects operate out of quite different frameworks. They use differing terms according to differing rhetorics; they ask different questions, even of the same event; and, not surprisingly, they sometimes arrive at distinctly divergent answers. The scientist often finds it hard to accept that his memories and his beliefs are simply data to the historian of science, not unquestioned truths.

To give a few examples illustrating the differing perspectives: I read with interest a recent newspaper account of a reception at the Smithsonian's exhibit on atom smashers. A reporter interviewing a pioneer in the field, M. Stanley Livingston, raised the matter of atom bombs, receiving this interesting reply: "We weren't responsible . . . I don't like to associate with that. Atom smashing explained and explored nuclear physics. . . . Let's not disturb the history of nuclear physics by bringing that into it."[8] Reading this reminded me of a seminar I attended at the Folger Library during the Copernicus celebrations. In a discussion of the impact on the Renaissance mind of natural and man-made wonders, I casually mentioned that Galileo was interested in the Arsenal in Venice. A physicist from one of our leading universities told me flatly that I was wrong. I shall never forget his words: "A good scientist is not interested in technology." It was a first principle he had undoubtedly discovered while helping to build the A-bomb at Los Alamos.

Having turned to contemporary history, I recently wrote to a distinguished emeritus professor of mathematics. I was working on a paper, since published, on how American mathematicians reacted to the coming of refugee colleagues in the 1930s.[9] I received a prompt, courteous response, concluding with an expression of puzzlement as to my purpose in undertaking such an arduous task. He could not see how my paper would aid mathematics. I replied with thanks, noting that I had no intention of providing a service to mathematics but hoped my article would contribute to the history of the sciences in the United States.

Few historians of science would identify with the viewpoints of the three scientists. I would hope that most scientists will think differently from their three colleagues. That, in fact, has been my experience. There is a point in having a history of science that attempts to answer time-, place-, and person-related questions about the growth of science; a history of science that is real history and not a counterfeit consisting of second-rate science; a history of science whose valid findings may someday provide the basis for an applied or clinical study of current science. We need good history of science for the same reason we need to learn about stars, chromosomes, radiations, and elements. It is important for our understanding of who and where we are. It is intellectually and morally right. It is intellectually and morally wrong to limit—implicitly or otherwise—the history of so important a part of man's past.

Notes

1. *Science* 107 (1960):389.
2. *Washington Star* Feb. 7, 1980.
3. Leonard Wilson, "Medical History Without Medicine," *Journal of the History of Medicine and Allied Sciences* 35 (1980):5–7; Lloyd G. Stevenson, "A Second Opinion." *Bulletin of the History of Medicine* 54 (1960):134–140. Stevenson cited the *Science* news story. But both he and Wilson were clearly reacting independently of the AAAS talk.
4. *Science* 107 (1980):934–935.
5. F. L. Holmes, "The Fine Structure of Scientific Creativity," *History of Science*, 19 (1981):60–70.
6. A number of the writings of Charles C. Gillispie come to mind.
7. Weart's book successfully interweaves group sociology and science policy with developments in nuclear physics. Precedents for such attempts are quite common in the histories of applied fields. A recent outstanding example is

Reese V. Jenkins, *Images and Enterprise: Technology and the American Photographic Industry* (Baltimore, 1976).

Perhaps more important as precedents in the United States are the many writings of general historians which matter-of-factly treated interrelations of thought and action, particularly in U.S. history. Except for some of the historians of science, products of American graduate schools were not uncomfortable with the idea that science could show influences from outside its domain.

8. *Washington Post* July 23, 1980.

9. "Refugee Mathematicians in the United States of America, 1933–1941: Reception and Reaction," *Annals of Science*, 38 (1981):313–338 [Paper No. 12 in the present volume].

17. Through Paradigm-land to a Normal History of Science

Somewhere in the mind of Thomas S. Kuhn, I would speculate, was the sense of a great problem faced by historians of science and philosophers of science. The transition from the classical physics of Newton and his successors to a modern physics of quantum mechanics apparently shattered long-held beliefs about the nature of scientific advance. No longer could one assume progress, either in the form of cumulative additions to "positive" knowledge or of successive approximations to *the* Truth. Impermanence is now a real possibility. Tentativeness has replaced certainty.

The transition to modern physics had previously immediately effected philosophers of science and some scientists given to reflecting on their calling. When a professional history of science emerged after World War II, Kuhn was one of the few who worried about the implications of modern physics for the new breed of scholars. The result was a striking and influential work, *The Structure of Scientific Revolutions* (1962). Kuhn's ideas generated much discussion, particularly among philosophers and sociologists. This paper is primarily about what happened among the historians of science in the English-speaking world after they encountered *Structure*. It was originally presented in 1979 at the convention of the American Studies Association in Minneapolis and appeared in *Social Studies of Science* in 1980.

For various reasons I never accepted Kuhn's vision. To begin with, I approached the history of the sciences as a problem in social interaction, whether within scientific groups or between scientists and nonscientists. Kuhn basically wanted to explain the concepts and data of science in each historical epoch. "Explanation" to him decidedly emphasized scientific explanation even if given in a time-bound framework. Because of my concern with social or external issues, I was greatly interested in scientific activities far more routine than those represented by Kuhn's puzzle-solving normal scientists. That produced a skepticism on my part about both his normal science and his account of scientific revolutions. My intuitive reaction was that the routine had a greater role than Kuhn allowed both in the generation of novelties and in their eventual acceptance.

My own research clearly played a role in the genesis of that viewpoint. But my career path also had an effect. The laboratory was Kuhn's model of the world. Mine was the library, the archive, the museum. I did not see impermanence. Despite massive losses, so much still survived from the past. Where he saw each new paradigm creating a new world for laboratory research, I saw the continued turbulent inflow of new cultural artifacts—conceptual, paper, and three-dimensional. Where he saw the investigator playing games within sets of arbitrary rules (backgammon in the betatron), I saw scientists as authors, hopefully of useful myths.

A strong sense of intellectual rectitude kept Kuhn from taking the easy way out of slipping into positions about ultimate reality and the like. Apparently, impermanence had a deep significance for him beyond its obvious place in the world of physics. My historical beliefs called for understanding programmatic assertions of both scientists and of the students of their history. Above all, somehow I arrived at a strongly held belief in a historical ideology. To my mind, nothing is ever completely lost; the past is always with us (even in our genes). There is every possibility old beliefs will reappear in new guises, fresh and shining for eager minds and hands in the laboratory.

But the history of science becomes real history almost in proportion to the degree that the historian is able to stand back from the technical content of the subject in order to assess it in relation to the general intellectual pattern at the time, and the social needs and status which this reflects.[1]

The lead book review in the *New York Times* on Sunday, 10 March 1974, asserted that "Thomas Kuhn, among others [believes] that the traditions of rationality and objectivity governing our sciences, social sciences and conceptions of self are in the service of death." The book reviewed was not in the history of science, and the statement, by a professor of English, is a complete distortion of Kuhn's viewpoint. *The Structure of Scientific Revolutions*[2] is widely cited outside of the history of science, perhaps because science is important in our culture, and a striking historiographic theory of its development necessarily attracts interest. Kuhn's views are usually misunderstood and misused outside his own specialty. An author is not responsible for how others use his works, although Kuhn has ruefully admitted problems caused by his mode of exposition: "Part of the reason for its success is, I regretfully conclude, that it can be too nearly all things to all people."[3] All too often that arises from a readership unaware of both the context and intent of the author.

Perhaps the wide response springs from a more general current in the intellectual atmosphere. Kuhn postulates how something preexisting—the "paradigm"—structures beliefs, perceptions, and actions. Analogous concepts occur in Piaget, Gombrich, Chomsky, Foucault, and E. O. Wilson's sociobiology. Kuhn's writings have greatly influenced the philosophy of science and the sociology of science. What is striking is the near absence of examples of Kuhnian history of science. By this I mean histories following the general scenario set forth by Kuhn's theory. A few pieces attempt to test the theory in its entirety. The most impressive, David Edge and Michael Mulkay's *Astronomy Transformed* is critical of Kuhn's basic views.[4] Relatively few pieces essay verification of subdivisions of Kuhn's theories, although one does encounter superficial uses of terms like "paradigm" or "normal science."[5]

Judging by his last book (published in 1978), Kuhn himself does not practice Kuhnian history of science. *Black-Body Theory and the Quantum Discontinuity, 1894–1912*[6] is in a genre familiar to historians of science. The book chiefly concerns the work of the German theoretical physicist Max Planck, who is widely credited with introducing discontinuity, a key step in the history of modern quantum physics. The story has all the conventional earmarks which might have caused Kuhn to invoke his concept of "revolution." Instead, however, he purports to show that the notion of discontinuity did not enter Planck's intentions or conclusions: that was retroactively imposed on the historical account (and a reluctant Planck) by the scientific community. Kuhn's book, if his version is correct, is a rewriting of a community-sanctioned myth. The journal *History of Science* has an analogous article by Robert Olby, a British historian of biology. His purpose is clear from the title: "Mendel No Mendelian?"[7] Historians of science working this genre attempt to show that the scientific community, even the very investigator involved, may significantly misconstrue the essentials of a past achievement.

Before surveying Kuhn's views let us consider why they have had only minor application to history of science, particularly to the kinds of questions he discusses. Why didn't his readers grind out Kuhnian scenarios?

If Kuhn is read carefully, he is quite conscientious about noting qualifications, complexities, and exceptions. That is, Kuhn is not as Kuhnian as he emerged from commentaries or his own polemics with philosophers of science. My hypothesis is that historians after 1962 went into old scientific publications and manuscripts well aware of Kuhn's theory. Maybe they found his cycle, maybe not. But they also found those qualifications, complexities, and exceptions—and perhaps more. Perhaps they even encountered external factors. And these extra-paradigm cycle phenomena seemed more interesting per se, perhaps more attractive as a means of establishing their independent visibility as historians of science. Would confirming a fine point in Kuhn's theory do as well in attaining a tenured academic position? More important, historians of science found Kuhn's theory less useful as a means of organizing their findings.[8]

Investigators in areas close to Kuhn's own research interests simply did not find confirmation of his views. At a conference in 1975, one writer concluded, "at no point in the process does it seem to make much sense to say that Galileo is now doing purely normal research, or that later he is doing purely revolutionary work."[9] Another author at the same meeting discussed Einstein's refusal to accept the notion of revolutions as central to scientific change, casting doubt that such notions explained what Einstein accomplished.[10]

Or consider three recent separate investigations of electricity and magnetism in the eighteenth century. The first rejected the implication of one doctrinaire paradigm, Newtonianism, replacing another, Cartesianism: "Not for them [the scientists] the dogmatism of the paradigm or the Lakatosian research programme."[11] The second politely concedes Kuhn's superiority to Ravetz and Bernal but simply doubts the applicability of his categories of "normal" and "revolutionary" to the events in question.[12] A third author writes, "the power and unexpectedness of the condenser, the distress even fear it initially provoked, guaranteed that it did not become one of those shy 'anomalies' that are supposed gradually to prepare the ground for scientific revolution."[13] From writings like these, and even from silences where nonspecialists might expect traces of influences, one can conclude that the doctrines associated with *The Structure* have not become paradigmatic (if I may lapse into Kuhnian) in the history of science, whatever their currency elsewhere. Why that is so is the concern of what follows.

One of the attractive characteristics of Kuhn's historiographical writings is the personal, autobiographical element it contains. At points it conveys almost a sense of Romantic angst. More than once readers are reminded that Kuhn started out as a physicist. During a reading of Aristotle he had a great moment of illumination—like Saul on the road to Damascus. How could such an intelligent man write such absurdities, absurd, that is, by our present conceptions? When Kuhn realized that Aristotle had a completely different mode of imposing conceptual order on the external world, he became a historian and set forth on the road to *The Structure*. This was around 1947. By 1957, he had reached the positions which appear in the work whose writing started in 1961.[14]

The *New York Times* reviewer was correct in sensing in Kuhn an element of repudiation of science—or, rather, of science as commonly understood. Although Kuhn left physics when times were good for that discipline and its prestige high, other signs of disaffection existed. Some were disturbed that the great advances of the preceding generation had culminated in an alliance of theory and destruction and with the reduction of theory to nuclear engineering, both placing the research community in the firm embrace of the State. The decade 1947–1957 was the era of F. A. Hayek's *The Counter-Revolution of Science* (1952) and of the writings of Michael Polanyi, both stressing political and intellectual misuse of science by Marxists and the heirs of positivism, and favoring a conservative, tradition-laden view of science. Polanyi in particular stressed the autonomy of science and the tacitness of its knowledge, themes Kuhn later elaborated. The sociologist of science, Bernard Barber, somewhat later wrote about scientists resisting innovation. And Kuhn would soon write of the role of dogma in science.[15]

In a narrow sense, in 1947 Kuhn was repudiating the history of science produced by practicing scientists and the pattern of education this entailed. What might be called practitioners' history still flourishes. From this tradition have come a number of works of great distinction, and valuable compilations, biographical and bibliographical. But most such writings are marred by a particularly acute form of the Whig interpretation—an inevitable cumulative progression to our present towering intellectual rectitude. Kuhn argues against practitioners' history. Again and again in *The Structure,* he rails against science textbooks which impose this false view of the past. Yet Kuhn accepts that this false educational strategy is a necessary way of advancing scientific knowledge.[16]

Kuhn's problem, around 1947–1957, was to find how to write history of science in such a way that its conclusions were meaningful in a sense comparable or parallel to science itself. That exercise quickly reduced to a problem of boundary definition. If the concepts and data of science are self-generating (by an "internal logic"), then this so-called internal history could follow the ahistorical lines of practitioners' history. If they were wholly or partly determined by external contexts, then history of science might become a Marxist scenario, a sub-branch of sociology of science, or merely part of general history. Influenced by the prevailing mode among English-speaking historians of science,[17] Kuhn expanded the boundary of internal history to include elements of general intellectual history. Only under rare circumstances, according to Kuhn, are external factors translatable into intellectual factors affecting the core or internal history. Otherwise the context of general intellectual history could determine not only the timing, location, and rate of scientific change but also the content of science itself.

Having so defined the boundaries, Kuhn then produced a formula which made the history of science possible, a cyclical law–like effect analogous to the products of science. Cyclical theories of history, however attractive, are inherently improbable. They also run afoul of the normal tendency of historical practice to stress particulars of time and place, producing distinctive, short-term generalizations of a sort that are atheoretical by the usual standards of the natural and social sciences. For that reason it was almost inevitable that a historian of science could write recently of "the Procrustean application of the terms 'paradigm,' 'normal science,' and 'scientific revolution.'"[18]

Kuhn's cycle (or more precisely, sequence of cycles) starts with a preparadigm phase which is prescientific in the sense that competing schools, each perhaps with its own paradigm, argue about fundamentals. Somehow—and Kuhn is unclear on how this occurs—a consensus emerges, making one paradigm the basis for the practice of a mature science. This practice Kuhn has named "normal science." It is the most controversial aspect of his argument. In normal science the dominant paradigm is possessed by an interacting group of investigators sometimes referred to as an "invisible college." Typically, the group numbers about one hundred but, in a significant later modification, Kuhn brings the possible total down to "fewer than twenty-five people."[19] Normal science is characterized by the great certainty animating its practitioners. Because the paradigm has settled basic questions and provided models for research, scientists can concentrate on narrower technicalities rather than waste time arguing about larger but unanswerable questions. What scientists engage in is "puzzle solving," finding ways of solving problems whose answers are already known from the paradigm. According to Kuhn, most of science is normal science, and most of the clichés about science he counters originate here. Normal science is cumulative and represents the most certain form of knowledge provided by science.

But the practice of normal science inevitably produces anomalies, deviant answers unpredicted by the paradigm. Initially, they are explained away or suppressed. As anomalies persistently recur, they inevitably generate a crisis—that is, a breakdown of faith in the paradigm. If not resolved, this, in turn, produces a revolutionary situation. This resembles the preparadigm phase in that scientists behave like philosophers and ask big questions, even becoming susceptible to influences outside of science. New paradigms compete for primacy, one eventually winning out and becoming the basis for a new round of normal science. As in the transition from the preparadigm to paradigm stage, Kuhn is not very clear as to just how one paradigm wins out in the revolutionary competition.

Kuhn's description of normal science as the norm outraged philosophers and others to whom science was an enterprise forever at the edge of knowledge, undertaken by individuals continually challenging existing concepts.[20] A tradition-bound community was anathema to those believing in scientists as a band of adventurous explorers of the unknown. But there is more than a little doubt that normal science is an accurate description of most of the research done by scientists. Do they always know the outcome in the sense given by Kuhn? If so, why are they forever surprised by anomalies? Kuhnian normal science bears a striking resemblance to Whig history. It is best seen as "normal" in retrospect from a comfortable armchair in a library. To the actual participants, Whigs and scientists, solutions and problems were often vexingly uncertain.

A few general characteristics of Kuhn's theory merit attention: the nature of paradigms; the question of the discontinuity, cumulativity, and relativity of science; the applicability of his theory outside of science; and the role of the group, or invisible college. As to paradigms, one ostensibly friendly critic discerned twenty-one senses in which Kuhn used the term in *The Structure*.[21] That and other criticisms from philosophers of science caused him to redefine the paradigm in two distinct ways: as "disciplinary matrix" and as "exemplar." The former, which includes a worldview, is far more comprehensive than the latter. Exemplars are the concrete puzzle-solving practices scientists learn from textbooks during their education. With that redefinition, Kuhn clarified one troublesome aspect of his views.[22]

The use of the word "revolution," plus a specific comparison to political revolutions, led many readers to believe Kuhn referred exclusively to great intellectual turning points (such as the Newtonian or Darwinian "revolutions") as analogous to great catacylsmic political upheavals, like the French and Russian Revolutions. But *The Structure* refers also (and more frequently) to another, or smaller "revolution"— the replacement of one exemplar by another (that is to say, less than twenty-five individuals agreeing on a new way of puzzle solving). Kuhn has flatly declared the exemplar a "more fundamental meaning of paradigm than disciplinary matrix."[23] One can conclude that little revolutions are, in a sense, more fundamental, and that big revolutions involving disciplinary matrices are less so. The latter provide more difficulties in restoring research to the state of normal science and are, one deduces, "abnormal" revolutions. All of this accords with Kuhn's view of a tradition-laden scientific community. Scientists, still enamored of a self-image as fearless innovators, often associate their youthful triumphs with revolution (in the big sense) and their later careers with a normal science forced on them by society. Rightly or wrongly, some discern in Kuhn a validation of the conventional wisdom of the scientific community.

Can the distinction between the little revolution and the big revolution, between the examplar and the disciplinary matrix, clarify some of the arguments found in the literature about Kuhn's work? For one thing, the outrage of Popper and others about the demeaning idea of normal science appears ill-founded. Kuhn's invisible college of twenty-five (or even one hundred) engage in science at the research front, even if his theory assigns them an unflattering term and implies a kind of mechanical routine. In fact, Kuhn is talking about such groups as the twenty-five leading theoreticians in quantum electrodynamics, or the one hundred leading experimental neurophysiologists, at particular time periods. "Normality" in the Kuhnian sense hardly refers to duffers or hacks. Furthermore, a small group of less than twenty-five in the grip of an exemplar is more likely to undergo revolutions at smaller time intervals than larger groups committed to more comprehensive disciplinary matrices. These smaller revolutions may come on so rapidly, in fact, that the effect may not differ much from the Popperian image of scientists as perpetual revolutionaries. Kuhn may approach Popper as a limit. Kuhn and Popper are really discussing the same people at the same locus, the research front—the former, thinking of his very internalistic exemplar; the latter really in the grip of an elusive disciplinary matrix.

Another genre in the history of science identifies an individual or a publication neglected in the historical literature of a science; it is sometimes sourly referred to as "precursoritis." The historian here demonstrates the "real" importance of that individual or publication. Perhaps the resurrection of Boscovich is a good example of the genre. This amounts to no more than showing that past accounts by scientists or by historians of science do not give the correct composition of past invisible colleges (to speak in Kuhnian). It gives us the names of the twenty-sixth or one-hundred-and-first member. There is no reason in principle why subsequent research should not turn up the twenty-seventh, the twenty-eighth, and so on, or the one-hundred and second, the one-hundred and third, and so on. As soon as science became a general activity in a society—not simply the doings, literally, of a few—the groups involved grew in size. The existence of priority squabbles and the recurrent rediscovery of precursors suggest that accounts of invisible colleges simply freeze into the literature one group's self-perception. Both Kuhn and Popper are really not interested in the possible larger group and the possible social processes they participate in. Nor are they more than marginally concerned with all these twenty-sixth individuals who play roles, however limited, in the intellectual processes they describe. From the standpoint of historical practice, it is probably impossible to distinguish between a conventional Kuhnian or Popperian interpretation once investigations expand beyond circumscribed bounds of some small body of investigators. In the modern period one always finds the twenty-sixth. Invariably, the twenty-sixth cannot unequivocally fit a preexisting philosophic scheme.

As a former modern physicist, Kuhn unsurprisingly made discontinuity a necessity of his view of the history of science. He firmly disavowed the widely held conventional interpretation of science as a cumulative progression to some ultimate truth or reality. Resolutely, Kuhn avoids any position implying that there is "something out there"—Nature or whatever. In his world there are only paradigms providing occasions for puzzle solving by invisible colleges. Each successive paradigm is equally valid as a successful organization of its materials for puzzle solving. More striking, each one places its practitioners in what is, to Kuhn, literally a different world.

That last point produced a counterbarrage charging him with subjectivity and relativism. Some philosophers of science (and others) wanted to preserve the notion of progress in science, the belief that certain theories are better than others, not somehow determined simply by a consensus of twenty-five or even one hundred scientists. Kuhn was forced to explain that successive paradigms represented higher logical states and that he held to certain permanent values which determined validity of paradigms. On examination these turn out closely to approximate the conventional wisdom of the scientific community. Simplicity, ability to predict, consonance with known facts, and the like are quite different than the uses of terms like "Gestalt switch" or of language implying parallels with religious conversions.[24] Kuhn's position, as revised, preserves a primacy not only of Western science, but of whatever are the current conceptions governing research.

Supporting this position and other specifics in Kuhn is his view of "mature sciences." Mature sciences are those in the grip of the Kuhnian cycle. Not all of the natural sciences are equally mature. The most mature (and, therefore, the earliest to leave the preparadigm phase) are mathematical in nature. Kuhn's implicit scale of maturity is very close to the Comtian hierarchy of fields. He is quite explicit in denying maturity and the benefits of the Kuhnian cycle to fields outside the natural sciences, except for a momentary wavering in the case of economics. Here he detected a group perhaps with consensus on both basics and research examplars. But that lapse was never repeated. The "internal logic" of Kuhn's theory simply rules out its applicability to all fields other than the mature sciences. Only here are groups in the grip of a paradigm.[25]

Such groups are essential to Kuhn's purposes. They define and determine the validity of successive paradigms. In the absence of a correspondence to an objective reality or to a logical or rational standard for validation of theories, the groups reign supreme. They are the sole measure and measurer of paradigms. Because of the importance of groups of paradigm-sharing, interacting scientists in his theory, Kuhn has statements about their collective and individual behavior. These sound and were interpreted (for example, by social scientists) as a call to introduce certain nontechnical or external factors into the analysis of the genesis of scientific exemplars.[26] Kuhn was seen as a liberator from both practitioners' history and dominant factions in the philosophy of science. But his intentions are much narrower.

As Kuhn repeatedly noted, "mature sciences are regularly more insulated from the external climate."[27] Except for periods of revolutionary crisis, Kuhn's groups are largely autonomous—or, to be more precise, should be for the optimum functioning of the cycle. External factors play a very limited role because of the necessity for relative autonomy. Some sense of these distinctions, the boundaries established by Kuhn's position, is in his statement about the ahistorical attitude, typical among scientists, toward early versions of published papers: they "illuminate only their author's intellectual biography, not the solution of his puzzle."[28] Intellectual biography and puzzle solving are somehow kept in separate niches in Kuhn's mind to protect both the autonomy of the mature sciences and to avoid practitioners' history.

To a large extent, practitioners' history is also a "great man" history. The post–World War II development of Anglo-American history of science as intellectual history tended toward a "great ideas" history. While part of this tradition, Kuhn aimed for a "great group" history. For the internal logic of science, Kuhn substituted the internal logic of the autonomous self-replicating specialist group. The older positivistic view complacently regarded science as the highest truth, a towering achievement of Western civilization which was the common heritage of all mankind. In contrast, Kuhn's cycle simply could go on forever, autonomously, without any teleological goal, existing for its own sake. Kuhn compared the process to natural selection, one of the rare references to biology in his works.[29] But Kuhn's description of science also bears a resemblance to a game like chess, with its groups of devotees. Kuhn can assert "that after a science has become thoroughly technical, particularly mathematically technical, its role as a force in intellectual history becomes relatively insignificant," in reference to mathematical physics, only because his system excludes such intellectual and political implications as those provoked by atomic weapons and nuclear power.[30]

By the latter half of the 1960s the intellectualist version of internal history appeared ascendant. In the United States both Marxist and sociological competition were apparently vanquished. In that period Kuhn's position had a kind of moderating quality, calling for the expansion of external history for its own sake and for whatever insight it could yield about the behavior of the group, particularly on the location, rate, and timing of scientific change. Once again his purposes were far narrower than many realized.

For one thing, Kuhn never wavered in his dedication to internal history as the core of the subject. His historical writing is largely devoid of the contextual treatment called for in his historiography. Having little practice in the genre, his historiographic views therefore appear more than a little condescending to external history. They are like the writings about science he deplores, the passages by general historians based on prefaces, rather than the actual practices embodied in the texts proper. Never does he acknowledge that good external history is both important and difficult. It presents more difficulties than the intense study of a relatively small number of sacred texts,[31] a great challenge because of the larger number of variables involved and the need to master a greater range and quantity of sources.

By 1971 an upsurge of interest in external history occurred, helping to crystallize the social history of science. Kuhn reacted uneasily:

Though I welcome the turn to the external history of science as redressing a balance which has long been seriously askew, its new popularity may not be an unmixed blessing. One reason it flourishes is undoubtedly the increasingly virulent antiscientific climate of these times.[32]

Obviously, this is a reaction to the turmoil associated with the Vietnam war. It is also an outrageous piece of guilt by association. It implies that external history means hostility to science. Kuhn is echoing the reaction of benighted scientists who think history of science should not stray from equations for fear of giving ammunition to critics. In 1977, Kuhn sadly noted that a disproportionate number of the younger historians (including many of the best) had turned to external history. Kuhn now worried about the future of his style of internalist history.[33] However, to the best of my knowledge, he has never in print challenged the validity of the newer social history of science. What Kuhn is expressing may be a form of distress accompanying generational succession, specifically centered on the apparent displacement of internal history as the core of the field. I think his concern is ill-founded. Internal history is alive and well but is often being practiced in a different mode. To explain that, I have to recount what I think happened to the history of science in the last decade or so in the English-speaking world.

For one thing, the late 1960s' sense of intellectualist triumph was illusory. The history of science was never that monolithic. Earlier, one of the founders of history of science as an academic field in America, Henry Guerlac of Cornell, had called for a social history of science and warned against an excessively intellectualized history of science.[34] The study of the sciences within the United States arose initially from American history, largely independent of academic history of science.[35] The practice of this particular specialty never recognized the internal–external distinction nor the Kuhnian canon. In 1955, at a conference from which came the NSF support program for history, philosophy and sociology of science (and later the National Institutes of Health support for history of biomedicine), Richard H. Shryock, a pioneer in the study of science in the United States, expressed a viewpoint widely present among certain sectors of the American historical community:

Some participants [in the Conference] held that distinct investigations of intellectual and social history . . . might be admirable in themselves but would not add up to the final word. Also needed . . . are studies which present the intellectual and social factors in close juxtaposition. For one thing, social presuppositions or interests often insinuate themselves into what seem at first glance to be the most objective ideas of the investigator, with the result that it may be misleading to omit the role of environmental factors even in tracing ideas as such. In other words, it is very difficult to isolate ideas in so "pure" a state that they contain no social ingredients.

More obvious is the fact that it is sometimes misleading to attempt any evaluation of ideas outside their social context. . . . It would seem more effective procedure . . . to deal with both ideas and circumstances in a common context.[36]

In Europe, such views had Marxist overtones or were associated with specific political positions. MacLeod has recently placed this position in a historiographic context. "Science has become . . . 'epistemologically privileged,' conferring cultural superiority on the ideas and on the representatives of the rational elite. The internal history of science has had as its object the defense of that privilege."[37] In Great Britain, more so than in the United States, this has stimulated a social history of science and a sociology of science repudiating both Robert K. Merton's functionalist sociology of science and Kuhn's theory of scientific revolutions.[38]

Another important strand is the influence of anthropology, a field seeking to "reconstruct individual cultural systems in their entirety." This influence is also stronger in Britain than America. Following Durkheim, anthropology placed science and its practitioners in an autonomous, epistemologically privileged position. At the same time, by reinforcing traditional interests in non-Western belief systems, anthropology gave new respectability to the sociology of knowledge. It also reinforced consideration, already underway among historians of Renaissance and early modern science, of those cultural activities, like astrology and the occult arts, once designated as "pseudo-science."[39] The expanded boundary of the history of science was redrawn to embrace all understanding and uses of nature.[40] Claims for autonomy, whether of scientific content or of scientific group, become part of their respective cultural systems. A sense of this interpretation was given by Roy Porter in 1973:

To understand science we must see it in the context of, and constitutive of, social structure, social change, and social consciousness. But it is simpleminded always to expect to find science responding, in any immediate way, to social conditions. For a mere glance will show that because men of science in East and West have always comprised some sort of mandarinate or clerisy, whether as an élite set apart by wealth, talent, privilege, or by state patronage, etc., they have been cushioned from some of the more obvious and potent social movements of their day, often isolated from economic and technical pressures from below, by systems of cultural mediations.[41]

Growing interest in the history of applied science also reflects this emerging consensus. Kuhn noted that technology (or the crafts) can influence science, but he has been equivocal on whether this is relevant to fully mature fields.[42] In his explanation of the course of the physical sciences from the Scientific Revolution of the seventeenth century until the emergence of modern physics, Kuhn distinguishes between Baconian fields and classical ones.[43] The latter go back to antiquity and are maturely mathematized. But he noted without elaboration the existence of other mature classical fields, like anatomy, associated with medicine. Never does he explore the possibility that these undergo a different course, perhaps because of the relationship to utility and to a specific social group (physicians). That is, if we look also at applied fields (like engineering) in conjunction with the basic disciplines, can we discern not one structure but many? Or even something utterly different from revolutions?

One early critic of Kuhn complained about the "absence of final technical validation."[44] A later writer decries the overemphasis on the distinction between science and technology, pointing out that most scientists now (and even in the past) had technological involvements.[45] Still another notes the characteristic "modern scientific fusion of the theoretical and the practical."[46] Clearly, those interested in science policy would favour concern with applications.

At the same time there is a vigorous tradition in the history of technology stressing the contributions of nonscientists or of nonscientific practices in technological advance. This has been countered by investigations purporting to show scientific contributions to technology in the seventeenth and eighteenth centuries, and to the Industrial Revolution.[47] In an earlier era, practitioners and positivists, Kuhn's favorite targets, handled the question of technology very simply. They lumped it with science, complacently citing its triumphs as proof of progress. Kuhn cannot look for "technical validation" because that implies that the purpose of science, of his cycle, is better technology. Kuhn is firm about keeping the cycle pure; it functions best when untrammeled by applied concerns.[48] Perhaps this is a faint echo of the response of some physicists to the coming of nuclear warfare. In contrast, historians of science are less and less inclined to disregard applied consequences.

Almost all of the early historians of science within the United States had interests in technology, medicine, or agriculture. Perhaps there is a connection here to one of them, A. Hunter Dupree of Brown University, raising another criticism against the older internal history. In 1966 he insisted that the history of science is properly and primarily concerned with human beings, not with things—even such abstract things as scientific concepts and data. That these are important elements in the study of the behavior of a specific group but are not the principal focus of concern is now a truism of the social history of science.[49]

This belief explicitly strikes at an assumption Kuhn shares with other internalists and with practitioners' history—the necessity of a detailed, even profound technical scientific competence.[50] It is extraordinary to someone of my vintage to read recently that "if you go back in time even half a century, for example to do research on early quantum mechanics, it is questionable whether a knowledge of modern quantum physics may not be more of a disadvantage than an advantage."[51] The contrary opinion holds that the historian has to enter vicariously into the creative scientific process both in order to re-create it and to analyze historically the great investigator. Retrospectively, the historian becomes a peer of the great past figures of science. Quite often this is accomplished not because of scientific talents of the historian but because the historian has the benefit of hindsight, of subsequent scientific developments. It is very hard for internalist historians to avoid the perils of Whig history.[52] There is a tendency to slide over the necessary distinction between the creator and the critic. When internal history is done well, there is no doubt of its value to every kind of analyst of the sciences.

In a few quarters there is an insistence that past science is not in principle barred to the historian. One of the great classics of the history of science is Herbert Butterfield's, *The Origins of Modern Science, 1300–1800.*[53] Butterfield, whose work on British history gave us the generalized concept of the Whig interpretation, was Regius Professor of History at Cambridge. The official histories of the atomic energy program of the United States and the United Kingdom were done by general historians. One of them, Margaret Gowing of Oxford, wrote:

Of course it is difficult enough for a non-scientist to learn enough science, but historians learn to handle all kinds of other subtle concepts and theories. . . . My own scientific education and scientific proclivity were minimal, but I find it easier to understand how a nuclear reactor works than to understand the finer points of theology or even superannuation schemes, and the scientists and engineers appear satisfied with my history. This experience enables me to exhort historians: "Be not afraid of science."[54]

Many still come to the field from the sciences. And *Astronomy Transformed* provides a paradigm for future team research practice: Edge is a radio astronomer turned historian; Mulkay is a sociologist of science.

The prevailing consensus very decidedly includes the content of science as an object of study. There are relatively few historians of science who are either militantly internalist or externalist; many now reject the distinction. There is a general recognition of the necessity to have knowledge of both and, ideally, to investigate the two together. Thomas Kuhn may have played a role in fostering this new consensus by stimulating discussion and research leading to the recognition of the difficulty of disentangling the internal from the external.[55]

The distinction between internal and external is handled by a firm desire to avoid all or nothing formulations. There is little profit in having not one, but two black boxes—one labeled "internal," for externalist historians; the other "external," for internalists. Again I quote Martin Rudwick, now on the purpose of history of science: "to relate the cognitive contents as closely as possible to the cultural circumstances in which they develop."[56] This emerging consensus does not regard history of science as part of science nor as an independent domain. As Rudwick has noted:

A substantial number of historians of science have begun to write about the history of science in the same way as other historians write about other aspects of past cultures—economic, social, political, religious, artistic, and so on. In other words, the contrasts of method and attitude between the history of science and the rest of history are being broken down.[57]

An American historian has recently noted: "We need a way to connect history of science with social and cultural history at the bottom rather than at the top." And a few pages later, he adds: "We feel that technical, dry, internalist history of science is sterile, but we also know that it forms an essential foundation for our interpretive studies."[58]

I am not implying a monolithic consensus. Divergences of purpose and practice exist. Perhaps the most consequential is the distinction between those historians of science who resolutely consider their task as primarily the "exposition and elucidation of substantial aspects of the scientific cultures,"[59] largely for their own sake, and those viewing their specialty as providing basic knowledge for application either by other historians or in such fields as science policy. Disagreement between the two viewpoints can become heated, but the products of both overlap considerably in style and in content.

What unites them, in the last analysis, is the rejection or minimization of the internal–external split and a resulting difference in the historiographic emphasis on causation. In the past historians of science considered historical causes to operate almost like mechanical impact or fluid pressure; at the most sophisticated level we were offered the equivalent of inductive action at a distance. What are being described and analyzed today are configurations, usually very complex. These configurations act both as exemplification and justification, and they change in involved patterns not adequately delineated by relatively simple formulas like Kuhn's. We are just beginning to discover what is involved.

Notes

1. A. J. Turner, "The History of Greenwich Observatory (Review)," *History of Science* 16 (1978):222–226; quote at p. 224.
2. T. S. Kuhn, *The Structure of Scientific Revolutions* (Chicago: University of Chicago Press, 1962; rev. ed., 1970). In this paper, references to *The Structure* are to the revised 2nd ed. of 1970.
3. T. S. Kuhn, "Second Thoughts on Paradigms," in Frederick Suppe (ed.), *The Structure of Scientific Theories* (Urbana, Ill.: University of Illinois Press, 1975), pp. 459–482; quote at p. 459. This and many other shorter pieces are reprinted in Kuhn, *The Essential Tension* (Chicago: University of Chicago Press, 1977). I am indebted to Ian Hacking of Stanford for his preprint, "The Essential Tension (Review Article)," now in *History and Theory*, 18 (1979):223–236, the best review of Kuhn by a philosopher of science. In a private communication to me, Kuhn has completely repudiated the statement in the *New York Times*.
4. David O. Edge and Michael J. Mulkay, *Astronomy Transformed: The Emergence of Radio Astronomy in Britain* (New York: Wiley-Interscience, 1976); see pp. 360, 386–395, and fn. *j* on p. 432, referring to text on p. 103. On p. 390 the authors write: "Thus the assimilation of novelties often requires, not a drastic revision of existing assumption accompanied by confused resistance, but the enthusiastic displacement of existing ideas and techniques into problem areas not previously encountered." A sociologist, H. Gilman McCann, recently published a monograph supporting Kuhn based on the revolution associated with Lavoisier: *Chemistry Transformed* (Norwood, N.J.: Ablex, 1978). This ostensibly supports Kuhn, but it is a better verification of McCann than Kuhn. Given the sweeping nature of Kuhn's theory and the imprecise, shifting language of its author, it is doubtful whether it is susceptible to either verification or falsification. I am concerned primarily with its fruitfulness for practicing historians of science.
5. For the former, see the dispute on Darwin between John C. Greene and Leonard Wilson in D. H. D. Roller (ed.), *Perspectives in the History of Science and Technology* (Norman: University of Oklahoma Press, 1971), pp. 3–37. For the latter, see E. J. Yoxen, "Where Does Schrödinger's 'What is Life?' Belong in the History of Molecular Biology?" *History of Science* 17 (1979):17–52; "paradigm" appears on p. 45, but its absence would make no difference whatsoever. I should add that my comments on the practice of historians of science is based on the English-language literature.
6. T. S. Kuhn, *Black-Body Theory and the Quantum Discontinuity, 1894–1912* (Oxford and New York: Oxford University Press, 1978). The September

1979 issue of *Isis* had a review symposium on this book which appeared after this passage was written. The three reviewers generally agreed with the position taken above. Two of the three were quite critical of the lack of consonance of the product with Kuhn's theoretical views.

7. R. C. Olby, "Mendel No Mendelian?" *History of Science* 17 (1979):53–72. See also Augustine Brannigan, "The Reification of Mendel," *Social Studies of Science* 9 (1979):423–454. To Kuhn, perhaps to others, it is not simply the received tradition which is at issue but the relationship of the historian of science to the contemporary ongoing scientific venture.

8. Put very simply, historians of science more and more act and think like general historians and less like scientists, philosophers of science, or a special breed of historical practitioners. See Tore Frangsmyr, "Science or History: George Sarton and the Positivist Tradition in the History of Science," *Lychnos* (1973/74):104–144.

9. T. B. Settle, "On Normal and Extraordinary Science," in Arthur Beer and K. A. Strand (eds.), *Yesterday and Today: Proceedings of a Commemorative Conference Held in Washington in Honor of Nicolaus Copernicus* (Oxford and Elmsford, N.Y.: Pergamon Press, 1975), pp. 105–111; quote at p. 109. The book is Vol. 17 of the *Vistas in Astronomy* series.

10. Martin J. Klein, "Einstein on Scientific Revolutions," ibid., pp. 113–120.

11. R. W. Home, "Newtonianism and the Theory of the Magnet," *History of Science* 15 (1977):256–266; quote at p. 263.

12. W. D. Hackmann, "The Relationship Between Concept and Instrument Design in Eighteenth-Century Experimental Sciences," *Annals of Science* 36 (1979):205–224, esp. 217ff.

13. J. L. Heilbron, *Electricity in the 17th and 18th Centuries . . .* (Berkeley: University of California Press, 1979), p. 309.

14. See the autobiographical preface to Kuhn, *The Essential Tension*, esp. pp. xiff, xvi–xvii.

15. Michael Polanyi's best known work is *Personal Knowledge: Towards a Post-Critical Philosophy* (London and Boston: Routledge & Kegan Paul, 1958), but see his *The Logic of Liberty . . .* (Chicago: University of Chicago Press, 1951). Bernard Barber's *Science and the Social Order* (New York: Free Press, 1952), and his "Resistance by Scientists to Scientific Discovery," *Science* 134 (Sept. 1, 1961):596–602, both show the influence of Robert K. Merton. Later, in the preface to *The Essential Tension*, p. xxi, Kuhn would repudiate British sociologists of science using him to attack Merton. They were misled by *The Structure*, and by such papers as his "The Function of Dogma in Scientific Research," in A. C. Crombie (ed.), *Scientific Change . . .* (New York: Basic Books, 1963), pp. 347–369, to see him as against the tradition-laden, conservative scientific community.

16. Kuhn, *Structure*, pp. 46ff, 80, 136ff, 165ff, 187ff.

17. Perhaps the best general survey of the development of the history of science is Arnold Thackray's chapter in Paul T. Durbin (ed.), *A Guide to the Culture of Science, Technology, and Medicine* (New York: Free Press, 1980). I appreciate Professor Thackray's courtesy in making available a draft for my use. For a superb statement of this position, see A. R. Hall, "Can the History of Science Be History?" *British Journal for the History of Science* 4 (1969):

207–220. On the last page Hall expressed significant skepticism about Kuhn's theory. My own observations are that many American historians of science similarly resisted Kuhn's cycle.

18. Martin Rudwick, *The History of the Natural Sciences as Cultural History* (Amsterdam: Rede-Vrije Universiteit, 1975), pp. 14–15; see also p. 21 and fn. 60 for other adverse comments. In the same year, in another inaugural lecture, *What's Science to History, or History to Science?* (Oxford and New York: Oxford University Press—Clarendon Press, 1975), Margaret Gowing also voiced a position counter both to Hall and to Kuhn.

19. Kuhn, *Structure*, p. 181.

20. Perhaps the best single source for this is Imre Lakatos and Alan Musgrave (eds.), *Criticism and the Growth of Knowledge* (Cambridge and New York: Cambridge University Press, 1970). For somewhat different perspectives, see R. H. Stuewer (ed.), *Historical and Philosophical Perspectives of Science* (Minneapolis: University of Minnesota Press, 1970), and Suppe (ed.), *Scientific Theories*. An interesting article on the changes in philosophy and history of science is Stephen Toulmin's "From Form to Function: Philosophy and History of Science in the 1950s and Now," *Daedalus* 106, no. 3 (Summer 1977):143–162.

21. Margaret Masterman, in Lakatos and Musgrave (eds)., *Criticism*, pp. 59–89.

22. Kuhn, "Second Thoughts."

23. Kuhn, *Essential Tension*, p. 463. The distinction between the two meanings and the two revolutions are forever being confused, especially by sociologists and philosophers. Even Edge and Mulkay, *Astronomy Transformed*, p. 392, stumble about here.

24. Hacking, "Essential Review," p. 228; Kuhn, *Structure*, pp. 97–98, refers to logically higher theories. Toulmin correctly notes that all these second thoughts make Kuhn less unconventional and less attractive, even to those leaning to his basic views: "Rediscovering History: New Directions in Philosophy of Science," *Encounter* 36 (Jan. 1971):53–64, esp. 60–63.

25. A convenient summary of Kuhn's developed position is in the entry "Science, history of," in the *International Encyclopedia of the Social Sciences*; reprinted in *Essential Tension*, pp. 105–126. See also *Structure*, pp. 165–168, 207–210.

26. Two notable examples are Barry Barnes, *Scientific Knowledge and Sociology Theory* (London and Boston: Routledge & Kegan Paul, 1974) and David Bloor, *Knowledge and Social Imagery* (London and Boston: Routledge & Kegan Paul, 1976).

27. Kuhn, *Essential Tension*, p. 148f.

28. Ibid., p. 347.

29. Kuhn, *Structure*, p. 205f.

30. Kuhn, *Essential Tension*, pp. 3, 133ff.

31. Ibid., the biographical introduction. One can say that Kuhn wants to study the texts intensely to climb inside the mind of the author. Only what is inside the mind is of interest; externalities have to come second. Note here that this attitude is quite like those of other "high culture" studies in which the products are all-important, almost existing independently of humans. Kuhn

cannot go that far because of the perils of practitioners' history—yet this mind-set prevents him from fully thinking out relations of ideas to human contexts.

32. Ibid., p. 16ff.

33. This mediation appears in Kuhn's contribution to the 1977 Reston Conference, "Critical Problems in the Philosophy of Science"; see also his "History of Science," in P. D. Asquith and Henry E. Kyburg, Jr. (eds.), *Current Research in Philosophy of Science: Proceedings of the PSA Critical Research Problems Conference* (East Lansing, Mich.: Philosophy of Science Association, 1979), pp. 121–128.

34. Frangsmyr, "Positivist Tradition," p. 108, called my attention to Guerlac's contribution to the Ninth International Congress of the History of Science, 1950. See Henry Guerlac's "Some Historical Assumptions of the History of Science," in Crombie (eds.), *Scientific Change*, pp. 797–812, and Alexandre Koyré's response, pp. 847–857. At Cornell and other American campuses historians committed to a history of ideas version of history of science imperceptibly drew closer in attitudes and practices to colleagues in other historical specialties. Perhaps they began to believe the words of R. H. Shryock quoted below (see n. 36).

35. My own recollections are in the introduction to the volume I edited, *Science in America Since 1820* (New York: Science History Publications, 1976).

36. Richard H. Shryock, "The Nature of the Conference [Conference on the History, Philosophy, and Sociology of Science]," *Proceedings of the American Philosophical Society* 99 (1955):329.

37. From Roy MacLeod's survey of "Changing Perspectives in the Social History of Science," in Ina Spiegel-Rosing and Derek de Solla Price (eds.), *Science, Technology and Society: A Cross-Disciplinary Perspective* (London and Beverly Hills, Calif.: Sage Publications, 1977), pp. 149–195; quote at p. 161.

38. Among sociologists, three examples: M. J. Mulkay, G. N. Gilbert, and S. Woolgar, "Problem Areas and Research Networks in Science," *Sociology* 9 (1975):187–203; Richard Whitley, "Types of Science, Organizational Strategies and Patterns of Work in Research Laboratories in Different Fields," *Social Science Information* 17 (1978):427–447; Ron Johnson, "Contextual Knowledge: A Model for the Overthrow of the Internal/External Dichotomy in Science," *Australian and New Zealand Journal of Sociology* 12 (1976):193–203.

39. A fascinating historiographic chain arise here. Keith Thomas's *Religion and the Decline of Magic* . . . (London: Weidenfeld & Nicolson, 1971), precipitated a debate with the anthropologist Hildred Geertz in the *Journal of Interdisciplinary History* 6 (1975):71–109—a marvelous illustration of the differing attitudes of the historian and the anthropologist. For historians of science, Thomas's book is arresting as his closing chapters deal with the period of the Scientific Revolution in England. But Thomas was also assaulted as a worshipper of science and technology from the vantage of the eighteenth century by E. P. Thompson, in *Midland History* 1 (1972):53–55. Although a Marxist, Thompson has great reservations about theory arbitrarily imposing order on the sources. This is discussed in Richard Johnson's "Edward Thompson, Eugene Genovese, and Socialist-Humanist History," *History Workshop* 6 (Autumn 1978):85ff, but the position is not always clear amid the sprawl of Thompson's *The Poverty of Theory & Other Essays* (New York: Monthly Review Press,

1978). Put another way, historians are prone to neglect the importunities of theorists in the face of authentic sources. Kuhn's theory might find it hard to assimilate the glorious diversity encountered by his colleagues in print and in manuscript.

40.   A phrase borrowed from Thackray in Durbin (ed.), *Guide.*

41.   Roy Porter, "The Industrial Revolution and the Rise of the Science of Geology," in M. Teich and R. Young (eds.), *Changing Perspective in the History of Science* . . . (London: Heinemann, 1973), pp. 320–343; quote at p. 342.

42.   Kuhn, *Structure*, pp. 15–16, 161; and *Essential Tension*, pp. 141–143, 157.

43.   Although published in full in English only in 1976 and reprinted in *Essential Tension*, pp. 31–65, these ideas appear earlier in Kuhn's work.

44.   Masterman, in Lakatos and Musgrave, (eds.), *Criticism*, pp. 59–89.

45.   In Gowing, *What's Science*, p. 19f.

46.   Rudwick, *Natural Sciences*, p. 8.

47.   Good introductions to these issues are in A. P. Molella and N. Reingold, "Theorists and Ingenious Mechanics: Joseph Henry Defines Science," *Science Studies* 3 (1973):323–351, and A. E. Musson's preface to *Science, Technology, and Economic Growth in the Eighteenth Century* (London: Methuen, 1972). For a differing view, see A. R. Hall, "What Did the Industrial Revolution in Britain Owe to Science?" in N. McKendrick (ed.), *Historical Perspectives* . . . (London: Europa, 1974), pp. 129–151.

48.   Kuhn, *Structure*, p. 96.

49.   A. Hunter Dupree, "The History of American Science: A Field Finds Itself," *American Historical Review* 81 (1966):863–874.

50.   In 1975 the History of Science Society commissioned a *Report on Undergraduate Education in the History of Science*, which noted (p. 23) that "Neither the contributions that have formed our contemporary understanding of the history of science, nor the practices of graduate programs considered together, nor the active composition of the History of Science Society justifies the view that every historian of science must have considerable knowledge of science." The same report noted that about half of the graduate students had science backgrounds. See also Richard French and Michael Gross, "A Survey of North American Graduate Students in the History of Science, 1970–71," *Science Studies* 3 (1973):161–179.

51.   Rudwick, *Natural Sciences*, p. 6.

52.   Toulmin, "Form to Function," p. 149.

53.   H. Butterfield, *The Origins of Modern Science, 1300–1800* (London: Bell, 1949).

54.   Gowing, *What's Science*, p. 11.

55.   MacLeod, "Changing Perspectives," pp. 157–158. I think this is an overstatement, but MacLeod is not the only one who believes so.

56.   Rudwick, *Natural Sciences*, p. 7. In 1971, the year Kuhn complained of the dangerous rise of external history, a survey of undergraduate students undertaken for the History of Science Society reported that more than two-thirds of those expressing an interest in social history also had a concern with the internal history of an area of science.

57.   Ibid., p. 5.

58. Thomas L. Hankins, "In Defence of Biography: The Use of Biography in the History of Science," *History of Science* 17 (1979):1–16; quotes at pp. 4, 13.

59. From an unpublished talk of my colleague Paul Forman, "Geneses of Scientific Ideas as a Historiographic Goal." I am grateful for his permission to use the work.

Index